Animal Applications of Research in Mammalian Development

SERIES EDITORS

John Inglis and Jan A. Witkowski

Cold Spring Harbor Laboratory

CURRENT COMMUNICATIONS

In Cell & Molecular Biology

---

1 *Electrophoresis of Large DNA Molecules: Theory and Applications*
2 *Cellular and Molecular Aspects of Fiber Carcinogenesis*
3 *Apoptosis: The Molecular Basis of Cell Death*
4 *Animal Applications of Research in Mammalian Development*

Forthcoming

5 *Molecular Biology of Oxygen-free Radical Scavenging*

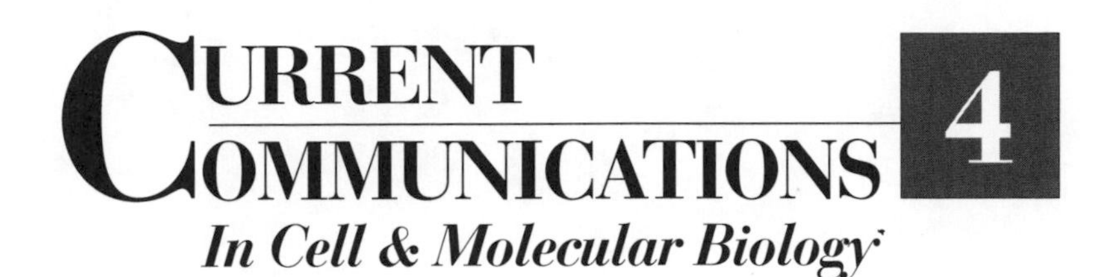

Animal Applications of Research in Mammalian Development

Edited by

Roger A. Pedersen
University of California
San Francisco

Anne McLaren
University College
London, England

Neal L. First
University of Wisconsin
Madison

Cold Spring Harbor Laboratory Press 1991

CURRENT COMMUNICATIONS 4
In Cell & Molecular Biology
Animal Applications of Research in Mammalian Development

Printed in the United States of America
Cover design by Leon Bolognese & Associates Inc.

Front Cover: (*Left*) A transgenic female Alpine goat that secreted tPA in her milk throughout lactation. (*Right*) Two of her offspring. (Courtesy of K. Ebert, Tufts University Schools of Veterinary Medicine and Medicine and Dental Medicine.)

Back cover: Schematic diagram outlining the methodologies used in the production of transgenic goats by microinjection of fusion genes. (Courtesy of K. Ebert and Thomas Smith. See related article by Ebert and Selgrath, this volume.)

Library of Congress Cataloging-in-Publication Data

Animal applications of research in mammalian development / edited by Roger A. Pedersen, Anne McLaren, Neal First.
p. cm. — (Current communications in cell & molecular biology ; 4)
Includes bibliographical references and index.
ISBN 0-87969-333-9
1. Veterinary embryology. 2. Livestock—Embryos. 3. Livestock—Development. 4. Livestock—Genetic engineering. I. Pedersen, Roger A. II. McLaren, Anne. III. First, Neal L. IV. Series: Current communications in cell and molecular biology ; 4.
[DNLM: 1. Animals, Domestic—embryology. 2. Gametogenesis. 3. Genetic Engineering. SF 767.5 A598]
SF767.5.A55 1991
636.089'264--dc20
DLC
for Library of Congress 91-10980
CIP

The articles published in this book have not been peer-reviewed. They express their authors' views, which are not necessarily endorsed by Cold Spring Harbor Laboratory.

All Cold Spring Harbor Laboratory Press publications may be ordered directly from Cold Spring Harbor Laboratory Press, 10 Skyline Drive, Plainview, New York 11803. Phone: 1-800-843-4388. In New York (516) 349-1930. FAX: (516) 349-1946.

A Special Tribute to Dr. John Dennis Biggers

In view of the dimension and importance of the issues addressed in this volume, it is entirely fitting that it arose from a meeting honoring Dr. John D. Biggers, whose own contributions have established the scope of the field of mammalian development and reproductive biology. Dr. Biggers, in collaboration with Dr. Anne McLaren, showed that mouse embryos cultured from the eight-cell stage to the blastocyst stage could progress successfully to term after transfer to foster mothers. The pioneering studies of John Biggers and his collaborators fueled the renaissance in experimental studies on mammalian embryos. In the intervening years, he has made major contributions in the culture of oocytes and embryos, their metabolism, physiology, and endocrinology, as well as biostatistical approaches for optimizing the use of embryos for experimental purposes. We are indebted to John Biggers not only for his research contributions, but also for his role in addressing ethical and philosophical implications of research in mammalian development, and for his perspective on the historical roots of our field. The scope of his contributions is so great that it is difficult to imagine anyone working in the field of mammalian developmental and reproductive biology who has not benefited from John Biggers' work. Nevertheless, some of us consider ourselves uniquely fortunate, in that we had him as a graduate or postdoctoral mentor, collaborator, or colleague. When a number of us gathered at the Banbury Center on October 15–18, 1989, for a meeting in honor of John Biggers, it was clear to what degree current and future work in embryos of domestic species benefited from his past contributions and current interests. Therefore, we celebrate our legacy by dedicating this volume to John Biggers, our mentor, colleague, and friend.

Contents

Preface

The application of experimental mammalian embryology to domestic species began with the earliest studies in our field and continues at a breathless pace. The objective of Heape's pioneering embryo transfer studies, published in 1891, was the transfer of embryos from one variety of rabbit to another. Now, a century later, we have witnessed the phenomenal successes of in vitro fertilization and embryo transfer, applied both to human infertility and to animal reproduction. The details of those two areas of accomplishment are fully covered elsewhere and are not considered here. Instead, we have looked toward future applications of the knowledge accumulated from research in early mammalian development and reproductive biology. We envision here the animal application of current and incipient biotechnology with several objectives that have potentially enormous impact. These applications include expanding gamete resources through use of cultured ovarian primordial oocytes or immature oocytes; cryopreservation of oocytes and embryos; in vitro production and cloning of embryos; use of domestic species as experimental models leading to a comparative understanding of early mammalian development; the introduction of exogenous genes (transgenes) that are subsequently expressed, yielding products that alter the physiology of the animal or can be harvested in substantial quantity; and rapid transfer of genes between stocks using DNA markers to assist introgression. These applications have the potential to transform animal genetics, biotechnology, and domestic food production. The extent to which this vision is realized will depend on many factors, including market forces, social and ethical concerns, and the level of support for these objectives from government and private sources. However, it is clear from the chapters included in this volume that the conceptual basis of these future applications already exists and is being rapidly extended by current work in both university and commercial institutions.

Important lessons from this volume can also be seen in areas that impact on the future of the field of mammalian developmental biology. First, basic research in mammalian development is increasingly being carried out in species other than mice, and this could ultimately provide a general understanding of the events of early mammalian embryogenesis. It is especially hopeful that these insights will extend to embryos of our own species. Second, there will be a potentially tremendous economic benefit from the application of embryo biotechnology to improving livestock and increasing food production. Similar gains may accrue from efforts to produce bioactive proteins in domestic species. These economic forces will lead to an increasing proportion of basic research being done in commercial organizations or with support from such sources. This "commercialization" of basic research could have a negative impact on future progress in the field unless researchers maintain the discipline of publishing their results and experimental detail. Even though timely publication may appear to conflict with the short-term economic interests of the private organization, it is nevertheless essential to the health and effective long-term progress in the field.

The applications envisioned here, and others that may occur to the reader, raise important ethical issues that deserve social debate. Although these issues are not addressed in this volume directly, they will inevitably impact on decisions that determine the future development and applications of mammalian embryological research.

This volume would not have been possible without the enduring encouragement from John Inglis, Director of the Cold Spring Harbor Laboratory Press; the generous support from Jan Witkowski, Director of the Banbury Center, and Jim Watson, in his role as Director of the Cold Spring Harbor Laboratory; the administrative effort of Bea Toliver of Banbury Center; the cheerful hospitality of Katja Davies and other staff members of Robertson House; the guidance of Dr. Betsey Williams; and the editorial and production expertise of Dorothy Brown, Virginia Chanda, Mary Cozza, and the entire staff of the Cold Spring Harbor Laboratory Press. Finally, one of us (R.A.P.) extends a special thanks to the authors and other participants for their personal support during the harrowing hours following the Loma Prieta earthquake that devastated the San Francisco Bay area while our Banbury Center meeting was in progress.

R.A.P.
A.M.
N.L.F.

The preparation of papers in this book was assisted by the authors' participation in a meeting at the Banbury Conference Center which was generously supported by Cold Spring Harbor Laboratory's Corporate Sponsors.

Alafi Capital Company
American Cyanamid Company
Amersham International plc
AMGen Inc.
Applied Biosystems, Inc.
Becton Dickinson and Company
Beecham Pharmaceuticals
Boehringer Mannheim Corporation
Ciba-Geigy Corporation/Ciba-Geigy Limited
Diagnostic Products Corporation
E.I. du Pont de Nemours & Company
Eastman Kodak Company
Genentech, Inc.
Genetics Institute
Glaxo Research Laboratories
Hoffmann-La Roche Inc.
Johnson & Johnson
Life Technologies, Inc.
Eli Lilly and Company
Millipore Corporation
Monsanto Company
Oncogene Science, Inc.
Pall Corporation
Perkin-Elmer Cetus Instruments
Pfizer Inc.
Pharmacia Inc.
Schering-Plough Corporation
Tambrands Inc.
The Upjohn Company
The Wellcome Research Laboratories,
  Burroughs Wellcome Co.
Wyeth-Ayerst Research

The preparation of papers in this book was assisted by the [illegible] in a meeting at the Banbury Conference Center, which was generously supported by Cold Spring Harbor Laboratory Corporate Sponsors.

New Advances in Reproductive Biology of Gametes and Embryos

N.L. First

Department of Meat and Animal Science
University of Wisconsin
Madison, Wisconsin 53706

INTRODUCTION

Animal reproduction has entered an exciting new era in which new ways for rapidly propagating or designing domestic animals are being developed. These new techniques take advantage of procedures in cellular and molecular biology, including the ability to transfer new genes into the genome of domestic animals and the identification of genes or chromosome restriction-fragment-length polymorphisms (RFLPs) that are associated with important production traits. Each technique assists in direct and rapid selection for a particular trait. The molecular approach complements the reproductive methods of in vitro production of embryos, multiplication of embryos by splitting and cloning, artificial insemination, and embryo transfer. As a result, our knowledge of the physiology of reproductive processes is enhanced, and chances for embryo survival and fecundity are increased.

The following chapters review the state of the art in each of these exciting areas of reproductive science. Much of the science presented in this book is beginning to reach application in the livestock industry. Commercial companies are already beginning to offer services such as in vitro production of embryos, embryo splitting, cloning, sexing, and genetic screening while continuing research to increase efficiencies and to develop useful new transgenic lines of cattle, sheep, and swine.

Although the major effort to develop these technologies has been with cattle, laboratory animals are equally important to the development of new tools for animal propagation and for understanding the physiology of reproduction. The laboratory species most widely used as a model for mammalian reproduction has been the mouse, and during the past years, the laboratory of John Biggers has been a leader in the study of the physiology of gametes and embryos in mice. Biggers' experiments with oocyte maturation, embryo culture, metabolism, and development, as well as blastocyst and implantation physiology, have provided the groundwork that has led to successful systems for maturing oocytes and culturing embryos in vitro for domestic species. These studies have provided the key components for in vitro production of embryos, cloning of embryos, gene transfer, and indeed any procedure performed in vitro and requiring in vitro culture.

TECHNIQUES FOR IN VITRO EMBRYO PRODUCTION

Oocyte Maturation

The physiology of oocyte maturation is reviewed by Racowsky, and cryopreservation of oocytes and embryos is presented by Van Blerkom (both this volume). In domestic species, oocytes are recovered from follicles matured in vivo either with or without superovulation by ultrasound-guided vaginal laparoscopy. These oocytes can then be fertilized and will proceed through embryo development with good success (cattle, Sirard et al. 1985; Lambert et al. 1986; Leibfried-Rutledge et al. 1987, 1989; swine, Cheng et al. 1986; Nagai et al. 1988; sheep, Crozet et al. 1987). Recovery from genetically valuable cows provides a supply of oocytes with high genetic value for use in gene transfer and production of founding embryos for embryo cloning. By preventing development of the dominant follicle, von der Schams et al. (1991) have recently shown that approximately eight oocytes could be harvested from a valuable cow every 2–3 days throughout the year. If these oocytes are fertilized in vitro, it is estimated that one genetically valuable cow could produce nearly 100 calves per year.

By coculture with granulosa cells from the ovarian follicle and rigorous oocyte quality selection, oocytes recovered at the abattoir from follicles more than 3 mm can also yield a large supply and nearly normal rate of fertilization and embryo development. Incidence of development to morula or blastocyst stage for in-vivo-matured and -fertilized oocytes was 55%, in vivo matured/in vitro fertilized oocytes was 45%, and in-vitro-matured and -fertilized oocytes was 23–63% (Leibfried-Rutledge et al. 1989; Gordon and Lu 1990). The bovine ovary contains thousands of growing oocytes in follicles of less than 1 mm. A major challenge to scientists is how to mature and harvest from this early period of oocyte growth, and some success has been achieved with mouse oocytes (Eppig and Schroeder 1989). Methods for maturing oocytes from the growing oocyte pool (see Racowsky, this volume) should also be applicable to fetal oocytes and would allow rapid genetic progress through marker-assisted selection and velogenesis as described by Georges (this volume).

In Vitro Fertilization

The second part of production of embryos in vitro is the sperm capacitation and fertilization system. In general, any agent that causes Ca^{++} entry into the sperm acrosome and a pH increase within the sperm causes capacitation (First and Parrish 1988; Parrish 1990). Numerous capacitation systems have been developed, including high-ionic-strength media, glycosaminoglycans (such as heparin and fucose sulfate), aging, pH shift, calcium ionophores, caffeine, and oviduct fluid (First and Parrish 1987, 1988; Parrish et al. 1989). With appropriate sperm capacitation and incubation in serum-free medium at body temperature, in vitro fertilization has been successful, with fertilization rates as high as 70–80% in cattle (Saeki et al. 1991), sheep (Fukui et al. 1988), swine (Nagai et al. 1988; Kikuchi et al. 1991), and goats (Jiang et al. 1991b).

Embryo Culture

Embryos of domestic animals can develop in surrogate oviducts of rabbits (Boland 1984; Sirard et al. 1985; Lambert et al. 1986) or sheep (Willadsen 1979; Eyestone et al. 1987). Whittingham and co-workers (Muggleton-Harris et al. 1982) showed that cytoplasm from a strain of mice that did not undergo a two-cell block of development in culture at the zygote stage would rescue a blocking strain, which suggested that the two-cell block was due to a cytoplasmic event or interaction of the culture media with the cytoplasm. It was later shown by Chatot et al. (1989) that the early mouse embryo could not metabolize glucose and that deletion of glucose from the culture media allowed embryo development through the blocked period. Glucose deletion, deletion of phosphate, addition of glutamine, and reduction of oxygen in the culture gas have all been shown to be beneficial to the progression of embryo development through the blocked period in mice, hamsters, and most domestic animals.

This information has resulted in optimization of media ingredients for mouse embryo culture (Lawitts and Biggers 1991) and development of defined media for culture of mouse (Chatot et al. 1989; Lawitts and Biggers 1991), hamster (Seshagiri and Bavister 1989a,b), and cattle (Rosenkrans and First 1991). It was also shown in mice (Biggers et al. 1962), sheep (Gandolfi and Moor 1987; Rexroad and Powell 1988), and cattle (Eyestone and First 1989) that embryos could be cultured through the period of blocked development by coculture with cells or cell-conditioned media derived from oviduct cells (Eyestone and First 1989) or by coculture with cells from a number of different tissues, including cumulus cells, uterine cells, and trophoblast cells (Camous et al. 1984; Heyman et al. 1987; Goto et al. 1988; Jiang et al. 1991a) and even peritoneal fluid (Duby et al. 1991). These studies illustrate the importance of cellular metabolism (and perhaps growth factors from helper cells) in preimplantation embryo development and indicate a need to understand better the physiological requirements of early embryos. Important beginnings to this understanding are found in this volume in Biggers et al.

Because the period of successful culture is limited to about 3 days for rabbit and 5 days in sheep oviducts, later development may be partially dependent on still undefined uterine factors. Indeed, late development of bovine embryos has been enhanced by culture with uterine cells (Marquant-le Guienne et al. 1989; G.P. Singh and N.L. First, unpubl.) or a uterine cell growth factor, transforming growth factor β (TGF-β) (Marquant-le Guienne et al. 1989), or embryonic cells derived from the trophoblast (Scodras et al. 1991) or placenta. When culture is attempted in vitro, embryos of most species including cattle, sheep, and swine are blocked in development at the stage when the transition from maternal to embryonic control of development occurs (cattle: 8–16 cells, Barnes 1988; Barnes and First 1991; sheep: 8–16 cells, Crosby et al. 1988; swine: 4–8-cell stage, Norberg 1973).

Further progress in defining culture requirements for early mammalian embryos is aided by the simplex optimization procedure (Lawitts and Biggers 1991), in which alternative media formulations are evaluated efficiently. A comparative approach to early mammalian development is likely to be the most effective way of understanding not only culture requirements, but also morphogenetic mechanisms during peri-implantation development, as described by Cruz and Pedersen (this volume).

The ultimate application for embryos produced in vitro is transfer to the uterus of a foster mother. The stages compatible with transfer in domestic species are the morula or blastocyst stages, and the development of the blastocyst and its ability to hatch from the zona pellucida and implant in the uterus are important to the successful development of the embryo. The importance of ion transport mechanisms in this preimplantation development is reviewed by Biggers et al. (this volume).

Pregnancy Recognition and Embryo Survival

In domestic animals, the interaction of the uterus and conceptus is of great importance to pregnancy recognition and embryo survival. During cleavage stages, the presence of an embryo causes an increase in maternal platelet-activating factor (Collier et al. 1990), which can be detected in some species as an early assay for pregnancy (Morton et al. 1983; Stock et al. 1990). In swine, early embryos are made synchronous or asynchronous by factors such as time of ovulation (Pope 1988; Pope et al. 1990) and embryo genotype (Ford et al. 1988), leading to conditions that favor the more advanced embryos in a competitive uterine environment (Pope 1988; Pope et al. 1990). Pregnancy recognition and many of the changes in porcine uterine protein secretion are induced by the developing embryo as it acquires the ability to produce estrogen (Bazer et al.

1986; Geisert et al. 1990). In sheep and cattle, pregnancy recognition is dependent on embryo expression of an αl-interferon (Roberts et al. 1990). Further understanding of the interaction of conceptus and uterus in pregnancy recognition, and of factors such as synchrony influencing the survival of embryos, is expected to lead to systems for prevention of embryo loss.

MULTIPLICATION OF EMBRYOS

The ability to produce multiple copies of an embryo would provide a powerful selection and propagation tool especially useful for traits of low heritability. A large number of genetically identical embryos would provide a means for embryo selection wherein clonal lines descendent from one embryo could be selected by progeny testing for further clonal multiplication. This system approaches that of phenotypic selection and could permit rapid changes in characteristics such as meat or milk production and, when combined with in vitro production of embryos, could provide a way to produce large numbers of high-quality embryos for frozen storage and commercial transfer. Two methods of embryo multiplication—embryo bisection and nuclear transplantation—are discussed here and by Prather and Robl (this volume).

Embryo Bisection

The bisection technique was developed originally by Willadsen (1979) with sheep embryos. The most effective technique uses only one micromanipulator; the embryo is held in place by attachment to an electrostatically charged culture dish while the sectioning knife, a fragment of razor blade, is forced through the embryo as a guillotine (Fig. 1). The cells left on the splitting knife can then be used to sex the embryo (see below; Herr et al. 1990b,d). Bisection reduces the number of cells by half, yet as long as there are portions of the inner cell mass in both fragments, there are sufficient cells for normal embryo development. Embryos of cattle, sheep, swine, horses, rabbits, and mice have been bisected at the morula or blastocyst stage to produce twin offspring, and in a few cases, trisected or quartered.

FIGURE 1 Bisection of a blastocyst-stage bovine embryo held at the bottom of an electrostatically charged culture dish. (Photo courtesy of Dr. Charles Herr, A.B. Technology, PT4 Limited, Australia.)

The genetic gain realized by incorporating embryo bisection into dairy cattle selection procedures has been estimated to be 1.8 times greater (Gearheart et al. 1989) than conventional procedures, and cloning of embryos to four or more identical individuals has been estimated to be four to five times more powerful (Smith 1989). In one large study, the pregnancy rate was 56.5% from transplantation of 515 intact blastocyst-stage embryos and 52.4% from transplantation of 842 half-embryos (Leibo 1988). Another study recently reported that the number of pregnancies produced per bisected embryo was 1.04 (Gray et al. 1991), whereas the usual rate for a single whole embryo is 0.5 to 0.6. When embryos were bisected, 27.7% produced twins, 44.8% resulted in a single pregnancy, and 27.5% failed to produce a pregnancy (Gray et al. 1991).

Pregnancy rates from bisected embryos are not affected by surgical versus nonsurgical transfer, method of bisection, or parity of the recipient (Kippax et al. 1991). Pregnancies have resulted from bisected embryos frozen by techniques that cause little damage and allow cell multiplication in agar blocks before transfer (Lehn-Jensen and Willadsen 1983), but the pregnancy rate is less than normal when embryos are frozen immediately after bisection. In addition, the pregnancy rate is not reduced when fewer than ten cells are removed by biopsy before freezing (Bondioli et al. 1989).

To date, there has been no evidence for increased incidence of birth defects or abnormal offspring from transfer of bisected embryos into recipient cows. The frequency of an additional splitting and spontaneous twinning has, however, been increased by bisection (Seidel 1985). In sheep and cattle, this technique is being combined with sexing of the embryo, wherein two to eight cells are removed from the trophoblast at the time of bisection and sexed before transferring 3–7 hours later (Herr et al. 1990b,d).

Nuclear Transplantation

The second method for producing multiple copies of an embryo is by nuclear transplantation. This procedure is a modification of a procedure originally developed and shown to be successful for the mouse (McGrath and Solter 1983; Barton et al. 1987) and has successfully produced viable embryos and offspring in cattle (Prather et al. 1987; Barnes et al. 1990; Bondioli et al. 1990), sheep (Willadsen 1986; Smith and Wilmut 1989; McLaughlin et al. 1990), rabbits (Stice and Robl 1988), and swine (Prather et al. 1990). The procedure (Fig. 2; described in Prather and Robl, this volume) involves transfer of a blastomere or nucleus from a valuable embryo at a multicellular stage (usually 20–64-cell stage) into an enucleated metaphase II oocyte. The oocyte then develops to a multicellular stage and is used as a donor in a serial recloning.

Nuclear transplantation is being developed in private industry as well as by university research. Thus far, nuclear transplantation in cattle has been successfully performed using low-cost in-vitro-matured oocytes from abattoir-recovered ovaries and with serial nuclear transfers (Barnes et al. 1990). Enucleation of the oocyte has been made more accurate by fluorescent visualization of the chromatin to be removed (Tsunoda et al. 1988; Westhuesin et al. 1990). However, the efficiencies are less than desired, with approximately 20–25% of the nuclear transplantations resulting in transferable embryos and approximately 30% of the embryos transferred into cows resulting in completed pregnancies. Throughout the United States and Canada, several hundred pregnancies have been produced in cattle by this procedure, and recloning has been

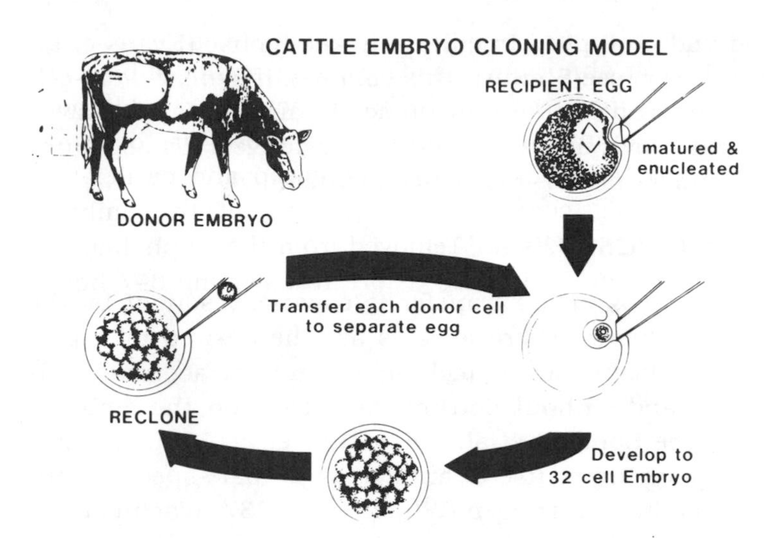

FIGURE 2 Procedures used to multiply bovine embryos by nuclear transfer. Late-stage multicellular embryos of high value are recovered by nonsurgical flush from the uterus of an inseminated cow. The individual cells or blastomeres are removed from the valuable donor embryo and transferred into enucleated oocytes, and the resulting new embryos are developed to morula stage when the process can be repeated (presented originally as a report to W.R. Grace and Company by N.L. First).

performed. To date, the largest number of calves cloned from one embryo has been 11 calves born at Granada Genetics in 1990.

The keys to a successful cloning system for the livestock industry are the ability to produce offspring from donor embryos of large cell number and the capacity for serial nuclear transplantations as the clones develop to advanced cell number. In sheep embryos, the frequency of development to blastocyst after use of donor cells from the blastocyst inner cell mass was 57% and only three pregnancies resulted (Smith and Wilmut 1989). In rabbits, blastocysts have been produced from inner-cell-mass cells, but at a lower frequency than from the 32-cell morula-stage blastomeres in nuclear transplantation (Collas and Robl 1991). In cattle, embryos of morula and early blastocyst stages recovered as late as day 6 have produced good results as donors in cloning (Bondioli et al. 1990). This is very close to a stage where embryonic stem cells can be recovered and multiplied in culture in the mouse (Evans et al. 1990; for review, see Stewart, this volume). If similar stem-cell multiplication can be done in domestic animals and if stem cells should prove useful in cloning by nuclear transplantation, the number of possible clones is unlimited.

SEXING OF EMBRYOS AND SPERM

Sexing of embryos before transfer is especially sought by the dairy cattle industry where females are the desired milk producers. To be useful, sexing techniques must be accurate, efficient, rapid, and without detrimental effects on the embryos (for review, see Bondioli et al. 1989; Herr et al. 1990a). Methods for karyotyping or use of antibodies to male-specific antigens such as the H-Y antigen (White et al. 1987; Wachtel et al. 1988; White 1988) have been used, but they require a large number of cells, are time-consuming assays, and are often inaccurate. When a fluorescent second antibody approach is used, the method is nondamaging to embryos and is approximately 85% accurate (White et al. 1987; Wachtel et al. 1988).

Recently, highly accurate methods for sexing embryos which need only a few cells and provide a quick answer have been developed. These methods involve the amplification of sex-specific sequences from a single blastomere by use of the polymerase chain reaction (PCR) and the subsequent visualization of these sequences on agarose or polyacrylamide gels (Kirzenbaum et al. 1989; Herr et al. 1990a). The cloning and sequencing of both repetitive and unique DNA elements from the Y chromosome have been reported for bovine by Khandekar et al. (1986), Leonard et al. (1987), Ellis et al. (1988), Popescu et al. (1988), Vaiman et al. (1988), Reed et al. (1988), Bondioli et al. (1989), Herr et al. (1989a,b), and Aasen and Medrano (1990) and for sheep and goats by Reed et al. (1988), Aasen and Medrano (1990), and Herr et al. (1990c,d). These sequences have been assayed by ^{32}P-labeled dot-blot hybridization to cells from bovine embryo biopsies (Ellis et al. 1988; Bondioli et al. 1989) or in situ hybridization (Leonard et al. 1987; Vaiman et al. 1988). However, these methods are not practical for field use because of the time required to obtain an answer genetically. PCR amplification is much quicker and provides a more concise answer concerning the sex of cattle or sheep embryos (Herr et al. 1989a,b, 1990a,b,c,d; Kirzenbaum et al. 1989). Even the PCR approach, however, requires delicate laboratory embryonic procedures to isolate embryonic material by blastomere biopsy or embryo bisections. However, when genes for which genetic screening is desirable are identified, sequenced, and cloned, it is expected that the cattle embryo transfer industry will utilize genetic screening for many traits other than sex.

Sexing procedures may be improved if an assay can distinguish between DNAs from the X and Y chromosomes, as published for domestic animals by Aasen and Medrano (1990). There are unpublished reports of similar systems being used with cattle embryos and swine embryos. Whereas the sexing of embryos provides a way to predetermine sex, the commercial production of offspring of a chosen sex would be greatly facilitated if sperm could be sexed and distributed through artificial insemination. X-bearing sperm could be used for dairy cattle production, and specific superior cows might be inseminated with Y-bearing sperm to produce beef or dairy bulls.

Until recently, however, the prospects for sexing sperm have been dim. It has been known for some time that stained DNAs from X and Y sperm could be separated by fluorescence-activated cell sorting but that the sperm did not survive the separation process. Recently, Johnson et al. (1990) reported a modification of this method wherein sperm remain alive. When female rabbits were inseminated with sorted sperm from the X-chromosome-bearing population, 94% of the offspring born were females. After insemination with the Y-chromosome-bearing sperm, 81% of the offspring were male. Because of the limited resolution of the sorting procedure, there was overlap between the base of the X- and Y-chromosome-bearing peaks. If the procedure is modified such that small sperm numbers are used, it should be possible to obtain sperm from only the completely separated peaks of X- and Y-chromosome-sorted populations, thereby increasing the accuracy of sexing to nearly 100%. Commercial use through artificial insemination will be limited until more efficient sorting systems are developed and damaging effects of the fluorescent staining on chromatin can be prevented.

GENE TRANSFER

Methods for transferring genes into embryos by microinjection of DNA into the pronucleus of an egg have been developed for cattle, sheep, swine, rabbits, mice, and rats (for review, see Ebert and Selgrath, this volume). Although large numbers of new strains of mice have been made, in general, the efficiency is relatively low for all species. More efficient and effective methods are being developed and include the infection of genes into embryos by replication-defective viral vectors (a method applied to mice and chickens but not yet to cattle, sheep, or swine) and the introduction of genes into cultured embryonic stem (ES) cells that are then used to form the germ cells of an embryo. The latter has been applied to mice and hamsters (Evans et al. 1990) as discussed by Stewart (this volume), but not yet to ES cells of cattle, sheep, or swine (Evans et al. 1990). Use of the ES cell gene transfer method and selection of cultured cells for homologous recombination between an introduced and native gene sequence allow gene insertion or deletion at specific chromosome sites, which could be used for correction of genetic defects or for adding new genes (see Stewart, this volume).

The time-dependent and tissue-specific expression of a gene is related to promoter-enhancer sequences associated with a given gene. A catalog of promoter-enhancer sequences able to provide gene expression only in specific tissues and at specific times (see Ebert and Selgrath, this volume) is being prepared. For example, muscle-specific promoters such as α-actin allow direction of gene expression specifically in skeletal muscle (Shani 1986, 1991). If the appropriate genes controlling physiology in domestic species were known, this could lead to control over meat quality traits such as the marbling of meat, the growth of the muscle, and perhaps its flavor and tenderness. Thus far, the principal trait targeted for change through use of transgenic technology has been that of growth. Even as reviewed by Ebert and Selgrath (this volume) for cattle, by Pinkert (1991) and Pursel et al. (1990) for swine, and by Rexroad (1990) for sheep, the appropriate combination of promoter and growth gene sequences for controlled growth of muscle mass (free of side effects) has not yet been discovered.

Another commercial application for the transgene technology is the potential for modifying milk both as a food and as a source of pharmaceutical protein (reviewed by Ebert and Selgrath, this volume; see also Wilmut et al. 1990; Henninghausen et al. 1991). Several pharmaceutical companies are developing transgenic animal programs for production of new proteins in milk, and the dairy and food industry has targeted research toward changing milk constituents to make new kinds of cheese, a higher cheese-yielding milk, antibacterial milk, and human milk from farm animals (Wilmut et al. 1990). Experiments targeting promoter sequences specific for mammary gland epithelial cells such as the acidic whey promoter, β-lactoglobulin, α-lactalbumin, or casein (all proteins produced exclusively by the mammary gland) could result in production of new proteins (perhaps pharmaceutical proteins) in milk and allow changing the protein, carbohydrate, or fat composition of milk. To date, recombinant clotting factor IX (Simons et al. 1987) and tissue plasminogen activator (Andres et al. 1987) have been produced in milk.

The exciting aspect of gene transfer is that it provides a way to create new strains of animals that are transgenic for useful genes never found in their species and it facilitates the introduction of allelic genes existing in a breed or strain at a low frequency. The application of this methodology will depend on the identification and understanding of potentially useful genes for transfer into an animal system.

Defining the genes associated with desirable traits such as milk production, disease resistance, reproduction, growth, and carcass quality will provide not only candidate genes for transfer, but also probes or chromosomal RFLPs, which can be used as powerful and accurate animal selection tools. Major efforts are needed in mapping the genome of domestic animals to identify gene sequences or RFLPs linked to production traits. Mapping the genome of domestic animals and the development of chromosomal markers for this purpose are discussed by Georges (this volume). Much of the progress in understanding the genome of domestic animals is coming from the large mapping efforts for the human and mouse genome through development of methodology and through similarities in chromosomal gene organization across these species.

FUTURE PROSPECTS

A likely scenario in animal breeding will be the production of sperm and embryos of precisely tailored genetic ability to resist disease and to produce a high-quality meat or milk product efficiently. Additionally, the sex of the offspring will be preselected and the resulting embryos will be from tested and highly selected clonal lines that are mass-produced to be cost-affordable. With herds of identical cows, breeders will be able to test sexed sperm on representative oocytes to identify favorable specific-combining ability of the sire and dam line and then to breed selected combinations in larger numbers to increase the level of genetically based performance further. The extent to which this scenario is fulfilled may well be determined more by market forces and social policy than by limitations in biotechnology.

REFERENCES

Aasen, E. and J.F. Medrano. 1990. Amplification of the ZFY and ZFX genes for sex identification in humans, cattle, sheep and goats. *Biotechnology* **8:** 1279.

Andres, A.-C., C.A. Schonenberger, B. Groner, L. Henninghausen, M. LeMeur, and P. Gerlinger. 1987. Ha-ras oncogene expression directly by a milk protein gene promoter: Tissue specificity, hormonal regulation, and tumor induction in transgenic mice. *Proc. Natl. Acad. Sci.* **84:** 1299.

Barnes, F.L. 1988. "Characterization of the onset of embryonic control and early development in the bovine embryo." Ph.D. thesis, University of Wisconsin, Madison.

Barnes, F.L. and N.L. First. 1991. Embryonic transcription in in vitro cultured bovine embryos. *Mol. Reprod. Dev.* (in press).

Barnes, F.L., M.E. Westhusin, and C.R. Looney. 1990. Embryo cloning: Principals and progress. In *Proceedings of the 4th World Congress on Genetics Applied to Livestock Production*, vol. XVI, p. 323. Edinburgh, Scotland.

Barton, S.C., M.L. Norris, and M.A.H. Surani. 1987. Nuclear transplantation in fertilized and partheongenetically activated eggs. In *Mammalian development: A practical approach* (ed. M. Monk), p. 235. IRL Press, Oxford, England.

Bazer, F.W., J.L. Vallet, R.M. Roberts, D.C. Sharp, and W.W. Thatcher. 1986. Role of conceptus secretory products in establishment of pregnancy. *J. Reprod. Fertil.* **76:** 841.

Biggers, J.D., R.B.L. Gwatkin, and R.L. Brinster. 1962. Development of mouse embryos in organ culture of fallopian tubes on a chemically defined medium. *Nature* **194:** 747.

Boland, M.P. 1984. Use of rabbit oviduct as a screening tool for the viability of mammalian eggs. *Theriogenology* **21:** 126.

Bondioli, K.R., M.E. Westhusin, and C.R. Looney. 1990. Production of identical bovine offspring by nuclear transfer. *Theriogenology* **33:** 165.

Bondioli, K.R., S.B. Ellis, J.H. Pryor, M.W. Williams, and M.M. Harpold. 1989. The use of male-specific chromosomal DNA fragments to determine the sex of bovine preimplantation embryos. *Theriogenology* **31:** 95.

Camous, S., Y. Heyman, W. Meziou, and Y. Menezo. 1984. Cleavage beyond the block stage and survival after transfer of early bovine embryos cultured with trophoblastic vesicles. *J. Reprod. Fertil.* **72:** 479.

Chatot, C.L., C.A. Ziomek, B.D. Bavister, J.L. Lewis, and R.I. Torress. 1989. An improved culture medium supports development of random-bred 1-cell mouse embryos in vitro. *J. Reprod. Fertil.* **86:** 679.

Cheng, W.T.K., R.M. Moor, and C. Polge. 1986. In vitro fertilization of pig and sheep oocytes matured in vivo and in vitro. *Theriogenology* **25:** 146. (Abstr.)

Collas, P. and J. Robl. 1991. Development of rabbit nuclear transplant embryos from morula and blastocyst stage donor nuclei. *Theriogenology* **35:** 190.

Collier, M., C.O. Neill, A.J. Ammit, and D.M. Saunders. 1990. Measurement of human embryo derived platelet activating factor (PAF) using a quantitative bioassay of platelet aggregation. *Hum. Reprod.* **5:** 323.

Crosby, I.M., F. Gandolfi, and R.M. Moor. 1988. Control of protein synthesis during early cleavage of sheep embryos. *J. Reprod. Fertil.* **82:** 769.

Crozet, N., D. Huneau, V. Desmedt, N.C. Theron, D. Szollosi, S.S. Torries, and C. Sevellec. 1987. In vitro fertilization with normal development in the sheep. *Gamete Res.* **16:** 159.

Duby, R.T., P. Collas, and J.M. Robl. 1991. Development of rabbit embryos to the hatched blastocyst stage in rabbit peritoneal fluid in vitro. *Theriogenology* **35:** 197. (Abstr.)

Ellis, S.B., K.R. Bondioli, M.E. Williams, J.H. Pryor, and M.M. Harpold. 1988. Sex determination of bovine embryos using male-specific DNA probes. *Theriogenology* **29:** 242a. (Abstr.)

Eppig, J.J. and A.E. Schroeder. 1989. Capacity of mouse oocytes from preantral follicles to undergo embryogenesis and development to live young after growth, maturation and fertilization in vitro. *Biol. Reprod.* **41:** 268.

Evans, M.J., S.L. Notarianni, and R.M. Moor. 1990. Derivation and preliminary characterization of pluripotent cell lines from porcine and bovine blastocysts. *Theriogenology* **33:** 125.

Eyestone, W.H. and N.L. First. 1989. Co-culture of early bovine embryos to the blastocyst stage with oviductal tissue or in conditioned medium. *J. Reprod. Fertil.* **85:** 715.

Eyestone, W.H., J. Vignieri, and N.L. First. 1987. Co-culture of early bovine embryos with oviductal epithelium. *Theriogenology* **27:** 288. (Abstr.)

First, N.L. and J.J. Parrish. 1987. In vitro fertilization of ruminants. Second International Symposium on reproduction in domestic ruminants. *J. Reprod. Fertil.* **34:** 151.

———. 1988. Sperm maturation and in vitro fertilization. In *The 11th International Congress on Animal Reproduction and Artificial Insemination*, vol. I. Dublin, Ireland.

Ford, S.P., N.K. Schwartz, M.F. Rothschild, A.J. Conley, and C.M. Warner. 1988. Influence of SLA haplotype on preimplantation embryonic cell number in miniature pigs. *J. Reprod. Fertil.* **84:** 99.

Fukui, Y., A.M. Glew, F. Gandolfi, and R.M. Moor. 1988. Ram specific effects on in-vitro fertilization and cleavage of sheep oocytes matured in vitro. *J. Reprod. Fertil.* **82:** 337.

Gandolfi, F. and R.M. Moor. 1987. Stimulation of early development in the sheep by co-culture with oviductal epithelial cells. *J. Reprod. Fertil.* **81:** 21.

Geisert, R.D., M.T. Zavy, R.J. Moffatt, R.M. Blair, and T. Yellin. 1990. Embryonic steroids and the establishment of pregnancy in pigs. *J. Reprod. Fertil.* **40:** 293.

Gearheart, W.W., C. Smith, and G. Teepher. 1989. Multiple ovulation and embryo manipulation in the improvement of beef cattle: Relative theoretical rates of genetic change. *J. Anim. Sci.* **67:** 2863.

Gordon, I. and K.H. Lu. 1990. Production of embryos in vitro and its impact on livestock production. *Theriogenology* **33:** 77.

Goto, K., Y. Kajihara, S. Kosalka, M. Koba, Y. Nakanishi, and K. Ogawa. 1988. Pregnancies after co-culture of cumulus cells with bovine embryos derived from in-vitro fertilization of in vitro matured follicular oocytes. *J. Reprod. Fertil.* **83:** 753.

Gray, K.R., K.R. Bondioli, and C.L. Bitts. 1991. The commercial application of embryo splitting in beef cattle. *Theriogenology* **35:** 37.

Henninghausen, L., C. Westphal, L. Sankaran, and C.W. Pittius. 1991. Regulation of expression of genes for milk proteins. In *Transgenic animals* (ed. N.L. First and F.P. Haseltine), p. 65. Butterworth-Heinemann, Boston.

Herr, C., K.I. Matthaei, and K.C. Reed. 1989a. Accuracy of a rapid Y-chromosome-detecting bovine embryo sexing assay. In *Australia and New Zealand Society for Cell Biology, 8th Annual Meeting*, p. 50. (Abstr.) University of Melbourne.

———. 1990a. Rapid, accurate sexing of livestock embryos. In *Proceedings of the 4th World Congress on Genetics Applied to Livestock Production*, vol. XVI, p. 323.

Herr, C., N.A. Holt, H.I. Matthaei, and K.C. Reed. 1990b. Sex of progeny that developed from bovine embryos sexed with a rapid Y-chromosome-detection assay. *Theriogenology* **33:** 247. (Abstr.)

Herr, C., K.I. Matthaei, N. Holt, and K. Reed. 1989b. Field implementation of a rapid Y-chromosome-detecting bovine embryos sexing assay. In *Australia and New Zealand Society for Cell Biology, 8th Annual Meeting*, p. 51. (Abstr.) University of Melbourne.

Herr, C., K.I. Matthaei, U. Petrzak, and K.C. Reed. 1990c. A rapid Y-chromosome-detecting ovine embryo sexing assay. *Theriogenology* **33:** 245. (Abstr.)

Herr, C., N. Holt, U. Petrzak, K. Old, and K. Reed. 1990d. Increased number of pregnancies per collected embryo by bisection of blastocyst stage ovine embryos. *Theriogenology* **33:** 244. (Abstr.)

Heyman, Y., Y. Menezo, P. Chesne, S. Camous, and V. Garnier. 1987. In vitro cleavage of bovine and ovine early embryos: Improved development using co-culture with trophoblastic vesicles. *Theriogenology* **27:** 57.

Jiang, H.S., W.L. Wang, K.H. Lu, I. Gordon, and C. Polge. 1991a. Roles of different cell monolayers in the co-culture of bovine embryos. *Theriogenology* **35:** 216. (Abstr.)

Jiang, S., X. Yang, S. Chang, W. Heuwieser, and R.H. Foote. 1991b. Effect of sperm capacitation and oocyte maturation procedures on fertilization and development of bovine oocytes in vitro. *Theriogenology* **35:** 218. (Abstr.)

Johnson, L.A., J.P. Flook, and H.W. Hawk. 1990. Sex preselection in rabbits: Live births from X and Y sperm separated by DNA and cell sorting. *Biol. Reprod.* **41:** 199.

Khandekar, P., G.P. Talwar, and R. Chaudhury. 1986. Cloning of Y derived DNA sequences of bovine origin in *Escherichia coli*. *J. Biosci.* **10:** 481.

Kikuchi, K., T. Nagai, and J. Motlik. 1991. Effect of follicle cells on in-vitro fertilization of pig follicular oocytes. *Theriogenology* **35:** 225. (Abstr.)

Kippax, I.S., W.B. Christie, and T.G. Rowan. 1991. Effects of method of splitting, stage of development and presence or absence of zona pellucida on foetal survival in commercial bovine embryo transfer of bisected embryos. *Theriogenology* **35:** 25.

Kirzenbaum, M., M. Vaiman, C. Cotinot, and M. Fellous. 1989. PCR sexing bovine embryos using Y chromosome specific sequences. *J. Cell Biochem.* (suppl.) **13E:** 293.

Lambert, R.D., M.A. Sirard, C. Bernard, R. Beland, J.E. Rioux, P. Leclerc, D.P. Menard, and M. Bedoya. 1986. In vitro fertilization of bovine oocytes matured in vivo and collected at laparoscopy. *Theriogenology* **25:** 117.

Lawitts, J.A. and J.D. Biggers. 1991. Optimization of mouse embryo culture using simplex methods. *J. Reprod. Fertil.* **91:** 543.

Lehn-Jensen, H. and S.M. Willadsen. 1983 Deep-freezing of cow "half" and "quarter" embryos. *Theriogenology* **19:** 49.

Leibfried-Rutledge, M.L., E.S. Critser, J.J. Parrish, and N.L. First. 1989. In vitro maturation and fertilization of bovine oocytes. *Theriogenology* **31:** 61.

Leibfried-Rutledge, M.L., E.S. Critser, W.H. Eyestone, D.L. Northey, and N.L. First. 1987. Development potential of bovine oocytes matured in vitro or in vivo. *Biol. Reprod.* **36:** 376.

Leibo, S.P. 1988. Bisection of mammalian embryos by micromanipulation. In *The American Fertility Society Regional Postgraduate Course. Hands-on IVF, Cryopreservation and Micromanipulation, April 25–29*. Madison, Wisconsin.

Leonard, M., M. Kirszenbaum, C. Cotinot, P. Chesne, Y. Heyman, M.G. Stinnakre, C. Bishop, C. Delouis, M. Vaiman, and M. Fellous. 1987. Sexing bovine embryos using chromosome specific DNA probes. *Theriogenology* **27:** 248. (Abstr.)

Marquant-le Guienne, B., M. Gerard, A. Solari, and C. Thibault. 1989. In vitro culture of bovine eggs fertilized either in vivo or in vitro. *Reprod. Nutr. Dev.* **29:** 559.

McGrath, J. and D. Solter. 1983. Nuclear transplantation in the mouse embryo by microsurgery and cell fusion. *Science* **220:** 1300.

McLaughlin, K.J., L. Davies, and R.F. Seamark. 1990. In vitro embryo culture in the production of identical Merino lambs by nuclear transplantation. *Reprod. Fertil. Dev.* **2:** 619.

Morton, H., D.J. Morton, and F. Ellendorff. 1983. The appearance and characteristics of early pregnancy factor in the pig. *J. Reprod. Fertil.* **69:** 437.

Muggleton-Harris, A., D.G. Whittingham, and L. Wilson. 1982. Cytoplasmic control of preimplantation development in vitro in the mouse. *Nature* **299:** 460.

Nagai, T., T. Takahashi, H. Masuda, Y. Shioya, M. Kuwayama, M. Fukushima, S. Iwasaki, and A. Hanada. 1988. In-vitro fertilization of pig oocytes by frozen boar spermatozoa. *J. Reprod. Fertil.* **84:** 585.

Norberg, H.S. 1973. Ultrastructural aspects of the preattached pig embryo cleavage and early blastocyst stages. *J. Anat. Entwickl-Gerch* **143:** 95.

Parrish, J.J. 1990. Application of in vitro fertilization to domestic animals. In *The biology and chemistry of mammalian fertilization* (ed. P.M. Wasserman), vol. II, p. 111. CRC Press, New York.

Parrish, J.J., J.L. Susko-Parrish, R.R. Handrow, M.M. Sims, and N.L. First. 1989. Capacitation of bovine spermatozoa by oviduct fluid. *Biol. Reprod.* **40:** 1020.

Pinkert, C.A., D.L. Kooyman, and T.J. Dyer. 1991. Enhanced growth performance in transgenic swine. In *Transgenic animals* (ed. N.L. First and F.P. Haseltine), p. 251. Butterworth-Heinemann, Boston.

Pope, W.F. 1988. Uterine asynchrony, a cause of embryonic loss. *Biol.*

Reprod. **39:** 999.

Pope, W.F., S. Xie, D.M. Broermann, and K.P. Nephew. 1990. Causes and consequences of early embryonic diversity in pigs. *J. Reprod. Fertil.* (suppl.) **40:** 251.

Popescu, C.P., C. Cotinot, J. Boscher, and M. Kirszenbaum. 1988. Chromosomal localization of a bovine male specific probe. *Ann. Genet.* **31:** 39.

Prather, R.S., M.M. Sims, and N.L. First. 1990. Nuclear transplantation in early pig embryos. *Biol. Reprod.* **41:** 414.

Prather, R.S., F.L. Barnes, M.L. Sims, J.M. Robl, W.H. Eyestone, and N.L. First. 1987. Nuclear transplantation in the bovine: Assessment of donor nuclei and recipient oocyte. *Biol. Reprod.* **37:** 859.

Pursel, V.G., R.E. Hammer, D.J. Bolt, R.D. Palmiter, and R.L. Brinster. 1990. Integration, expression, and germline transmission of growth-related genes in pigs. *J. Reprod. Fertil.* (suppl.) **41:** 77.

Reed, K.C., M.E. Matthews, and M.A. Jones. 1988. Sex determination in ruminants using Y-chromosome specific polynucleotides. Published Patent Application. Patent Cooperative Treaty No. WO88/ 01300.

Rexroad, C. 1990. Transgenic ruminants. In *Proceedings OTA Workshop on Biotechnology, May 1990.* Washington, D.C.

Rexroad, C.E., Jr. and A.M. Powell. 1988. Coculture of ovine ova with oviductal cells in medium 199. *J. Anim. Sci.* **66:** 947.

Roberts, R.M., P.V. Malathy, T.R. Hansen, C.E. Farin, and K. Imakawa. 1990. Bovine conceptus products involved in pregnancy recognition. *J. Anim. Sci.* (suppl.) **2:** 28.

Rosenkrans, C.F. and N.L. First. 1991. Culture of bovine zygotes to the blastocyst stage: Effects of amino acids and vitamins. *Theriogenology* **35:** 266. (Abstr.)

Saeki, K., M. Hoshi, M.L. Leibfried-Rutledge, and N.L. First. 1991. In vitro fertilization and development of bovine oocytes matured in serum-free medium. *Biol. Reprod.* **44:** 256.

Scodras, J.M., J.W. Pollard, and K.J. Betteridge. 1991. Effect of somatic cell type on bovine embryonic development in coculture. *Theriogenology* **35:** 269. (Abstr.).

Seidel, G.E., Jr. 1985. Are identical twins produced from micromanipulation always identical? In *Proceedings of the Annual Conference on Artificial Insemination and Embryo Transfer in Beef Cattle (Denver)*, p. 50. National Association of Animal Breeders, Columbia, Missouri.

Seshagiri, P.B. and B.D. Bavister. 1989a. Glucose inhibits development of hamster 8 cell embryos in vitro. *Biol. Reprod.* **40:** 599.

———. 1989b. Phosphate is required for inhibition by glucose of development of hamster 8 cell embryos in vitro. *Biol. Reprod.* **40:** 607.

Shani, M. 1986. Tissue specific and developmentally regulated expression of a chimeric actin/globin gene in transgenic mice. *Mol. Cell Biol.* **6:** 2624.

———. 1991. Application of germline transformation to the study of myogenesis. In *Transgenic animals* (ed. N.L. First and F.P. Haseltine), p. 55. Butterworth-Heinemann, Boston.

Simons, J.P., M. McClenaghan, and A.J. Clark. 1987. Alteration of the quality of milk by expression of sheep beta-lactoglobulin in transgenic mice. *Science* **328:** 530.

Sirard, M.A., R.D. Lambert, D.P. Menard, and M. Bedoya. 1985. Pregnancies after in vitro fertilization of cow follicular oocytes, their incubation in rabbit oviducts and their transfer to the cow uterus. *J. Reprod. Fertil.* **75:** 551.

Smith, C. 1989. Cloning and genetic improvement of beef cattle. *Anim. Prod.* **49:** 49.

Smith, L.C. and T. Wilmut. 1989. Influence of nuclear and cytoplasmic activity on the development in vivo of sheep embryos after nuclear transplantation. *Biol. Reprod.* **40:** 1027.

Stice, S.L. and J.M. Robl. 1988. Nuclear reprogramming in nuclear transplant rabbit embryos. *Biol. Reprod.* **39:** 657.

Stock, A.E., R. Axtell, H. Jones, G. Jones, and W. Hansel. 1990. Evaluation of human and bovine embryos by measurements of PAF in the culture medium. *Theriogenology* **33:** 332. (Abstr.)

Tsunoda, Y., Y. Shioda, M. Onodera, K. Nakamura, and T. Uchida. 1988. Differential sensitivity of mouse pronuclei and zygote cytoplasm to Hoechst staining and ultraviolet irradiation. *J. Reprod. Fertil.* **82:** 173.

Vaiman, M., C. Cotinot, M. Kirszenbaum, M. Leonard, P. Chesne, Y. Heyman, M.-G. Stinnakre, C. Bishop, and M. Fellous. 1988. Sexing of bovine embryo using male-specific nuclei acid probes. In *Proceedings of the World Sheep and Cattle Breeding Congress*, p. 93.

von der Schams, A., L.A.J. van der Westerlaken, A.A.C. deWit, W.H. Eyestone, and H.A. de Boer. 1991. Ultrasound-guided transvaginal collection of oocytes in the cow. *Theriogenology* **35:** 288.

Wachtel, S.S., D. Nakamura, G. Wachtel, W. Fenton, M.G. Kent, and V. Jaswaney. 1988. Sex selection with monoclonal H-Y antibody. *Fertil. Steril.* **50:** 355.

Westhusin, M.E., M.J. Levanduski, R. Scarborough, C.R. Looney, and K.R. Bondioli. 1990. Utilization of fluorescent staining to identify enucleated demi-oocytes for utilization in bovine nuclear transfer. *Biol. Reprod.* (suppl. 1) **41:** 407. (Abstr.)

White, K.L. 1988. Identification of embryonic sex by immunological methods. In *The American Fertility Society Workshop, Hand-on IVF, Cryopreservation and Micromanipulation, April 25–29*. Madison, Wisconsin.

White, K.L., G.B. Anderson, and R.H. Bon Durant. 1987. Expression of a male-specific factor on various stages of preimplantation bovine embryos. *Biol. Reprod.* **37:** 867.

Willadsen, S.M. 1979. A method for culture of micromanipulated sheep embryos and its use to produce monozygotic twins. *Nature* **227:** 298.

———. 1986. Nuclear transplantation in sheep embryos. *Nature* **320:** 63.

Wilmut, I., A.L. Archibald, S. Harris, M. McClenaghan, J.P. Simons, C.B.A. Whitelaw, and A.J. Clark. 1990. Modification of milk composition. *J. Reprod. Fertil.* (suppl.) **41:** 135.

Gamete Resources: Origin and Production of Oocytes

C. Racowsky
Department of Obstetrics and Gynecology, University of Arizona
Tucson, Arizona 85724

I. INTRODUCTION

Classically, the major purpose underlying experimental manipulation of mammalian oocytes has been to understand the physiological processes that control their origin, meiotic status, and developmental competency. Today, oocyte experimentation has additional applications both in the clinical field of infertility and in the agricultural industry for enhancement of the reproductive capacity of genetically desirable stock. Progress with these applications, however, has been only moderate compared with that in other areas of biotechnology. The reasons for this relate not only to a relative paucity of material for experimentation, but also to our incomplete knowledge regarding the factors that regulate the growth, meiotic competency, and fertilizability of mammalian oocytes.

In this chapter, a comparative approach will be taken to review our current understanding of these processes, particularly as such processes impact upon the production of developmentally competent zygotes in vitro. Attention will be focused on research with laboratory and domestic species, with limited reference being made to human reproduction. Extensive reviews have recently been published on several aspects of the physiology of human oocytes (Frydman et al. 1988; Janson et al. 1988; Mashiach et al. 1988).

II. PHYSIOLOGICAL PRODUCTION OF DEVELOPMENTALLY COMPETENT OOCYTES

A. Origin of Female Gametes

Mammalian germ cells derive from primordial cells which have an embryonic origin and that in mice, at least, arise from the

 0-87969-333-9/91 $3.00 + 00

epiblast (Gardner and Rossant 1979). Initially by passive movement, and subsequently by active locomotion, the primordial germ cells reach the undifferentiated gonadal ridge from which develops the fetal ovary (for review, see McLaren 1988). During migration and for some time after they have reached the genital ridge, the germ cells proliferate by many mitotic divisions to give rise to several thousand gonia within the primitive gonad (Mintz and Russell 1957). In contrast to the male germ line, however, that of the female is finite by virtue of the cessation of mitosis, and the entrance into meiosis, of all of the oogonia (for review, see Byskov 1987). The natural reproductive capacity of the female mammal is further limited as a consequence of atresia, in which a large number of the germ cells degenerate throughout oogonial and oocyte development (Baker 1972).

The meiotic process in female mammals is believed to be similar in all species (Byskov 1987), although the timing and duration of the various stages differ from one species to another (Peters 1970). In most species, the oogonia enter meiosis prenatally so that by the time of birth, or shortly thereafter, most oocytes have reached the diplotene (dictyate) stage of meiotic prophase (Borum 1961; Baker 1979). However, in some mammals such as the hamster, neonate ovaries contain oogonia, the meiotic process being initiated during the first few weeks postpartum (for review, see McLaren 1988).

Regardless of the timing of the onset of oogenesis, oocytes of all mammalian species become arrested in their meiotic progression at the dictyate, germinal vesicle (GV) stage of development (Fig. 1A). This extended period of arrest, the duration of which is species-dependent, is terminated sometime after the animal has reached sexual maturity when a mature antral follicle is exposed to preovulatory gonadotropin (Ayalon et al. 1972). Following such follicular gonadotropic stimulation, the oocyte becomes irreversibly committed to resume meiosis and subsequently undergoes breakdown of the GV (GVBD), which is followed by overt changes in chromatin configuration. These changes involve condensation of the chromatin at diakinesis (Fig. 1B), progression through metaphase I (Fig. 1C), completion of the first meiotic division (Fig. 1D), and transformation of the oocyte to a mature oocyte

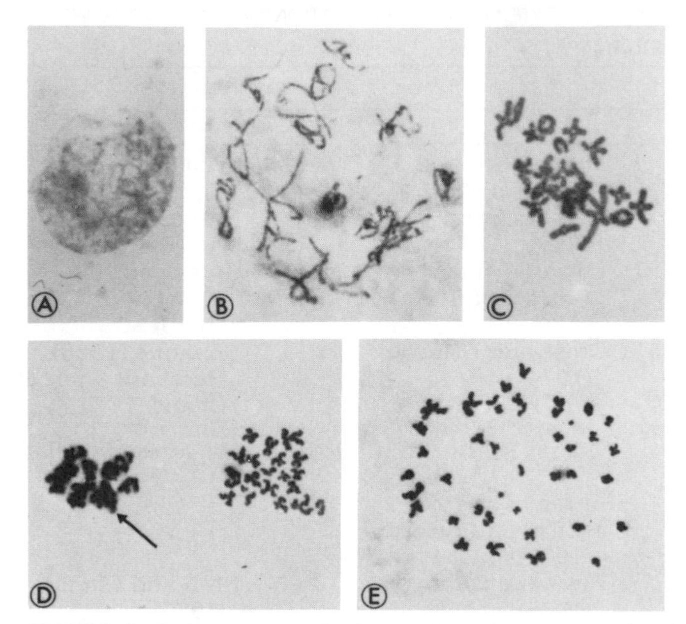

FIGURE 1 Progression of chromatin changes in the hamster oocyte during meiotic maturation. (*A*) Fibrous configuration of a germinal vesicle in the meiotically arrested oocyte; magnification, 933x. (*B*) Condensed chromatin of a meiotically maturing oocyte at early diakinesis; magnification, 1066x. (*C*) Bivalents present at metaphase I; magnification, 1024x. (*D*) Telophase I, following segregation of homologous chromosomes at anaphase I. Polar body chromosomes are undergoing degeneration (arrow); magnification, 1226x. (*E*) Typical chromosome configuration of an oocyte in which homolog segregation was perturbed at metaphase I, giving rise to a diploid metaphase II oocyte; magnification, 960x.

at metaphase of the second division. These events, inclusively, are referred to as oocyte meiotic maturation (for review, see Donahue 1972). The duration for completion of oocyte meiotic maturation for any one species varies considerably, although it falls within a species-specific range (Table 1). In addition to the nuclear changes that accompany meiotic maturation, cytoplasmic maturation renders the oocyte competent to undergo normal fertilization and embryonic development (Chang 1955a,b; for review, see Thibault et al. 1987).

In any natural reproductive cycle, the average number of such developmentally competent, mature ovarian oocytes is

TABLE 1 *DURATION OF OOCYTE MEIOTIC MATURATION IN VIVO AND IN VITRO IN VARIOUS MAMMALIAN SPECIES*

Species	Duration of meiotic maturation (hr)			
	in vivo	study	in vitro	study
Mouse	9	McGaughey and Chang (1969)	8	Zeilmaker 1978
	12	Merchant and Chang (1971)	12	Neal and Baker (1973)
	12.5–13	Edwards and Gates (1959)	12–13	Cross and Brinster (1970)
	17–18	Cross and Brinster (1974)	17	Donahue (1968); Merchant and Chang (1971)
Rat	7–8	Odor (1955)	7–10	Magnusson et al. (1977)
	9	Ahren et al. (1978)		
	9	Tsafriri and Kraicer (1972)		
	13	Niwa and Chang (1975)	13	Niwa and Chang (1975)
Hamster	11	Yanagimachi and Chang (1961); Ward (1948)	12	Racowsky et al. (1989a); Edwards (1962)
Rabbit	8	Pincus and Enzmann (1935)		
	9–9.5	Thibault (1972)	9–9.5	Thibault (1972)
	10	Chang (1955a,b)	12	Van Blerkom and McGaughey (1978)
Sheep	18–24	Hay and Moor (1973)	17–25	Quirke and Gurdun (1971)
	21–24	Dziuk (1965)		
Pig	36–37	Hunter and Polge (1966)	32–42	Motlik (1972)
			43–44	McGaughey and Polge (1971)
Cow	19	Hyttel et al. (1986)	24–25	Younis et al. (1989)
	27–29	Callesen et al. (1986)	25–28	Xu et al. (1988)
			48	Hensleigh and Hunter (1985)
			48	Homa (1988)

Time zero corresponds to the time of hCG injection in vivo and to the time of follicular release of the oocytes in vitro. (Expanded, with permission, from Skoblina 1988.)

TABLE 2 *NUMBER OF NATURAL OVULATIONS IN EACH CYCLE IN VARIOUS MAMMALIAN SPECIES*

Species	Number of ovulations	Study
Mouse	9	Edwards and Fowler (1960)
	9	McLaren (1967)
Rat	10	Blandau (1952)
	12	Nuti et al. (1975)
	14	Yoshinaga et al. (1979)
Hamster	10	Greenwald (1962)
	10–14	Moore and Greenwald (1974)
Rabbit	10	Kennelly and Foote (1965)
	11	Boving (1956)
Sheep	1–2	Hulet et al. (1969)
	1–2	Dermody et al. (1970)
Pig	13	Bazer et al. (1969)
	13	Dzuik (1968)
	11–19	Longenecker and Day (1968)
Cow	1–2	reviewed by Dzuik (1973)

species-specific (Table 2) and small compared with the plethora of sperm in a typical semen sample. Thus, to obtain the optimal in vitro production of developmentally competent oocytes, considerable efforts have been focused on identifying factors within the follicular environment, and within the oocyte itself, that are responsible for acquisition of nuclear meiotic competency, on the one hand, and for cytoplasmic maturation and full developmental competency, on the other.

B. Acquisition of Nuclear Meiotic Competency

The postnatal ovary is a dynamic structure in which vesicular follicles are continually growing from those of the primordial type (for review, see Baker 1987). With the onset of sexual maturity, growth of a cohort of vesicular follicles proceeds in each reproductive cycle, under the influence of a precise hormonal milieu, to give rise to a group of fully expanded, mature antral follicles. Each antral follicle is composed of several layers of theca cells, separated from the underlying membrana

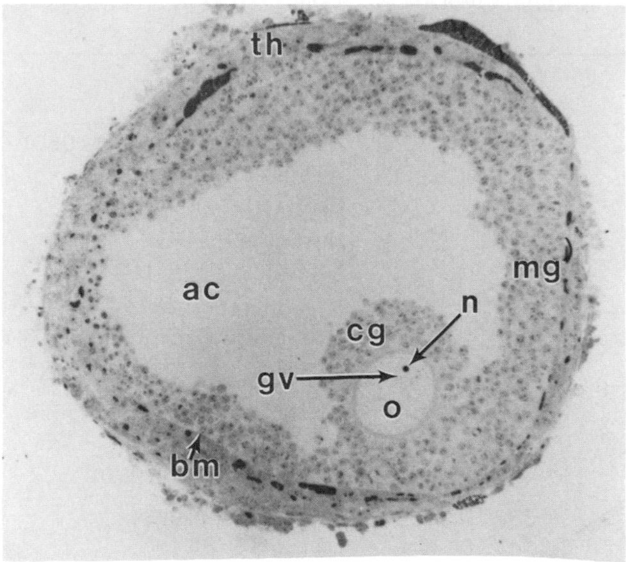

FIGURE 2 Cross-section of a mature hamster follicle dissected from an ovary prior to the endogenous surge of LH. The outside thecal layers (th) are separated from the underlying membrana granulosa cell layers (mg) by a vascularized basement membrane (bm). The innermost layer of membrana granulosa cells bounders the fluid-filled antral cavity (ac) in which is suspended the oocyte (o), surrounded by several layers of cumulus granulosa cells (cg). The oocyte is at the germinal vesicle (gv) stage, and a nucleolus (n) is visible. Magnification, 100x.

granulosa cells by a vascularized basement membrane (Fig. 2). The innermost layer of membrana granulosa cells encloses the antrum, a fluid-filled cavity in which the oocyte-cumulus complex is suspended while remaining attached to the membrana granulosa cells by a cumulus cell stalk. The oocyte-cumulus complex is composed of an oocyte encompassed by a thin glycoprotein coat, the zona pellucida, which is surrounded by several layers of cumulus cells (Fig. 3A). Those cells in the innermost layer extend numerous processes into the zona pellucida (Fig. 3B), some of which traverse the zona to interact with the planar surface of the oolemma among the microvilli. Prior to the gonadotropic surge, gap junctions are usually present at these points of contact (Fig. 4A,B). In addition to

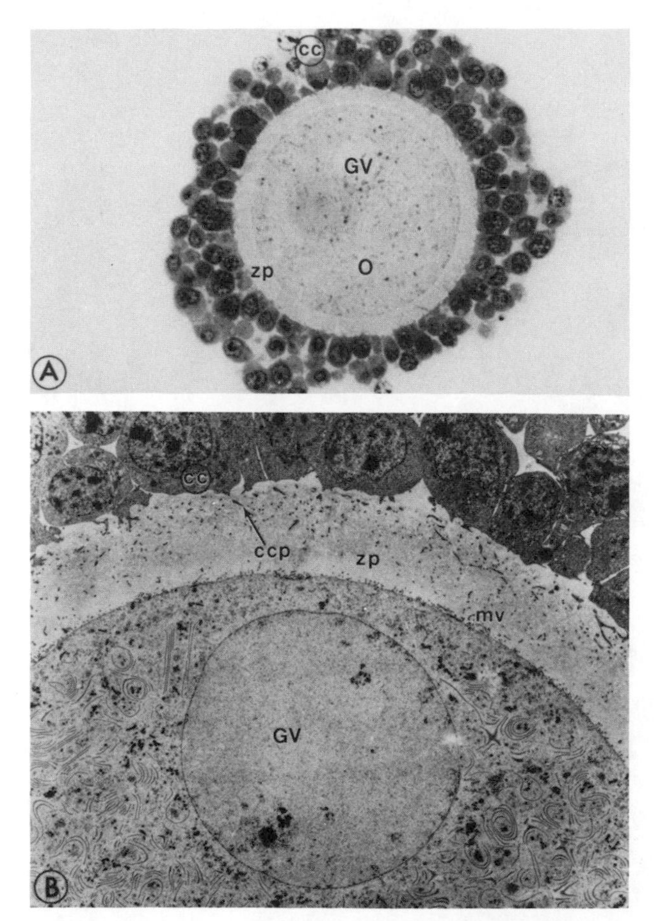

FIGURE 3 Transmission electron micrographs of a hamster oocyte-cumulus complex collected prior to the LH surge. (*A*) The germinal vesicle (GV) of this oocyte is clearly visible and the cumulus cell (cc) layers completely surround the oocyte. The oocyte is encompassed by a thin, glycoprotein envelope, the zona pellucida (zp); magnification, 215x. (*B*) Innermost layer of cumulus cells (cc) extends numerous processes (ccp) into the zona pellucida (zp). The surface of the oocyte exhibits large numbers of much smaller protrusions, the microvilli (mv). Magnification, 1553x.

these heterologous gap junctions, gap junctions also exist among the cumulus (Fig. 5A) and membrana (Fig. 5B) granulosa cells.

Within the mammalian follicle there is an extensive syncyti-

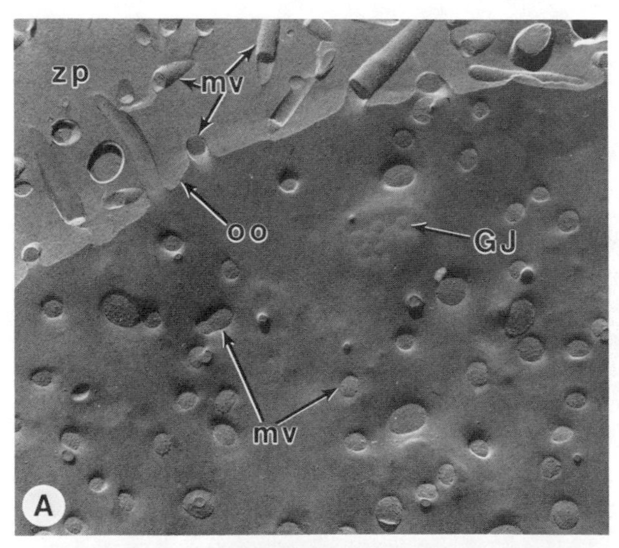

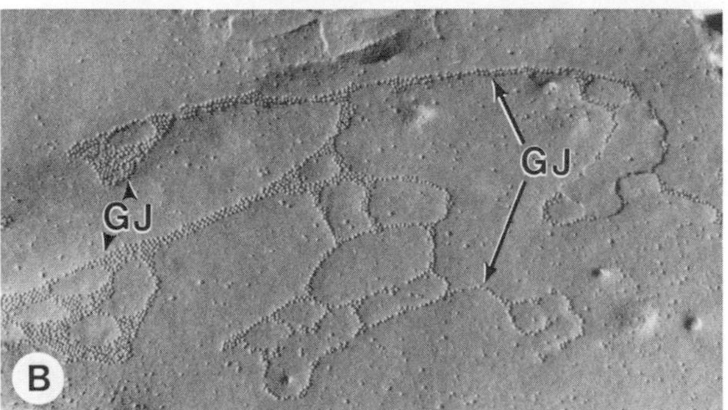

FIGURE 4 Freeze-fracture replicas of hamster oocyte–cumulus complexes immediately after follicular release prior to the ovulatory gonadotropin surge. (*A*) Planar surface of a hamster oocyte revealing gap junction (GJ) and sheared off microvilli (mv) on the surface of the oocyte, with microvilli extending from the oolemma (oo) into the zona pellucida (zp). Magnification, 15,675x. (*B*) Oocyte-cumulus gap junctions (GJ) organized as particle aggregates in linear arrays (arrows) and in circular arrays (arrowheads). Magnification, 37,908x. (Reprinted, with permission of John Wiley & Sons, Inc., from McGaughey, R.W., C. Racowsky, V. Rider, K.V. Baldwin, A.A. DeMarais, and S.W. Webster. 1990. Ultrastructural correlates of meiotic maturation in mammalian oocytes. *J. Electron Microsc. Tech.* **16:** 257.)

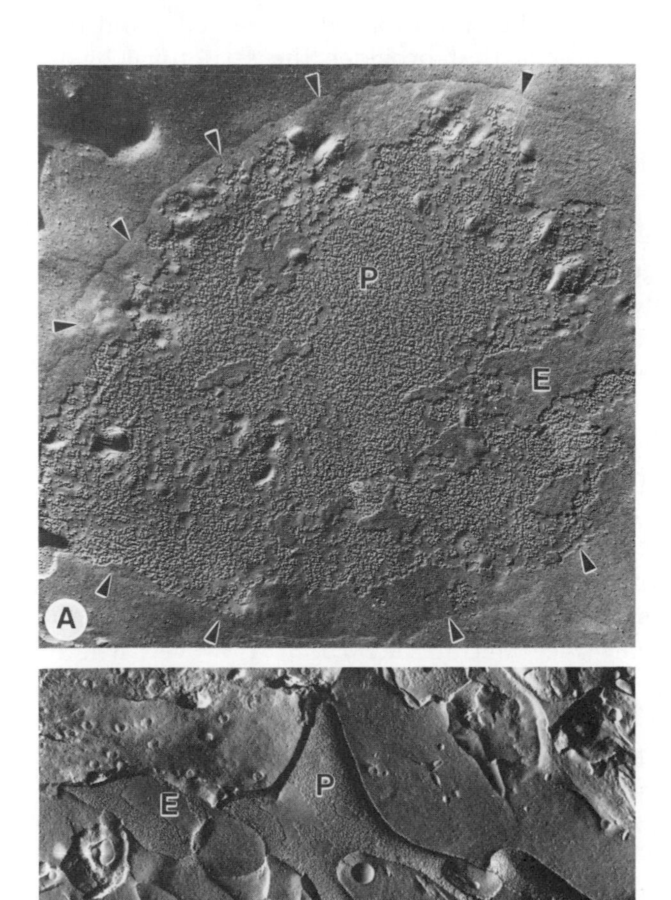

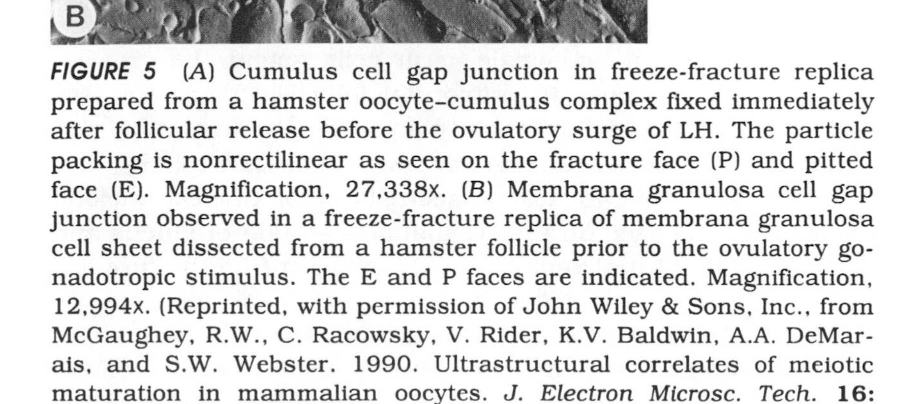

FIGURE 5 (*A*) Cumulus cell gap junction in freeze-fracture replica prepared from a hamster oocyte–cumulus complex fixed immediately after follicular release before the ovulatory surge of LH. The particle packing is nonrectilinear as seen on the fracture face (P) and pitted face (E). Magnification, 27,338x. (*B*) Membrana granulosa cell gap junction observed in a freeze-fracture replica of membrana granulosa cell sheet dissected from a hamster follicle prior to the ovulatory gonadotropic stimulus. The E and P faces are indicated. Magnification, 12,994x. (Reprinted, with permission of John Wiley & Sons, Inc., from McGaughey, R.W., C. Racowsky, V. Rider, K.V. Baldwin, A.A. DeMarais, and S.W. Webster. 1990. Ultrastructural correlates of meiotic maturation in mammalian oocytes. *J. Electron Microsc. Tech.* **16:** 257.)

um involving the oocyte, the cumulus granulosa cells, and the membrana granulosa cells that provides the structural basis for transfer of nutritional and regulatory molecules from the somatic cells into the oocyte (for review, see Larsen and Wert 1988). As follicular growth proceeds, the oocyte undergoes a concomitant increase in size that is dependent on this functional follicular gap junctional network. The absolute dependency of oocyte growth on such oocyte-granulosa coupling has been demonstrated in vitro (Eppig 1977, 1979; Bacharova et al. 1980; Herlands and Schultz 1984) and has been shown to be directly related to the extent of heterologous intercellular coupling (Herlands and Schultz 1984).

Under normal circumstances, it is only with the establishment of the correct hormonal balance of follicle-stimulating hormone (FSH) and luteinizing hormone (LH) during each reproductive cycle, after the animal has reached sexual maturity, that growth of the follicle and its associated oocyte is permitted to result in meiotic maturation and subsequent ovulation (for review, see McGaughey 1983). However, it is of considerable practical interest to determine whether an oocyte acquires meiotic competency prior to expression of such competency in vivo and whether such acquisition relates to the size of the cell itself, to the size of the follicle in which it is enclosed, and/or to an intrinsic time-related developmental program. It is only after such determinations have been made that one can attempt to exploit the considerable pool of germ plasm present in the oocytes of primordial and small preantral follicles. Such exploitation has potentially far-reaching implications, since most female germ cells spend the greatest part of their existence in primordial follicles (for review, see McLaren 1988), and the majority are normally wasted in vivo through the process of atresia (for review, see Baker 1987).

The assay used to assess acquisition of nuclear meiotic competency is based on the original observation of Pincus and Enzmann (1935) that rabbit oocytes released from mature antral follicles will resume meiosis (i.e., undergo spontaneous GVBD) and complete oocyte maturation if maintained under appropriate conditions in vitro. Thus, lack of meiotic competency may relate to inadequate culture conditions, rather than to an absence of meiotic competency in the oocyte per se. This possibility must be seriously considered in light of a dependence on specific medium supplements for meiotic resumption in some species (Gwatkin and Haidri 1973) and of established differences in nutritional requirements of oocytes derived from follicles of varying size (Haidri and Gwatkin 1973).

In all mammals studied, the oocyte acquires the ability first to resume meiosis and then to complete meiotic maturation (Sorensen and Wassarman 1976; Motlik et al. 1984). However, the dependence of each of these processes on the size of either the oocyte or the follicle and/or on some time-related developmental program remains to be defined for most mammalian species. Nevertheless, in those species examined to date, acquisition of competency to resume meiosis occurs when the oocyte has attained a minimum size, which is species-specific. Mouse oocytes, for example, must be greater than 60 μm in diameter (Sorensen and Wassarman 1976), and hamster oocytes must be at least 80 μm (Iwamatsu and Yanagimachi 1975), in order for spontaneous GVBD to occur. In the case of the mouse (Szybec 1972; Sorensen and Wassarman 1976) and, indeed, the rat (Erickson and Ryan 1976), this minimum size may be considerably less than the maximum size attained by the fully grown oocyte. Therefore, in some species, incompletely grown oocytes are capable of resuming meiosis, provided they have grown to the limiting size associated with this capability.

In other species, however, such as the pig (Motlik et al. 1984) and the hamster (Iwamatsu and Yanagimachi 1975), follicle-free, incompletely grown oocytes cannot undergo GVBD in vitro. In these species, this ability is only acquired with an increase in follicular size and a concomitant increase in oocyte size, which ultimately results in maximum oocyte diameter (Iwamatsu and Yanagimachi 1975; Tsafriri and Channing 1975). Since there is a direct correlation between increasing follicle size and attainment of maximal oocyte size, not all studies have identified which of these two parameters is the limiting factor. Nevertheless, where such an identification has been made, the evidence is in favor of a stringent control exerted by oocyte size upon acquisition of meiotic competency (Iwamatsu and Yanagimachi 1975).

In the mouse, acquisition of the ability to resume meiosis is also apparently dependent on an intrinsic, time-related developmental program, since oocytes maintained on fibroblasts in vitro acquired this ability in the absence of any measurable growth (Canipari et al. 1984). In this study, the duration of culture required for in vitro acquisition of the capacity to mature was inversely related to the initial size of the oocyte at the time of isolation from granulosa cells and closely corresponded to developmental timing of acquisition of such ability in vivo. Although the basis for the onset of such a developmental program remains to be elucidated, it is likely triggered by the initiation of oocyte growth as the oocyte leaves its resting state, coincidental to the start of growth of the next cohort of growing follicles. One consequence of such a developmental program may be to ensure a minimum size below which oocytes are incompetent to resume meiosis (Eppig and Schroeder 1989). Although this type of program has only been demonstrated in the mouse (Canipari et al. 1984), the interesting possibility exists that dependence of meiotic competency on such a time-related developmental program may be universal in oocytes across all mammalian species.

The factors that regulate the ability of the oocyte to reach metaphase II and thereby complete meiotic maturation may contrast with those involved in regulating the competency of the oocyte to resume this process. Thus, a correlation between follicular size and frequency of completion of meiotic maturation has been demonstrated in many species (rabbit, Smith et al. 1978b; sheep, Moor and Trounson 1977; pig, Tsafriri and Channing 1975; Channing and Tsafriri 1977; Motlik et al. 1984; hamster, Iwamatsu and Yanagimachi 1975). Interestingly, however, this correlation seems to be absent in other species (cow, Thibault et al. 1976; Liebfried and First 1979; macaque, Smith et al. 1978a), suggesting that there may be species differences in this regard.

C. Regulation of Meiotic Maturation

Although oocytes nearing completion or having just completed their growth phase have acquired meiotic competency, they do not express such competency in vivo until their follicles receive stimulation by the preovulatory surge of gonadotropin (for review, see Lindner et al. 1974). Thus, the oocyte in vivo must be maintained in meiotic arrest by the action of a follicular "arrester," the effect of which is somehow overcome after exposure of the mature follicle to the preovulatory surge of LH (for review, see Thibault et al. 1987). Events mediating this gonadotropic action remain ill-defined, although several hypotheses have been proposed. One hypothesis suggests that LH induces a qualitative change in the arrester that renders it inactive (Moor et al. 1980b). Another proposes a gonadotropic-mediated effect that results in the production of a stimulator such as a growth factor, a steroid, or a prostaglandin, which in turn modulates the activity of maturation promoting factor (MPF), as is the case for steroid modulation of MPF in nonmammalian species such as the frog (Schuetz 1967) and starfish (Kanatani et al. 1969). A third hypothesis proposes that LH effects a quantitative change by causing a physical disruption of the follicular gap junctional network. This hypothesis further proposes that such a disruption interrupts transfer of cyclic adenosine monophosphate (cAMP) into the oocyte from its cellular source, the granulosa compartment, to reduce the level of arrester within the oocyte to a level below the minimum threshold required for maintenance of meiotic arrest (Dekel and Beers 1978, 1980).

In mammalian studies, attention has been focused on the last of these hypotheses. Available evidence supports a granulosa cell origin for the arrester (Foote and Thibault 1969; Leibfried and First 1980; Racowsky and Baldwin 1989a). In addition, it is quite clear from fluorescent dye studies (Racowsky and Satterlie 1985) that, in contrast to the functional gap junctional network that exists prior to the LH surge (Fig. 6A,B), exposure of the oocyte-cumulus complex to LH ultimately leads to disruption of this network (Fig. 6C,D) concomitant with cumulus mass expansion. However, inconsistent data exist regarding the timing of disruption of this network as it relates to release of the oocyte from meiotic arrest (for review, see Racowsky et al. 1989b). The majority of studies using radioactively labeled metabolites of either uridine or choline have shown no reduction in oocyte-cumulus cell coupling until after GVBD in vivo (Moor et al. 1981; Eppig 1982; Salustri and Siracusa 1983; Eppig and Downs 1988).

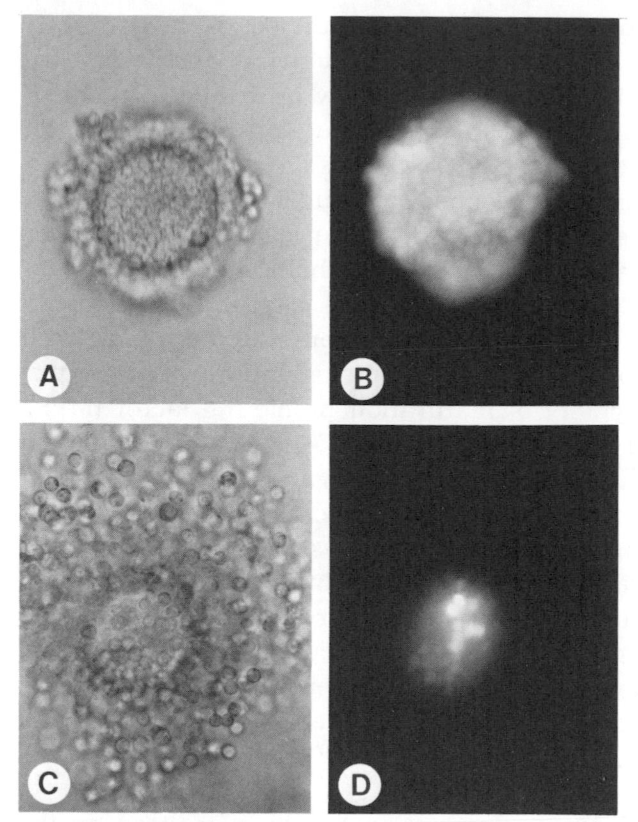

FIGURE 6 Fluorescent dye studies showing clear association between heterologous dye transfer within the oocyte-cumulus complex and the extent of cumulus mass expansion. Movement of lucifer yellow injected iontophoretically into a cumulus cell in a cumulus oophorus isolated immediately before (*A–B*) and 7 hr after (*C–D*) exposure to preovulatory gonadotropin. Note the markedly reduced movement of dye among the cumulus cells in the expanded cumulus mass. Magnification, 165x. (Reprinted, with permission, from Racowsky and Satterlie 1985.)

However, one study using such a metabolic coupling assay (Racowsky and Satterlie 1987) and other studies employing freeze-fracture electron microscopy (Larsen et al. 1986; Racowsky et al. 1989b) have indicated a tight relationship between oocyte meiotic status and down-regulation of follicular granulosa cell gap junctions in vivo.

In support of an arresting function for cAMP, some studies have demonstrated the maintenance of meiotic arrest in vitro concomitant with an increase in oocyte cAMP when levels of cumulus cell cAMP have been elevated artificially (Racowsky 1985a,b,c). However, although cAMP is capable of exerting an arresting action on all mammalian oocytes so far examined (for review, see Eppig and Downs 1988), this effect is only transient in many species (for review, see Homa 1988). In addition, although a significant drop in intra-oocyte cAMP levels occurs with GVBD in the rat (Racowsky 1984) and mouse (Schultz et al. 1983), no such decrease in cAMP levels has been observed in several other species including the sheep (Moor and Heslop 1981), hamster (Racowsky 1985a,b; Hubbard 1986), and pig (Racowsky 1985c). It is therefore likely that at least in some species, some component(s) in addition to cAMP is involved in the regulation of meiotic status. Such a component may act either as an arrester, which thus undergoes a decrease in intra-oocyte concentration after the LH surge, or as a maturation stimulator, which may bypass elevated levels of cAMP to induce GVBD.

Several recent studies support the involvement of a maturation stimulator in the release of the oocyte from meiotic arrest. Available evidence indicates a number of possible candidates for such a stimulator, including meiosis-stimulating factor (Westergaard et al. 1985) and prostaglandins (Murdoch 1988). In addition, several growth factors have been shown to exhibit GVBD-stimulating activity (for review, see Downs 1989). Epidermal growth factor (EGF) has been found to stimulate meiotic resumption in explanted rat follicles (Dekel and Sherizly 1985) and in follicle-free rat (Feng et al. 1987; Ueno et al. 1988) and mouse (Downs 1989) oocytes when the cumulus cells are present and elevated levels of intra-oocyte cAMP are maintained. Although less effective than EGF, insulin has also been found to antagonize cAMP-dependent meiotic arrest in mouse oocytes (Downs 1989). Transforming growth factor (TGF-β), insulin-like growth factors (IGF-I and IGF-II), and erythroid differentiation factor (activin-A) have all been shown to accelerate the rate of GVBD in spontaneously maturing rat oocytes (Feng et al. 1988; Itoh et al. 1990), but none of these peptides have been able to stimulate GVBD when elevated levels of intra-oocyte cAMP were maintained. Furthermore, a recent study by O et al. (1989) has revealed a marked inhibitory action of inhibin, a granulosa-cell-derived hormone, on GVBD in follicle-free rat oocytes. Collectively, these observations suggest that although EGF may play a central role in the stimulation of meiotic resumption, it probably acts in concert with other growth factors.

Whether EGF acts as a meiotic stimulator in a classically endocrine fashion, or whether it functions in an autocrine or a paracrine manner, is unknown. Any one of these alternatives is possible, since the peripheral concentration of this growth factor is sufficiently high, at least in mouse, to affect meiosis (Byyney et al. 1974). In addition, cultured ovarian cells in a variety of species have been shown to synthesize growth factors (cow, Neufeld et al. 1987; pig, Hammond et al. 1985; rat, Skinner et al. 1987). Porcine follicular fluid contains significant levels of EGF (Hsu et al. 1987), and there are binding sites for EGF within the ovary (for review, see Feng et al. 1987) that fluctuate in relation to the gonadotropic status of the follicle (St-Arnaud et al. 1983; Feng et al. 1987). An increase in the number of ovarian EGF receptors occurs during the afternoon of proestrus in the rat (St-Arnaud et al. 1983) and is concomitant with the onset of meiotic resumption.

Steroids may mediate the action of EGF in the regulation of meiotic resumption, since EGF stimulation of GVBD is dependent on the presence of cumulus granulosa cells (Downs et al. 1988), which are steroidogenically active (Racowsky 1983). EGF modulates granulosa cell steroidogenesis and has been shown to inhibit FSH stimulation of granulosa cell estrogen production (Hseuh et al. 1981; Jones et al. 1982) while enhancing FSH stimulation of progesterone synthesis (Jones et al. 1982). In this context, it is of interest that many studies have implicated an arresting role for estradiol and a releasing role for progesterone in the control of meiotic maturation in mammalian oocytes (for review, see Racowsky and Baldwin 1989b). Of particular note is a recent in vivo study in the golden Syrian hamster which revealed that the proportion of oocytes irreversibly committed to mature was positively correlated with follicular tissue progesterone levels and negatively correlated with follicular tissue estradiol levels (C. Racowsky et al., in prep.). The possibility therefore exists that following the preovulatory gonadotropin-induced increase in ovarian EGF receptors, elevated EGF binding causes a switch in granulosa cell steroidogenic activity in favor of progesterone, which, in turn, participates in the stimulation of meiotic resumption. However, other possible mechanisms of action of EGF in the control of oocyte meiotic status must be considered, given the dependence of the oocyte on granulosa cell metabolism and the variety of known effects of EGF on granulosa cell function, including its mitogenic action (Gospodarowicz et al. 1977; Gospodarowicz and Bialecki 1978, 1979) and its modulation of FSH-induced differentiation (Hseuh et al. 1981; Mondschein and Schomberg 1981; Jones et al. 1982).

The possible involvement of progesterone in the regulation of meiotic maturation in mammalian oocytes is strengthened by its proven role in stimulating GVBD in *Xenopus* oocytes through a cytoplasmic MPF (Masui and Markert 1971; Smith and Eckert 1971) and by the established ubiquity of MPF in maturing oocytes of all mammalian and nonmammalian species so far investigated (for review, see Hashimoto and Kishimoto 1988). Since MPF activity has also been identified in a wide variety of somatic cells entering mitosis (for review, see Cyert and Kirschner 1988), it is probable that MPF is a universal regulator of the G_2/M transition. In a series of recent studies, Sagata and his colleagues have shown that progesterone-induced oocyte maturation in *Xenopus* stimulates the synthesis of the c-*mos* proto-oncogene product, pp39*mos* (Sagata et al. 1988, 1989a), which is required for activation of MPF (Sagata et al. 1988, 1989b), probably through stabilization of this factor (Sagata et al. 1989a). Further support for the likelihood that common molecular mechanisms regulate oocyte meiotic maturation in mammalian and nonmammalian species is the recent demonstration that expression of the Mos protein is necessary for normal meiotic maturation in mouse oocytes (Paules et al. 1989). The collective evidence therefore suggests that common molecular mechanisms underlie regulation of oocyte maturation in mammalian and nonmammalian species.

D. Acquisition of Cytoplasmic Maturation

In preparation of the oocyte for fertilization, not only must meiotic maturation occur, but also the cytoplasm of the oocyte must undergo some critical changes in order to achieve competency to support sperm chromatin decondensation and subsequent male pronuclear formation (for review, see Thibault et al. 1987). For some years, it has been known that the ability of the oocyte to decondense the sperm nucleus is directly dependent on the meiotic maturational state of the oocyte. Thus, earlier in vitro fertilization studies (Thibault and Gerard 1970; Iwamatsu and Chang 1972; Barros and Munoz 1973; Niwa and Chang 1975), and more recent sperm microinjection experiments (Thadani 1979; Balakier and Tarkowski 1980), have shown that sperm decondensing activity is maximal in mature metaphase II oocytes, but minimal or absent in immature GV stage oocytes. Furthermore, this activity declines after fertilization and is either substantially reduced or absent in pronuclear stage oocytes (Usui and Yanagimachi 1976; Komar 1982).

Although an association between nuclear meiotic stage and cytoplasmic sperm decondensing activity is present in oocytes of all mammalian species so far investigated, the requirement for membrana granulosa cells in this process appears to vary among species. In some species, including the mouse and rat, the appearance of this activity is independent of the presence of membrana granulosa cells, since cumulus-enclosed oocytes matured in vitro in the absence of membrana granulosa cells are fertilizable and can proceed to full term upon transfer to recipient animals (mouse, Schroeder and Eppig 1984; Downs et al. 1986a; Eppig and Schroeder 1989; rat, Fleming et al.1985; Vanderhyden and Armstrong 1989). However, in other species, including the sheep and pig, acquisition of oocyte cytoplasmic competency is dependent on physical contact between the oocyte-cumulus complex and the underlying membrana granulosa cells (sheep, Staigmiller and Moor 1984; Crozet et al. 1987; pig, Iritani et al. 1978; Pavlok 1981; Mattioli et al. 1988a). Although results of earlier work with the cow oocyte indicated a similar membrana granulosa cell dependency (for review, see Thibault et al. 1987), a recent study reports that bovine oocytes matured with cumulus cells alone are capable of supporting normal fertilization and development to term (Fukuda et al. 1990).

Regardless of whether induction of cytoplasmic maturation

is dependent on the presence of membrana granulosa cells, this process is enhanced in all species by association of the oocyte with its adherent cumulus cells (for review, see Vanderhyden and Armstrong 1989). Thus, if the cumulus mass is removed before meiotic maturation in vitro, the percentage of such cumulus-free oocytes that support normal male pronucleus formation is substantially reduced (mouse, Schroeder and Eppig 1984; rat, Vanderhyden and Armstrong 1989; cow, Fukui and Sakuma 1980; Sirard et al. 1988; sheep, Staigmiller and Moor 1984). In addition, only those cultured pig oocytes that remain effectively coupled to their surrounding cumulus cells acquire the ability to decondense sperm nuclei and form male pronuclei (Mattioli et al. 1988a).

It remains to be determined whether such granulosa-cell-induced enhancement of oocyte developmental competency reflects an overall increase in oocyte metabolism resulting from the established nutritional support provided by both the cumulus mass and the membrana granulosa cells (Biggers et al. 1967; Zamboni 1974; Eppig 1977) or whether specific instructional signals arising from this somatic compartment are involved. In addition, the possibility exists that the mere presence of these somatic cells with their associated receptors enables hormones and/or growth factors to elicit responses that would otherwise be impossible due to an absence of such receptors on the oolemma. In support of this possibility, the documented LH-induced changes in the follicular steroid profile (Eiler and Nalbandov 1977; Hubbard and Greenwald 1982) arise largely from gonadotropic control of steroidogenic activity in cumulus cells (pig, Hillensjo and Channing 1980; Racowsky 1983, 1985c; rat, Hillensjo et al. 1981) and membrana granulosa cells (cow, Lacroix et al. 1974; rat, Fortune and Armstrong 1978; pig, Armstrong and Dorrington 1976; hamster, Makris and Ryan 1975, 1977). Moreover, evidence in several domestic species indicates a role for steroids in the induction of cytoplasmic maturation. Thus, as early as 1975, Soupart (1975) showed that normal male pronucleus formation only occurred in human oocytes when estradiol and then estradiol and 17α-OH progesterone were added sequentially to the culture medium. Consistent with this observation, high concentrations of follicular fluid progesterone in the cow corresponded to a high proportion of metaphase II oocytes and a high incidence of fertilization in vitro (Bousquet et al. 1988). These high fluid progesterone levels were accompanied by elevated levels of testosterone and androstenedione but a lower level of estradiol. Further support for steroidogenic control of developmental competency has been provided by Moor and his co-workers (Moor and Trounson 1977; Moor et al. 1980a; Osborn and Moor 1983), who reported that balanced steroid profiles in explanted sheep follicles are mandatory for the subsequent formation of a normal male pronucleus.

Since no steroid supplementation is required in medium used to promote acquisition of cytoplasmic maturation in either mouse (for review, see Eppig et al. 1988) or rat (Vanderhyden and Armstrong 1989) oocytes, there may be fundamental differences between short-cycling and long-cycling species in the mechanisms involved in regulating cytoplasmic maturation and developmental competence. However, as discussed by Vanderhyden and Armstrong (1989), intrafollicular hormone exposure in vivo before oocyte collection from gonadotropin-primed animals may initiate the processes responsible for expression of such competence in vitro. In this context, it is noteworthy that attainment of a high incidence of normal fertilization in mouse oocytes collected from gonadotropin-primed juveniles (Schroeder et al. 1988) and in rat oocytes collected from nongonadotropin-primed sexually mature animals (Zhang and Armstrong 1989) is dependent on the presence of FSH in the medium, which stimulates cumulus cell progesterone synthesis (Racowsky 1983). In addition, the incidence of developmental competence in cumulus-enclosed oocytes in vitro has been increased by LH in rat (Shalgi et al. 1979), mouse (Jinno et al. 1989), and cow (Younis et al. 1989) and by FSH in the mouse (Downs et al. 1986a; Jinno et al. 1989).

Whatever the mechanism(s) by which granulosa cells exert their beneficial effects on oocyte developmental competency, oocyte-cumulus junctional communication could mediate either a direct influence via membrana granulosa–cumulus junctional communication (for review, see Thibault et al. 1987) or an indirect influence by soluble factors secreted into the follicular fluid. In support of the latter mechanism, medium previously used to mature follicle-enclosed pig oocytes significantly increased the incidence of sperm nuclear decondensation in follicle-free, cumulus-enclosed pig oocytes compared with either unconditioned medium or medium previously conditioned with kidney explants (Mattioli et al. 1988a).

Although the metabolic changes that underlie acquisition of oocyte cytoplasm maturation are ill-defined, new protein synthesis within the oocyte is involved (for review, see Thibault et al. 1987). These newly synthesized proteins may be associated with modulating cytoplasmic glutathione levels, since such levels increase during meiotic resumption in hamster oocytes and are significantly higher in mature metaphase II oocytes (8 mM) than in either GV stage or pronuclear oocytes (4 mM and 6 mM, respectively; Perrault et al. 1988). Glutathione, the major intracellular free thiol (Meister and Anderson 1983), has been implicated as playing a key role in protamine disulfide bond reduction (Perrault et al. 1984; Calvin et al. 1986), which in turn is a prerequisite for sperm chromatin decondensation, and thus for male pronucleus formation (Perrault et al. 1984). Indeed, when glutathione synthesis is blocked during the early stages of oocyte maturation in vitro, the cytoplasm of chromosomally normal metaphase II oocytes is incapable of supporting sperm nuclear decondensation (Perrault et al. 1988). This observation indicates that the synthesis of glutathione during initial stages of oocyte maturation is essential for the acquisition of sperm nuclear decondensing ability.

E. Conclusions

The following conclusions can be drawn from the above discussion regarding the physiological production of developmentally competent mammalian oocytes:

1. Depending on the species and the study undertaken, acquisition of nuclear meiotic competency has been correlated with attainment of a minimum size of either the oocyte or the follicle. However, since oocyte size is directly related to follicle size, it is probable that the size of the oocyte is of paramount importance in the production of meiotically competent eggs in all mammalian species. Depending on the species, the critical oocyte size is either less than or equal to that of a fully grown oocyte. Although proven only

in the mouse oocyte, acquisition of meiotic competency in vivo may be dependent on a time-related developmental program, which coincidentally ensures a threshold size above which the oocyte is competent to complete meiotic maturation.

2. In all species, cAMP appears to be involved in the maintenance of meiotic arrest, and in some species at least, a significant drop in intra-oocyte cAMP may be responsible for meiotic resumption. Furthermore, gap junctional downregulation between the oocyte and the somatic compartment occurs concomitantly with events associated with meiotic resumption in some species, thus indicating an involvement of follicular gap junctions in the regulation of meiotic status. In addition, stimulation of the follicle by LH may promote the synthesis of a maturation promoter that overrides the effect of follicular arrester. The mechanisms that mediate this gonadotropic action are ill-defined, but they may involve EGF, various other growth factors, and progesterone.
3. Expression of meiotic competency by completion of spontaneous meiotic maturation in cultured follicle-free oocytes does not ensure normal cytoplasmic maturation and support of male pronuclear formation. Acquisition of such developmental competence is dependent on the presence of the cumulus mass during oocyte growth. Furthermore, this process is enhanced, at least in some species, by factors arising from the membrana granulosa compartment. The mechanisms responsible for these granulosa cell effects are unknown but may involve an overall increase in nutritional support for the oocyte and/or a specific instructional signal such as a steroid, arising in the somatic cell compartment in response to gonadotropin or growth hormone stimulation. Regardless of the nature of the granulosa cell signal, an increase in the synthesis of hamster oocyte glutathione occurs during the early stages of nuclear maturation and is prerequisite for attainment of acquisition of cytoplasmic maturation. The possibility therefore exists that increased glutathione synthesis is required in oocytes of all mammalian species in order for the cytoplasm to decondense the sperm nucleus.

4. The average number of mature oocytes produced in a single cycle for a given species is relatively small. Thus, attempts to increase the yield of oocytes having full potential for developmental competence must either target the population of growing antral follicles or tap the pool of pre-antral follicles.

In the following section, these concepts are considered in the nonphysiological production of meiotically mature, developmentally competent oocytes.

III. THE NONPHYSIOLOGICAL PRODUCTION OF DEVELOPMENTALLY COMPETENT OOCYTES

A. Production of Developmentally Competent Oocytes from Antral Follicles

1. Developmental Maturity of Animal. Both sexually mature and immature animals of laboratory and domestic species have been used to obtain oocytes from antral follicles. For laboratory animals, the antral follicles constitute the pool destined for ovulation in the next cycle, whether it be the immature animal's first cycle at puberty or the sexually mature animal's forthcoming cycle. Regardless of the species, there is a minimum age of donor animal consistent with ovaries containing oocytes of the minimum size compatible with meiotic competency (rat, Cole 1937; mouse, Szybek 1972; hamster, Iwamatsu and Yanagimachi 1975). Although this approach provides a relatively small number of follicles, it has the obvious advantage that both the follicle cells and the germ cells are exposed only to physiological levels of gonadotropin, thus avoiding any aberrancy induced by follicular exposure to the abnormally high levels of hormones associated with injection of exogenous gonadotropin (see Section III.A.2). Furthermore, in some species, the availability of developmentally competent oocytes is independent of the stage of the estrous cycle for oocyte collection. In the mature mouse, equivalent numbers of GV-stage oocytes were isolated at each stage of the estrous cycle, with the exception of metestrus, and the frequency of spontaneous maturation was similar regardless of the day of collection (Schroeder and Eppig 1989). In addition, the oocytes isolated from each stage of the estrous cycle exhibited similar developmental competencies as assessed by subsequent fertilization, rate of blastocyst formation, and development to term after transfer to pseudopregnant foster mothers (Eppig and Schroeder 1989). Although species differences may exist regarding the maturational capacity of oocytes isolated at different stages of the reproductive cycle (Leibfried and Bavister 1983), it is likely that at least some developmentally competent oocytes can be obtained at stages of the estrous cycle other than the optimum stage of proestrus.

The population of antral follicles collected from domestic animals is composed of all growing follicles whose diameters fall within some specified range. The lower value in this range must be above that consistent with acquisition of oocyte meiotic competency, unless the oocyte is allowed to grow in culture (see Section II.B; for review, see Tsafriri and Channing 1975), whereas for the sexually mature animal, the upper value must be well below the average diameter of a normal preovulatory follicle. Such a strategy thereby includes all follicles containing meiotically competent oocytes but avoids any follicles that may have been stimulated by endogenous preovulatory gonadotropin. This approach is well illustrated in the cow, where several reports demonstrate that, with attention to follicle size, it is possible to collect a pool of oocytes with a high level of developmental competency after maturation in vitro, regardless of the stage of the donor's estrous cycle at collection (Newcomb et al. 1978; Leibfried and First 1979; Critser et al. 1986; First and Parrish 1987). This approach is appropriate where the stage of estrous cycle of domestic animals at slaughter is unknown.

2. Hormonal Status of the Animal. Although sexually immature animals are capable of providing a small number of meiotically competent and developmentally competent (Schroeder and Eppig 1989) oocytes, the practical and economical importance to both researchers and stock breeders of obtaining a much increased number of oocytes has led to the development of superovulatory techniques. These techniques involve injection of FSH (for review, see Callesen et al. 1986) or the FSH-like gonadotropin, pregnant mares' serum gonadotropin (PMSG; for review, see Yun et al. 1989), for production of a large number of growing antral follicles in immature animals. This hormone injection is followed by injection of human chorionic gonadotropin (hCG) when in-vivo-matured oocytes are required. Substantial increases in numbers of ovulated oocytes have been demonstrated in the mouse (Wilson and Zarrow 1962), rat (Wilson and Zarrow 1962; Miller and Armstrong 1981a), hamster (Bodemer et al. 1959), and pig (Gibson et al. 1963) using immature animals of optimal age and injections of hormones at optimal doses, with discrete intervals between injections (Table 3). Furthermore, exposure of oocytes of immature mice to PMSG has a beneficial effect on developmental capacity (Schroeder and Eppig 1989). In the immature rat, although a low dose of PMSG results in a marginal increase in the number of viable oocytes (Miller and Armstrong 1981b), high doses are associated with a substantial increase in oocyte number, but severely impair several aspects of fecundity (Miller and Armstrong 1981b). Such impairments include variability in ovulatory response (Walton et al. 1983; Yun 1987), a reduction in fertilization rate (Walton et al. 1983), a substantial loss in preimplantation embryos (Betteridge 1977; Miller and Armstrong 1981b; Yun et al. 1988), and/or a failure to establish implantation (Miller and Armstrong 1981a). It may therefore be concluded that although the immature rat responds to PMSG in a manner similar to that of the mouse in terms of number of oocytes ovulated (Wilson and Zarrow 1962), the susceptibility of the reproductive system to PMSG-induced perturbation is much greater in the rat.

In addition to these disadvantages specific to some species, there are differences in response among strains of the same species (McLaren 1967) and differences in quality of PMSG from different commercial sources, as evidenced both by the production of different numbers of embryos obtained from immature mice and by variability in the capacity of these embryos to survive in culture. In addition, the source of PMSG may affect the development in culture of previously frozen embryos, with strain-specific differences in developmental response (Schmidt et al. 1989). Nevertheless, there are several advantages associated with induction of superovulation in sexually immature animals: (1) Such premature induction of folli-

TABLE 3 *OPTIMAL CONDITIONS FOR INDUCTION OF SUPEROVULATION IN IMMATURE LABORATORY MAMMALS*

Species	*Optimal dose (IU)*		*Optimal hours between PMSG and hCG*	*Average number of ova/animal*	*Study*
	PMSG	*hCG*			
Mouse	5–10	5–10	40	45	Wilson and Zarrow (1962)
Rat	30	20	26	56	Rowlands (1944)
	20	20	56	37	Austin (1950)
	30	5–10	56	50	Wilson and Zarrow (1962)
	20	0	—	23[a]	Yun et al. (1989)
	16	0	—	16[b]	Miller and Armstrong (1981b)
Hamster	20	20	54–56	35	Bodemer et al. (1959)

[a]Oocytes recovered from oviducts at 72 hr post-PMSG.
[b]Number of live fetuses.

cular growth reduces the time, and thus the cost, of maintaining the animals until they reach sexual maturity, and (2) the pool of available oocytes in an immature animal is much greater than that in a mature animal, since fewer oocytes have been lost to atresia over time and the numbers have not been reduced by prior ovulations.

Gonadotropin injection techniques have also been successfully applied in sexually mature laboratory and domestic species, resulting in the ovulation of supernumerary oocytes. In this case, the population of growing follicles includes a group of reserve follicles capable of responding to elevated FSH levels, in addition to those normally destined to ovulate in the cycle during which gonadotropin injection occurred. In several studies, this increase in yield of oocytes from sexually mature animals has been associated with an increased production of normal developmentally competent oocytes, as evidenced by the production of "super litters" in mice (Fowler and Edwards 1957) and hamsters (Fleming and Yanagimachi 1980). However, other studies have indicated some important disadvantages associated with superovulation with gonadotropins (for review, see Yun et al. 1989). First, superovulatory hormones are expensive, and given the high doses of PMSG required for optimal responses, particularly in domestic animals (pig, 1200 IU PMSG/animal [Longenecker and Day 1968]; sheep, 1250 IU PMSG/animal [Moor and Heslop 1981]; cow, 1500–3000 IU PMSG/animal [Callesen et al. 1986]), the costs can be considerable. Second, since there are species differences with respect to the optimum dose of gonadotropin required for production of the maximal number of developmentally competent oocytes, extensive trials are necessary for any species not characterized in this regard. In addition, since the season (Moore and Greenwald 1974) and both day of the estrous cycle and time of day at which gonadotropin is administered (Greenwald 1962; Fleming and Yanagimachi 1980) may affect the cycle length and the number of oocytes ovulated, the trials also have to define these parameters. Third, exposure of ovaries of both sexually mature and immature animals to such elevated levels of exogenously introduced gonadotropin may result in follicular hyperstimulation (Miller and Armstrong 1981a; Yun et al. 1989) and subsequent abnormal profiles of steroid output (for review, see Yun et al. 1987). Such aberrations in steroid synthesis have been correlated with disruption in the priming of the genital tract for normal embryo transport or implantation in some species (for review, see Yun et al. 1988). However, in several other species, the use of superovulatory doses of exogenous gonadotropins results in a disruption in development as indicated by abnormalities in the meiotic progression of cow oocytes (Callesen et al. 1986), the fertilizability of both sheep (Moor et al. 1980a) and rat (Yun et al. 1987) oocytes, and the number of postimplantation embryos (Cole 1937; Longenecker and Day 1968) or births (Fleming and Yanagimachi 1980) relative to the number of corpora lutea present. Discrepancies in the number of corpora lutea and embryos or live births may be accounted for in part by inadequate uterine space for the support of development of extraordinary numbers of fetuses (Parkes 1943; Hunter 1966), but other factors could also be involved.

The mechanisms underlying the detrimental effects of PMSG injection remain ill-defined, although the long half-life of the gonadotropin in circulation has been implicated in both laboratory (Sasamoto et al. 1972) and domestic (Bindon and Piper 1977) species. In addition, since PMSG possesses both LH and FSH activities inherent in the same protein molecule (Papkoff 1974), the possibility has been raised that the LH component may be responsible for adverse effects of PMSG (for review, see Murphy et al. 1984). In support of such a possibility, administration of various gonadotropin preparations containing different activity ratios of LH:FSH in a variety of species (sheep and goat, Armstrong and Evans et al. 1984; cow, Murphy et al. 1984; human, Jones et al. 1985; Russell et al. 1986) has shown that the FSH preparation with the lowest LH activity produced the maximum number of developmentally competent oocytes. Collectively, these observations indicate that an optimum LH:FSH ratio exists for each species. Additional corroborating data have been obtained from studies of superovulation in immature rats that were continuously infused with FSH preparations of varying purity. Superior rates of ovulation and of oocyte developmental potential were obtained when FSH preparations that were lowest in LH activity were used (Armstrong and Opavsky 1988).

3. Systems Used to Obtain Developmentally Competent Oocytes. Following Pincus and Enzmann's original observation (1935) that rabbit oocytes released from mature antral follicles will undergo GVBD and complete their meiotic maturation in vitro, numerous studies in a wide variety of species have resulted in the production of meiotically mature mammalian oocytes (for review, see Thibault et al. 1987). However, it is now well known that an oocyte which exhibits nuclear meiotic maturation in vitro has not necessarily undergone the cytoplasmic changes associated with acquisition of developmental competency (Thibault 1977). Therefore, where possible, it is desirable to obtain mature eggs from animals after induction of ovulation in vivo. Although this is a relatively simple undertaking in laboratory species (for review, see Gates 1971) and has also been achieved in domestic animals (sheep, Dziuk 1965; pig, Buttle and Hancock 1967; cow, Graves and Dzuik 1968; Callesen et al. 1986; Bousquet et al. 1988), the frequent use of slaughterhouse material in the latter often precludes this possibility. Systems for production of developmentally competent oocytes are needed in all species, not just when utilizing oocytes from preantral follicles (see Section III.B), but also to enable exploitation of the large pool of antral follicles present in slaughterhouse ovaries.

Since many studies have shown that in-vitro-matured oocytes cultured in the absence of follicular supplements have a much reduced developmental capability compared with those matured in vivo (Leibfried-Rutledge et al. 1987), follicular hormones and/or follicular granulosa cells may be involved in promotion of cytoplasmic competency. Numerous culture studies have thus rigorously assessed the effects of various factors on the ability of the oocyte to support normal male pronucleus formation and, in some cases, embryonic development resulting in a live birth. These factors include steroids (sheep, Moor and Trounson 1977; cow, Callesen et al. 1986; Bousquet et al. 1988; Fukui and Ono 1989; pig, Mattioli et al. 1988b), gonadotropins (for review, see Fukui and Ono 1989), and both membrana granulosa cells (for review, see Mattioli et al. 1988a) and cumulus granulosa cells (for review, see Sirard et al. 1988).

Meiotically mature oocytes can be obtained from oocytes

maintained within explanted follicles exposed to gonadotropin in vitro (Moor and Trounson 1977; Moor et al. 1980a,b) or, alternatively, from follicle-free oocytes cultured either devoid of (cumulus-free) or enclosed within (cumulus-enclosed) their cumulus cells. The follicle-enclosed system keeps the oocyte in its normal immediate environment, and therefore has the potential to provide the oocyte with those physiological factors that normally support acquisition of cytoplasmic competency. Indeed, where follicle-free systems have been found deficient, the possibility exists that such deficiencies may be accounted for by the absence of follicle-specific factors. Alternatively, such culture systems may lack components, not necessarily specific to the follicle, that are required for the acquisition of cytoplasmic competency. Despite the possible advantages of using explanted follicles, there are two major disadvantages associated with their use. First, a rapid enzyme digestion procedure for follicular collection has only been established in hamsters (Roy and Greenwald 1985). In other species follicle dissection is necessary, which is more time-consuming than is collection of follicle-free oocytes and also involves a certain attrition rate resulting from follicular puncture. Second, because the follicle is a complex three-dimensional structure, it has unique gas exchange requirements when maintained as an explant in culture (Baker et al. 1974), requiring careful assessment of cellular viability after incubation.

Compared with cumulus-enclosed oocytes cultured after follicular release, oocytes cultured cumulus-free exhibit an increased rate of GVBD in all species studied (hamster, Racowsky 1986; rat, Dekel and Beers 1980; Vanderhyden and Armstrong 1989), although the rate of polar body emission is slower in the absence of cumulus cells (Vanderhyden and Armstrong 1989). As discussed in Section II.D, there appears to be an absolute requirement for the presence of cumulus cells during meiotic maturation in order for the oocyte to acquire the ability to support male pronucleus formation. Furthermore, in two recent studies, oocytes selected for the compactness of their cumuli exhibited higher incidences of both meiotic maturation and subsequent sperm chromatin decondensation than control, unselected oocytes (rat, Vanderhyden and Armstrong 1989; cow, Younis et al. 1989; Fukuda et al.

1990). In addition to the effect of the compactness of the cumulus mass on developmental competency, meiotic competency is further affected by the overall beneficial effects of cumulus cells, as reflected by the improved development obtained when a maximum number of cumulus complexes are cultured in a minimum volume of medium (Sirard et al. 1988). Furthermore, the timing of insemination of mature oocytes is of critical importance in achieving maximal levels of developmental competency. In cases where oocytes must be held in culture prior to insemination, it is desirable to maintain them in the meiotically arrested GV stage (e.g., the presence of dibutyryl cAMP or a cAMP phosphodiesterase inhibitor such as 3-isobutyl-1-methylxanthine; IBMX), rather than in the metaphase II stage (Downs et al. 1986a). This has been demonstrated in in-vitro-matured mouse oocytes in which the developmental potential in follicle-free oocytes maintained in meiotic arrest for 24 hours was strikingly enhanced compared with that in oocytes incubated for the same duration but permitted to mature immediately after follicular release (Downs et al. 1986a). Such observations are consistent with significant reductions in fertilization rates with increasing age of matured oocytes both in vitro and in vivo (for review, see Austin 1970; Szollosi 1975).

Beneficial effects of either membrana granulosa cells or follicular fluid have been demonstrated in some studies with domestic species, in contrast to laboratory species, where neither of these follicular components is required for the acquisition of developmental competence (for review, see van de Sandt et al. 1990). The requirement for follicular somatic cells for male pronucleus formation has been demonstrated by Moor and Trounson in the sheep (1977). In this study, oocytes matured in either atretic or nonatretic explanted follicles were equally capable of undergoing blastocyst formation upon transfer to inseminated recipients. From randomly selected embryos derived from the atretic and nonatretic groups, 73% of recipient ewes gave birth to live lambs, as compared with 71% in the control group. Although not conclusive, these data might suggest that oocytes derived from atretic follicles retain their capability to support normal development.

Consistent with these observations in the sheep, the presence of membrana granulosa cells has been shown to enhance significantly the developmental capability of bovine oocytes (Lutterbach et al. 1987; Fukui and Ono 1989). However, such coculturing is clearly not an absolute requirement in this species, since developmentally competent oocytes have resulted following maturation in vitro in the absence of membrana granulosa cells (Sirard et al. 1988; Fukuda et al. 1990). In the pig, oocyte developmental competency is enhanced by the presence of a soluble factor within the follicular fluid, since oocytes matured either in medium supplemented with follicular fluid (Naito et al. 1988) or in medium previously conditioned by follicles in which oocytes had been induced to mature (Mattioli et al. 1988a) acquired cytoplasmic competency. The enhancing effect of follicular fluid was increased by the presence of FSH (Naito et al. 1988), suggesting that FSH has a beneficial effect on the induction of oocyte developmental competency in domestic species. Consistent with this possibility in domestic species, inclusion of FSH in medium for in vitro maturation of mouse oocytes enhances the incidence of developmental competency after oocyte collection from PMSG-primed sexually immature mice (Schroeder et al. 1988). Furthermore, addition of FSH to the maturation medium for oocytes removed from immature, non-PMSG-primed rats (i.e., oocyte-cumulus complexes containing low levels of FSH) is required in order to achieve a high level of fertilizability (Zhang and Armstrong 1989).

In addition to investigations of the effects of FSH on acquisition of oocyte developmental competency in vitro, considerable efforts have focused on determining the regulatory roles of steroids in this process. Several recent studies have provided support for a steroidal control of oocyte developmental competency. Pig oocytes matured in the presence of progesterone in vitro exhibited a higher incidence of fertilization compared with those oocytes matured in the absence of this steroid (Mattioli et al. 1988b). As discussed in Section II.D, high levels of progesterone accompanied by low levels of estradiol in follicular fluid have been associated with enhanced fertilizability in cow oocytes triggered to mature in vivo (Bousquet et al. 1988). These results are consistent with the established switch from a predominantly estrogenic to a primarily progestagenic follicular environment following the surge of preovulatory gonadotropin (for review, see Eiler and Nalbandov 1977; Hubbard and Greenwald 1982). However, other studies have shown strikingly beneficial effects of estradiol on developmental competency. Estradiol increased both the incidence of blastulation in sheep oocytes matured in explanted follicles exposed to LH and FSH (Moor and Trounson 1977) and the incidence of male pronucleus formation in in-vitro-matured cow oocytes (Fukushima and Fukui 1985; Stubbings et al. 1988; Younis et al. 1989). In contrast to these observations, no estradiol-induced enhancement of maturation, fertilization, or development of resultant cow embryos was achieved in other studies (Sirard et al. 1988; Fukui and Ono 1989). One group, in fact, has obtained normal, live calves from oocytes matured in the presence of FSH, but in the absence of estradiol, with subsequent fertilization using sperm capacitated with the divalent ionophore A23187 (Fukuda et al. 1990).

The reasons for the conflicting results obtained from the cow oocyte experiments remain unidentified, although several differences exist among the various studies regarding the nature and doses of gonadotropins and estradiol used and the media and the culture methods employed. For example, inclusion of serum as a protein supplement will inevitably introduce both gonadotropins and steroids into the culture system, which may or may not have an impact on the final outcome. It is essential that consideration be given to all aspects of a published culture system in any attempt to improve it. This point is further illustrated by the disparate results obtained when differences existed between the times of addition of FSH and LH to medium supplemented with estradiol. When the gonadotropins were added 6 hours after the onset of culture, so as to mimic the temporal changes that occur in vivo, a significantly higher incidence of developmental competency was observed (Stubbings et al. 1988) than when the gonadotropins were present throughout incubation (Fukui and Ono 1989).

The importance of serum in both the meiotic maturation and the fertilizability of cultured oocytes in vitro has been extensively studied in both laboratory (for review, see Vanderhyden and Armstrong 1989) and domestic (for review, see Fukui and Ono 1989) species. In addition, numerous studies have attempted to identify particular sera that have optimal effects on oocyte developmental competency, whether they be porcine-, fetal-calf-, or bovine-estrous-derived sera at a specific stage of the cycle. Interpretation of the results of these studies has been complicated by species differences in response (for review, see Racowsky 1985c) and by the considerable variation in commercial lots of fetal calf serum (FCS; Shiigi and Mishell 1975). Furthermore, sera added to culture media may play a role in maintaining cumulus-oocyte coupling, and thus in the enhancement of heterologous transfer to the oocyte of nutrients, hormones, and other growth factors that may be involved in controlling maturation (Leibfried-Rutledge et al. 1986) and/or in stabilizing the oolemma (Downs et al. 1986b).

In the two laboratory species studied extensively (mouse and rat), there are species differences in the degree of dependence on serum as a supplement during meiotic maturation in order for high rates of fertilization to be achieved. In the mouse, serum is required in medium used to mature oocytes to prevent zona hardening (Downs et al. 1986b) and for acquisition of cytoplasmic competency and subsequent expression of the developmental program (Schroeder and Eppig 1984). This requirement is considerably greater than that for the rat, since less than 10% of mouse oocytes matured in serum-free medium underwent subsequent fertilization in vitro (Downs et al. 1986a), whereas in the rat, this figure was closer to 30% (Vanderhyden and Armstrong 1989). Nevertheless, the presence of serum in the maturation medium significantly enhanced the fertilizability of in-vitro-matured rat oocytes and their subsequent progression to two-cell embryos (Vanderhyden and Armstrong 1989). However, serum is not required for either fertilization or subsequent preimplantation development in vitro in either species, since medium containing bovine serum albumin (BSA) as the protein supplement promotes both of these phenomena to the same extent as does serum.

The incidence of fertilization in oocytes matured in vitro is not affected by the type of serum in the oocyte maturation medium (mouse: FCS, newborn calf serum, or murine serum; Schroeder and Eppig 1984; rat: FCS, rat, goat, pig, or human serum; Vanderhyden and Armstrong 1989). However, it is affected by the composition of the commercial medium used for the maturation of oocytes (Schroeder and Eppig 1984; van de Sandt et al. 1990). In the former study, Whitten's medium was found to be superior to Eagle's minimal essential medium (MEM), whereas in the latter, Waymouth's MB 752/1 promoted the greatest embryonic development compared with eight other media tested. Furthermore, the subsequent development of in-vitro-matured oocytes both through the blastocyst stage in vitro and, after transfer to recipients to the production of live young, is affected by the commercial medium used for oocyte maturation (van de Sandt et al. 1990).

The effects of FCS and cow serum on developmental competency of in-vitro-matured cow oocytes have also been extensively studied. Meiotic maturation in this species is enhanced by the presence of FCS as a supplement, compared with BSA (Leibfried-Rutledge et al. 1986). An extension of this earlier study has shown that serum obtained from cows at any stage of the estrous cycle further enhances the incidence of maturation compared with FCS supplementation (Younis et al. 1989). In addition, when the maturation medium contained serum collected from cows on either day 0 or day 20 of the estrous cycle, the incidence of fertilization was increased, and early conditioning of oocytes with day-20 serum produced superior viability following in vitro fertilization: 53.1% of the two-cell embryos reached four- to eight-cell stages, whereas only 28.6–38.4% were capable of undergoing further cleavage after previous exposure to either FCS or sera obtained at other stages of the estrous cycle (Younis et al. 1989). In contrast to these results, Fukui and Ono (1989) found that serum obtained from estrous cows (Lu et al. 1987) supported a lower rate of fertilization than did FCS (34–52% vs. 57–71%, respectively), although the incidence of polyspermy was significantly lower with estrous cow serum than with FCS. Collectively, the results of these studies question the advantage of supplementing maturation medium with estrous cow serum, as compared with FCS, for enhancement of the acquisition of oocyte developmental competency. It therefore appears that estrous cow serum is not absolutely required for developmental competency, as further shown in a recent study in which a high incidence of in vitro fertilization was obtained in oocytes matured in medium supplemented with FCS (Fukuda et al. 1990).

Aside from the beneficial effect of serum in the maturation medium on the maintenance of the structural and/or chemical properties of the zona pellucida (Downs et al. 1986b), serum supplements may enhance fertilizability by maintaining normal cumulus cell metabolism and/or the coupling between the oocyte and the surrounding cumulus cells. Not only do cumulus cells, in the presence of serum, improve the incidence of completion of the first meiotic division (Leibfried-Rutledge et al. 1986) and control the rate of nuclear maturation as discussed above, but their presence appears to be necessary for the promotion of normal cytoplasmic maturation (see Section IID). In support of these possibilities, serum deprivation and/or cumulus cell removal during oocyte maturation in vitro results in an alteration in the zona pellucida that causes an increased resistance to proteolytic digestion and sperm penetration (Downs et al. 1986b) known as zona hardening (DeFelici and Siracusa 1982). The beneficial effects of serum on oocyte fertilizability can be duplicated by supplementation of culture medium with follicular fluid (DeFelici et al. 1985; Downs et al. 1986b), suggesting that some component of follicular fluid in vivo inhibits changes in the zona associated with hardening. It is noteworthy that fertilizability was progressively reduced with increasing time of exposure of mouse oocytes to serum-free medium after follicular release (Downs et al. 1986b). Although species differences may exist regarding factors that affect zona hardening, it seems prudent to eliminate any exposure of oocytes, regardless of species, to serum- or follicular-fluid-free medium when attempting to produce developmentally competent oocytes in vitro.

4. Conclusions. The following conclusions can be drawn from the above discussion regarding systems used to obtain developmentally competent mammalian oocytes:

1. The judicious use of hyperstimulatory hormones substantially increases the number of oocytes obtained from antral follicles for the in vitro production of developmentally competent oocytes. When using hyperstimulatory hormones, it is essential to assess the ability of oocytes to support both

normal early embryonic development and fetal development to birth. Such stringent criteria will exclude all oocytes in which hyperstimulatory hormone-induced perturbation of developmental function is manifested after the oocyte has reached metaphase II.

2. Satisfactory systems for obtaining in-vitro-matured, developmentally competent oocytes collected from immature PMSG-primed mice and rats are currently available. Both systems utilize cumulus-enclosed oocytes, but do not depend on the presence of membrana granulosa cells. In the mouse, FSH and either FCS or follicular fluid are required in the maturation medium. In addition, optimal results are obtained when the base maturation medium consists of Waymouth's MB 752/1 medium. When oocytes must be held in culture prior to insemination, it is beneficial to maintain them at the GV stage, rather than at the metaphase II stage. Using this system and all other methodologies described by van de Sandt et al. (1990), a normal developmental program can be obtained in mouse oocytes matured in vitro, as evidenced by fertilization and subsequent development to live offspring at levels comparable to oocytes matured in vivo. In the rat, the oocyte maturation system composed of FCS-supplemented MEM with Earle's salts, with the additional components as described by Vanderhyden and Armstrong (1989), produces developmentally competent oocytes capable of giving rise to viable fetuses in proportions similar to those in ovulated oocytes after in vitro fertilization.
3. In domestic species, the follicle-free systems currently available for the in vitro production of developmentally competent oocytes are still lacking in effectiveness. Although the explanted follicle system allows in vitro production of developmentally competent sheep oocytes at a level comparable to that in vivo, there are several disadvantages associated with it, making it an undesirable system for large-scale production. The best of the follicle-free approaches provides high levels of maturation and normal fertilization in vitro, but they are clearly deficient, as evidenced by the low percentages of oocytes forming morulae or blastocysts prior to transfer. Nevertheless, research to date indicates that if

certain conditions are met, this technology can result in the production of a small number of live births. When medium used to mature cumulus-enclosed oocytes is supplemented with membrana granulosa cells and/or follicular fluid, substantially increased incidences of developmental competency have been obtained in most studies. Whether the beneficial effects of these follicular supplements can be explained by their steroidal and/or gonadotropin properties is not definitively established. However, some of the studies on three of the most commonly used domestic species (pig, sheep, and cow) have demonstrated beneficial effects of estradiol, progesterone, FSH, and/or LH under certain circumstances. In the pig, oocyte cytoplasmic competency appears to be regulated by a soluble follicular fluid factor, the action of which is dependent on cumulus cell–oocyte coupling. In addition, progesterone appears to promote sperm decondensation in this species. In the sheep, balanced steroidal and gonadotropic profiles are required for promotion of oocyte developmental competency; specifically, estradiol in the presence of FSH and LH confers a beneficial effect upon this process. In the cow, as in the other two species examined, serum is required in the maturation medium in order to obtain developmentally competent oocytes. However, the need for utilizing estrous cow serum is controversial. Furthermore, supplementation of the maturation medium with membrana granulosa cells may not be absolutely necessary for the production of developmentally competent cow oocytes. Systems that do not require the use of estrous cow serum and membrana granulosa cells have the obvious advantage that they avoid the need for serum and granulosa cell collection from cows at specific stages of the estrous cycle. Utilization of the more simplified systems thereby substantially reduces the expense of this technology.

B. Production of Developmentally Competent Oocytes from Preantral Oocytes

1. Systems Used to Support Oocyte Growth In Vitro. The majority of oocytes in the population of small preantral follicles are of suboptimal size for support of normal development and generally undergo atresia in vivo, but with appropriate conditions for maintenance and growth in culture, there is the potential for utilization of the germ plasm represented by these oocytes. Although oocyte growth is not necessary for the acquisition of competency to undergo spontaneous maturation (see Section II.B), as oocytes increase in size, they acquire sequentially the ability to undergo GVBD, to produce a polar body, and, after fertilization, to support cleavage to the two-cell stage and subsequent development to the blastocyst stage (Eppig and Schroeder 1989). Therefore, since oocytes collected from preantral follicles are of a diameter less than the minimum consistent with acquisition of meiotic competency, any attempts to exploit the pool of oocytes in the preantral follicles must involve systems that support oocyte growth in vitro. The growth systems must also be capable of maintaining the oocyte in meiotic arrest, since optimal levels of normal embryonic development are obtained when insemination occurs at a specific time after the onset of maturation (see Section III.A.3) (Schroeder et al. 1988). To date, a follicle-free system for oocyte growth in culture has only been developed for the mouse (Eppig and Schroeder 1989). Although systems that support growth of preantral follicles have been developed for both the mouse (Qvist et al. 1990) and the rat (Daniel et al. 1989; Gore-Langton and Daniel 1990), only the latter is currently capable of supporting acquisition of oocyte meiotic competence (Gore-Langton and Daniel 1990).

The greatest number of preantral follicles are derived from sexually immature animals. In the case of the mouse, in vitro follicular growth is optimal when the cultures are begun with follicles recovered from 12–16-day-old pups (Qvist et al. 1990). In the case of the rat, 10–11-day-old pups provide isolated preantral follicles fully capable of growth under defined conditions (Daniel et al. 1989). The important determinants for growth of cultured mouse preantral follicles are the inclusion of stroma with the primary follicles, the use of culture dishes with hydrophobic membranes, supplementation of the medium with 50% postmenopausal human serum, and the presence of FSH (Qvist et al. 1990). In contrast, the isolated rat preantral follicle system does not require FSH to support both oocyte growth and the acquisition of meiotic competency of oocytes in vitro (Daniel et al. 1989), although cytoplasmic maturation of the meiotically mature oocytes is limited, as judged from the reduced normal fertilizability of such oocytes in vitro (Daniel et al. 1989). In addition, consistent with the known roles of FSH in the formation of antral follicles in vivo (for review, see Gore-Langton and Daniel 1990), antrum formation in preantral rat follicles in vitro never occurs in the absence of FSH, but may be induced by FSH and further enhanced by additional supplementation with estradiol (Gore-Langton and Daniel 1990).

In the case of the follicle-free system available for the growth of mouse oocytes, most of the work has been undertaken using 10–12-day-old pups in which the preantral follicles comprise the oocyte surrounded by about two layers of granulosa cells and the oocytes are in the mid-growth phase, measuring about 48–56 μm in diameter (Eppig and Downs 1987; Eppig and Schroeder 1989). Such oocytes are incompetent to undergo GVBD, but if cultured as cumulus-enclosed oocytes for 10–12 days in MEM with Earle's salts and 5% FBS, supplemented with either 2 mM hypoxanthine (Eppig and Downs 1987) or 0.05 mM IBMX (Eppig and Schroeder 1989), the oocytes will grow to diameters consistent with meiotic competency. In the case of the hypoxanthine system in which oocytes collected from 10–11-day-old mice were cultured for 12 days, the oocytes increased in diameter from 48 μm to 68 μm. In the case of the IBMX system in which oocytes collected from 12-day-old mice were cultured for 10 days, the oocytes increased in diameter from 56 μm to 68 μm. In each system, cultures were fed every other day by exchanging 1 ml of spent medium with 1 ml of fresh medium, and the phosphodiesterase inhibitor was used to maintain meiotic arrest, in the face of attainment of meiotic competency, as oocyte size increased. Although addition of progesterone, dihydrotestosterone, testosterone, and estradiol failed to enhance significantly the sizes of oocytes grown in vitro, estradiol in the growth medium significantly increased the frequency of GVBD in response to FSH in the maturation medium (Eppig et al. 1988; see Section III.B.2).

The presence of functional cumulus-oocyte coupling is of

critical importance in achieving maximum growth of the oocytes in vitro (Eppig 1977; Brower and Schultz 1982), as compared to oocyte growth which occurs in vivo (Eppig 1979; Herlands and Schultz 1984). Oocytes devoid of their cumulus granulosa cells, but cocultured with them (Bachvarova et al. 1980) or with fibroblasts (Canipari et al. 1984), remained viable in culture but exhibited a negligible increase in diameter. Although oocytes can establish metabolic cooperativity with other cell types in vitro, they exhibit a specific requirement for communication with granulosa cells in order to grow and to acquire developmental competency (Buccione et al. 1987).

2. Systems Used to Support Oocyte Maturation In Vitro. After oocytes are grown in vitro until they are large enough to mature, the growth medium containing phosphodiesterase inhibitor must be replaced by maturation medium lacking this inhibitor. In-vitro-grown rat oocytes undergo completion of maturation in such medium, but the incidence of both normal fertilization and development to the two-cell stage is low (8.3% and 7.6%, respectively; Daniel et al. 1989). In contrast, fertilization of in-vitro-grown mouse oocytes matured in the absence of phosphodiesterase inhibitor, but in the presence of FSH (1 μg/ml), results in production of normal two-cell embryos at levels comparable to that for either in-vivo- or in-vitro-matured oocytes from superovulated, 22-day-old mice (Eppig and Schroeder 1989). The basis of this beneficial action of FSH has not yet been established, but it may relate to a delaying action of FSH on the progression of meiotic resumption, since it retards the onset of GVBD in mouse oocytes (Salustri and Siracusa 1983). In addition, FSH can enhance the developmental potential of in-vivo-grown oocytes matured in vitro (Downs et al. 1986a) and increase the production of mature in-vitro-grown mouse oocytes (~80% exhibiting the first polar body as compared with ~40% in the absence of FSH; Eppig and Schroeder 1989). The possibility therefore exists that the reduced rate of normal fertilization in the in-vitro-grown rat oocytes (Daniel et al. 1989) may be accounted for by the absence of FSH in the maturation medium.

Despite the fact that production of normal two-cell mouse embryos from in-vitro-grown mouse oocytes is similar to that for in-vitro-matured oocytes obtained from 22-day-old mice, the transition from the two-cell stage to the expanded blastocyst stage was reduced. In addition, the numbers of cells within a blastocyst are significantly greater in embryos derived from oocytes isolated from 22-day-old mice, as compared to 18-day-old mice. The latter observation suggests that the oocytes of older mice are qualitatively different from oocytes of younger mice and that this difference may reflect the reduced developmental potential of the embryos obtained from in-vitro-grown oocytes, since only 5.1% live young were born following transfer of embryos obtained from in-vitro-grown oocytes to pseudopregnant foster mothers (Eppig and Schroeder 1989).

3. Conclusions. The following conclusions can be drawn from the above discussion regarding information currently available on obtaining developmentally competent mouse and rat oocytes from preantral follicles.

1. Although systems that support growth of isolated preantral mouse and rat follicles are available, only that for the rat provides meiotically competent oocytes, and such oocytes exhibit a low incidence of cytoplasmic competence.
2. A system is available that supports growth of preantral mouse oocytes, which enables the production of live young in foster mothers following meiotic maturation of grown oocytes in the presence of FSH, and after fertilization and embryonic development in vitro. However, the frequency of live young is low (5.1%), thus indicating the need for considerable improvements in this system. In light of the absolute requirement for functional granulosa cells in order for oocyte growth to occur, such improvements should include modifications to enhance granulosa cell metabolism, in addition to refinements targeting the oocyte itself.

IV. SUMMARY

This chapter has addressed the physiological parameters that regulate the production of mature, developmentally competent mammalian oocytes in vivo and has considered the application of these parameters to in vitro production of mature, developmentally competent oocytes. Since the size of the female germ line is finite and the natural reproductive capacity of the female mammal is further limited by the process of atresia, it is desirable to enhance the number of oocytes obtained from an-tral follicles in any single female by using superovulatory techniques. Significant progress has been made in laboratory species, particularly mousc and rat, regarding the production of developmentally competent oocytes from oocytes grown in vivo but matured in vitro. Thus, equivalent numbers of live births have been obtained following fertilization and transfer to pseudopregnant foster mothers with oocytes matured in vitro, as compared with those matured in vivo. Although progress in domestic species has not been as great, currently available systems have the capacity to support the production of oocytes capable of development to term, although the frequency of success is low.

To improve the follicle-free systems currently used to maintain oocytes of domestic species, several refinements could be adopted. These include (1) redefining the ranges of follicular size for those follicles chosen for oocyte donation; (2) establishing the maximum number of cumulus-enclosed oocytes that can be cultured in the smallest volume of medium; (3) applying careful selection criteria to the oocytes to exclude those lacking a complete, compact cumulus mass; (4) eliminating any exposure of oocytes to serum- or follicular-fluid-free medium; and (5) attempting to mimic in vitro the steroidogenic and gonadotropic profiles that accompany meiotic maturation in vivo.

To exploit the large pool of germ plasm present in preantral follicles, much of which would otherwise be lost to atresia, it will be necessary to develop systems for the in vitro growth and maturation of oocytes that support developmental competency and thus the capability for normal fertilization and embryo transfer. To date, such systems have only been developed for the rat and mouse, and they still require improvements because the frequencies of normal fertilization (rat) and live young (mouse) obtained from such in-vitro-grown oocytes are still low. There are considerable advantages to exploiting the pools of preantral follicles in other species. Perhaps the most important of these advantages relates to the possibility that a substantially increased number of developmentally competent ova can be obtained from a single female, compared with that derived solely from the antral follicles. In addition, this technology could have a further application in attempts to save endangered species, since the pool of preantral follicles of any recently deceased female could be utilized as a source of oocytes to be grown in culture for subsequent freezing at the GV stage (see Van Blerkom, this volume). Such a bank of meiotically competent oocytes could then be used at a later date for embryo transfer to suitable recipients following application of fertilization in vitro. Finally, the availability of such a large supply of developmentally competent oocytes would allow the development of techniques for the production of transgenics (see Ebert and Selgrath, this volume).

ACKNOWLEDGMENTS

The author thanks Kendall V. Baldwin and Alyce A. DeMarais for discussions and editorial assistance, Jennifer Ashley for printing the document, Gregory Hendricks for preparing Figure 2 for light microscopy, and Rachelle Dermer for printing the plates.

REFERENCES

Ahren, K., N. Dekel, L. Hamberger, T. Hillensjo, R. Hultborn, C. Magnusson, and A. Tsafriri. 1978. Metabolic and morphological changes produced by gonadotropins and cyclic AMP in the oocyte and cumulus oophorus of preovulatory rat follicles. *Ann. Biol. Anim. Biochim. Biophys.* **18:** 409.

Armstrong, D.T. and J.H. Dorrington. 1976. Androgens augment FSH-induced progesterone secretion by cultured rat granulosa cells. *Endocrinology* **99:** 1411.

Armstrong, D.T. and G. Evans. 1984. Hormonal regulation of reproduction: Induction of ovulation in sheep and goats with FSH preparations. *Proc. Int. Congr. Anim. Reprod. Artif. Insem.* **VII:** 8.

Armstrong, D.T. and M.A. Opavsky. 1988. Superovulation of immature rats by continuous infusion of follicle-stimulating hormone. *Biol. Reprod.* **39:** 511.

Austin, C.R. 1950. The fecundity of the immature rat following induced superovulation. *J. Endocrinol.* **6:** 293.

———. 1970. Ageing and reproduction: Post-ovulatory deterioration of the egg. *J. Reprod. Fertil.* (suppl.) **12:** 39.

Ayalon, D., A. Tsafriri, H.R. Lindner, T. Cordova, and A. Harell. 1972. Serum gonadotropin levels in proestrous rats in relation to the resumption of meiosis by the oocytes. *J. Reprod. Fertil.* **31:** 51.

Bachvarova, R., N.M. Baran, and A. Tejblum. 1980. Development of naked growing mouse oocytes *in vitro*. *J. Exp. Zool.* **211:** 159.

Baker, T.G. 1972. *Reproductive biology* (ed. H. Balin and S.R. Glasser), chapter 10. Excerpta Medica, Amsterdam.

———. 1979. The control of oogenesis in mammals. In *Ovarian follicular development and function* (ed. A.R. Midgley, Jr. and W.A. Sadler), p. 353. Raven Press, New York.

———. 1987. Oogenesis and ovulation. In *Germ cells and fertilization*, 2nd edition (ed. C.R. Austin and R.V. Short), p. 1. Cambridge University Press, Cambridge.

Baker, T.G., R.H.F. Hunter, and P. Neal. 1974. Studies on the maintenance of porcine Graafian follicles in organ culture. *Experentia* **31:** 133.

Balakier, H.G. and A.K. Tarkowski. 1980. The role of germinal vesicle karyoplasm in the development of male pronucleus in the mouse. *Exp. Cell. Res.* **128:** 79.

Barros, C. and G. Munoz. 1973. Sperm-egg interaction in immature hamster oocytes. *J. Exp. Zool.* **186:** 73.

Bazer, F.W., O.W. Robison, A.J. Clawson, and L.C. Ulberg. 1969. Uterine capacity at two stages of gestation in gilts following embryo superinduction. *J. Anim. Sci.* **29:** 30.

Betteridge, K.J. 1977. Superovulation. In *Embryo transfer in farm animals: A review of techniques and applications* (ed. K.J. Betteridge), p. 1. Canadian Department of Agriculture (Monograph 16), Ottawa.

Biggers, J.D., D.C. Whittingham, and R.P. Donahue. 1967. The pattern of energy metabolism in the mouse oocyte and zygote. *Proc. Natl. Acad. Sci.* **58:** 560.

Bindon, B.M. and L.R. Piper. 1977. Induction of ovulation in sheep and cattle by injections of PMSG and ovine anti-PMSG immune serum. *Theriogenology* **8:** 171. (Abstr.)

Blandau, R.J. 1952. The female factor in fertility and infertility. I. The effects of the pronuclei in rat ova. *Fertil. Steril.* **3:** 349.

Bodemer, C.W., R.E. Rumery, and R.J. Blandau. 1959. Studies on induced ovulation in the intact immature hamster. *Fertil. Steril.* **10:** 350.

Borum, K. 1961. Oogenesis in the mouse: A study of the meiotic prophase. *Exp. Cell. Res.* **24:** 495.

Bousquet, D., A. Goff, W.A. King, and T. Greve. 1988. Fertilization *in vitro* of bovine oocytes: Analysis of some factors affecting the fertilization rates. *Can. J. Vet. Res.* **52:** 277.

Boving, B.G. 1956. Rabbit blastocyst distribution. *Am. J. Anat.* **98:** 403.

Brower, P.T. and R.M. Schultz. 1982. Intercellular communication between granulosa cells and mouse oocytes: Existence and possible nutritional role during oocyte growth. *Dev. Biol.* **90:** 144.

Buccione, R., S. Cecconi, C. Tatone, F. Mangia, and R. Colonna. 1987. Follicle cell regulation of mammalian oocyte growth. *J. Exp. Zool.* **242:** 351.

Buttle, H.L. and J.L. Hancock. 1967. The control of ovulation in the sow. *J. Reprod. Fertil.* **14:** 485.

Byskov, A.G. 1987. Primordial germ cells and regulation of meiosis. In *Germ cells and fertilization*, 2nd edition (ed. C.R. Austin and R.V. Short), p. 1. Cambridge University Press, England.

Byyney, R.L., D.N. Orth, and S. Cohen. 1974. Epidermal growth factor: Effects of androgens and adrenergic agents. *Endocrinology* **95:** 776.

Callesen, H., T. Greve, and P. Hyttel. 1986. Preovulatory endocrinology and oocyte maturation in superovulated cattle. *Theriogenology* **25:** 71.

Calvin, H.I., K. Grosshans, and E. Blake. 1986. Estimation and manipulation of glutathione levels in prepubertal mouse ovaries and ova: Relevance of sperm nucleus transformations in the fertilized egg. *Gamete Res.* **14:** 265.

Canipari, R., F. Palombi, M. Riminucci, and F. Mangia. 1984. Early programming of maturation competence in mouse oogenesis. *Dev. Biol.* **12:** 519.

Chang, M.C. 1955a. Fertilization and normal development of follicular oocytes in the rabbit. *Science* **121:** 867.

———. 1955b. The maturation of rabbit oocytes in culture and their maturation, activation, fertilization, and subsequent development in the fallopian tube. *J. Exp. Zool.* **128:** 378.

Channing, C.P. and A. Tsafriri. 1977. Mechanism of action of luteinizing hormone and follicle-stimulating hormone on the ovary *in vitro*. *Metabolism* **26:** 413.

Cole, N.H. 1937. Superfecundity in rats treated with mare gonadotropic hormone. *Am. J. Physiol.* **119:** 704.

Critser, E.S., M.L. Leibfried-Rutledge, W.H. Eyestone, D.L. Northey, and N.L. First. 1986. Acquisition of developmental competence during maturation in vitro. *Theriogenology* **25:** 150. (Abstr.)

Cross, P.C. and R.L. Brinster. 1970. In vitro development of mouse eggs. *Biol. Reprod.* **3:** 298.

———. 1974. Leucine uptake and incorporation at three stages of mouse oocyte maturation. *Exp. Cell Res.* **86:** 43.

Crozet, N., D. Huneau, V. Desmedt, M.C. Theron, D. Szollosi, S. Torres, and C. Sevellec. 1987. *In vitro* fertilization with normal development in the sheep. *Gamete Res.* **16:** 159.

Cyert, M.S. and M.W. Kirschner. 1988. Regulation of MPF activity in vitro. *Cell* **53:** 185.

Daniel, S.A.J., D.T. Armstrong, and R.E. Gore-Langton. 1989. Growth and development of rat oocytes *in vitro*. *Gamete Res.* **24:** 109.

De Felici, M. and G. Siracusa. 1982. Spontaneous hardening of the zona pellucida of mouse oocytes during *in vitro* culture. *Gamete Res.* **6:** 107.

De Felici, M., A. Salustri, and G. Siracusa. 1985. *Spontaneous* hardening of the zona pellucida of mouse oocytes during *in vitro* culture. II. The effect of follicular fluid and glycosaminoglycans. *Gamete Res.* **12:** 227.

Dekel, N. and W.H. Beers. 1978. Rat oocyte maturation *in vitro*: Relief of cyclic AMP inhibition by gonadotropins. *Proc. Natl. Acad. Sci.* **75:** 4369.

———. 1980. Development of the rat oocyte in vitro: Inhibition of maturation in the presence or absence of the cumulus oophorus. *Dev. Biol.* **75:** 247.

Dekel, N. and I. Sherizly. 1985. Epidermal growth factor induces maturation of rat follicle-enclosed oocytes. *Endocrinology* **116:** 406.

Dermody, W.C., W.C. Foote, and C.V. Hulet. 1970. Effects of season and progesterone synchronization on ovulation rate in mature western range ewes. *J. Anim. Sci.* **30:** 214.

Donahue, R.P. 1968. Maturation of the mouse oocyte in vitro. I. Sequence and timing of nuclear progression. *J. Exp. Zool.* **169:** 237.

———. 1972. The relation of oocyte maturation to ovulation in mammals. In *Oogenesis* (ed. J.D. Biggers and A.W. Schuetz), p. 413. University Park Press, Baltimore, Maryland.

Downs, A.C., S.M. Schroeder, and J.J. Eppig. 1986a. Developmental capacity of mouse oocytes following maintenance of meiotic arrest *in vitro*. *Gamete Res.* **15:** 305.

———. 1986b. Serum maintains the fertilizability of mouse oocytes matured *in vitro* by preventing hardening of the zona pellucida. *Gamete Res.* **15:** 115.

Downs, S.M. 1989. Specificity of epidermal growth factor action on maturation of the murine oocyte and cumulus oophorus *in vitro*. *Biol. Reprod.* **41:** 371.

Downs, S.M., S.A.J. Daniel, and J. Eppig. 1988. Induction of maturation in cumulus cell-enclosed mouse oocytes by follicle-stimulating hormone and epidermal growth factor: Evidence for a positive stimulus of somatic cell origin. *J. Exp. Zool.* **245:** 86.

Dzuik, P.J. 1965. Timing of maturation and fertilization of the sheep egg. *Anat. Rec.* **153:** 211.

———. 1968. Effect of number of embryos and uterine space on embryo survival in the pig. *J. Anim. Sci.* **26:** 673.

———. 1973. Occurrence, control and induction of ovulation in pigs, sheep and cows. In *Handbook of physiology* (ed. R.O. Greep), vol. II, p. 143. American Physiological Society, Washington, D.C.

Edwards, R.G. 1962. Meiosis in ovarian oocytes of adult mammals. *Nature* **196:** 446.

Edwards, R.G. and A.H. Gates. 1959. Timing of the stages of the maturation divisions, ovulation, fertilization, and the first cleavage of eggs of adult mice treated with gonadotropins. *J. Endocrinol.* **18:** 292.

Edwards, R.G. and R.E. Fowler. 1960. Superovulation treatment of adult mice; their subsequent natural fertility and response to further treatment. *J. Endocrinol.* **21:** 147.

Eiler, H. and A.V. Nalbandov. 1977. Sex steroids in follicular fluid and blood plasma during the estrous cycle of pigs. *Endocrinology* **100:** 331.

Eppig, J.J. 1977. Mouse development *in vitro* with various culture systems. *Dev. Biol.* **60:** 371.

———. 1979. A comparison between oocyte growth in co-culture with granulosa cells and oocytes with granulosa cell-oocyte junctional contact maintained in vitro. *J. Exp. Zool.* **209:** 345.

———. 1982. The relationship between cumulus-oocyte coupling, oocyte meiotic maturation, and cumulus expansion. *Dev. Biol.* **89:** 268.

Eppig, J.J. and S.M. Downs. 1987. The effect of hypoxanthine on mouse oocyte growth and development *in vitro*: Maintenance of meiotic arrest and gonadotropin-induced oocyte maturation. *Dev. Biol.* **119:** 313.

———. 1988. Gonadotropin-induced murine oocyte maturation in vivo is not associated with decreased cyclic adenosine monophosphate in the oocyte-cumulus cell complex. *Gamete Res.* **20:** 125.

Eppig, J.J. and A.C. Schroeder. 1989. Capacity of mouse oocytes from preantral follicles to undergo embryogenesis and development to live young after growth, maturation, and fertilization *in vitro*. *Biol. Reprod.* **41:** 268.

Eppig, J.J., S.A.J. Daniel, and A.C. Schroeder. 1988. Growth and development of mouse oocytes *in vitro*. *Ann. N.Y. Acad. Sci.* **541:** 205.

Erickson, G.F. and K.J. Ryan. 1976. Spontaneous maturation of oocytes isolated from ovaries of immature hypophysectomized rats. *J. Exp. Zool.* **195:** 153.

Feng, P., K.J. Catt, and M. Knecht. 1988. Transforming growth factor β stimulates meiotic maturation of the rat oocyte. *Endocrinology* **122:** 181.

Feng, P., M. Knecht, and K.J. Catt. 1987. Hormonal control of epidermal growth factor receptors by gonadotropins during granulosa cell differentiation. *Endocrinology* **120:** 1121.

First, N.L. and J.J. Parrish. 1987. *In vitro* fertilization of ruminants. *J. Reprod. Fertil.* (suppl.) **34:** 151.

Fleming, A.D. and R. Yanagamachi. 1980. Superovulation and superpregnancy in the golden hamster. *Dev. Growth Differ.* **22:** 103.

Fleming, A.D., G. Evans, E.A. Walton, and D.T. Armstrong. 1985. Developmental capability of rat oocytes matured *in vitro* in defined medium. *Gamete Res.* **12:** 255.

Foote, W.D. and C. Thibault. 1969. Recherches experimentales sur la maturation *in vitro* des oocytes de truie et de veau. *Ann. Biol. Anim. Biochim. Biophys.* **9:** 329.

Fortune, J.E. and D.T. Armstrong. 1978. Hormonal control of 17β-estradiol biosynthesis in proestrous rat follicles: Estradiol production by isolated theca vs. granulosa. *Endocrinology* **102:** 227.

Fowler, R.E. and R.G. Edwards. 1957. Induction of superovulation and pregnancy in mature mice by gonadotrophins. *J. Endocrinol.* **15:** 374.

Frydman, R., R.G. Forman, J. Belaisch-Allart, A. Hazout, J.O. Rainhorn, N. Fries, and J. Testart. 1988. Improvements in ovarian stimulation for *in vitro* fertilization. *Ann. N.Y. Acad. Sci.* **541:** 30.

Fukuda, Y., M. Ichikawa, K. Naito, and Y. Toyoda. 1990. Birth of normal calves resulting from bovine oocytes matured, fertilized, and cultured with cumulus cells *in vitro* up to the blastocyst stage. *Biol. Reprod.* **42:** 114.

Fukui, Y. and H. Ono. 1989. Effects of sera, hormones and granulosa cells added to culture medium for *in-vitro* maturation, fertilization, cleavage and development of bovine oocytes. *J. Reprod. Fertil.* **86:** 501.

Fukui, Y. and Y. Sakuma. 1980. Maturation of bovine oocytes cultured *in vitro*: Relation to ovarian activity, follicular size and the presence or absence of cumulus cells. *Biol. Reprod.* **22:** 669.

Fukushima, M. and Y. Fukui. 1985. Effects of gonadotropins and steroids on the subsequent fertilizability of extrafollicular bovine oocytes cultured *in vitro*. *Anim. Reprod. Sci.* **9:** 323.

Gardner, R.L. and J. Rossant. 1979. Investigation of the fate of 4.5 day post-coitum mouse inner cell mass cells by blastocyst injection. *J. Embryol. Exp. Morphol.* **52:** 141.

Gates, A.H. 1971. Maximizing yield and developmental uniformity of eggs. In *Methods in mammalian embryology* (ed. J.C. Daniel, Jr.), p. 64. W.H. Freeman, San Francisco.

Gibson, E.W., S.C. Jaffe, J.F. Lasley, and B.N. Day. 1963. Reproductive performance in swine following superovulation. *J. Anim. Sci.* **22:** 858. (Abstr.)

Gore-Langton, R.E. and S.A.J. Daniel. 1990. Follicle stimulating hormone and estradiol regulate antrum-like reorganization of granulosa cells in rat preantral follicle cultures. *Biol. Reprod.* **43:** 65.

Gospodarowicz, D. and H. Bialecki. 1978. The effects of the epidermal and fibroblast growth factors on the replicative lifespan of cultured bovine granulosa cells. *Endocrinology* **103:** 854.

———. 1979. Fibroblast and epidermal growth factors are mitogenic agents for cultured granulosa cells of rodent, porcine, and human origin. *Endocrinology* **104:** 757.

Gospodarowicz, D., R. Ill, and C.R. Birdwell. 1977. Effects of fibroblast and epidermal growth factors on ovarian cell proliferation *in vitro*. I. Characterization of the response of granulosa cells

to FGF and EGF. *Endocrinology* **100:** 1108.
Graves, C.N. and P.J. Dzuik. 1968. Control of ovulation in dairy cattle with human chorionic gonadotrophin after treatment with 6-methyl-17-acetoxyprogesterone. *J. Reprod. Fertil.* **17:** 169.
Greenwald, G.S. 1962. Analysis of superovulation in the adult hamster. *Endocrinology* **71:** 378.
Gwatkin, R.B.L. and A.A. Haidri. 1973. Requirements for the maturation of hamster oocytes *in vitro*. *Exp. Cell. Res.* **76:** 1.
Haidri, A.A. and R.B.L. Gwatkin. 1973. Requirements for the maturation of hamster oocytes from preovulatory follicles. *J. Reprod. Fertil.* **35:** 173.
Hammond, J.M., J.L.S. Baranao, D. Skaleris, A.B. Knight, J. Romanus, and M.M. Rechler. 1985. Production of insulin-like factors by ovarian granulosa cells. *Endocrinology* **117:** 2553.
Hashimoto, N. and T. Kishimoto. 1988. Regulation of meiotic metaphase by a cytoplasmic maturation-promoting factor during mouse oocyte maturation. *Dev. Biol.* **126:** 242.
Hay, M.P. and R.M. Moor. 1973. The Graafian follicle of the sheep: Relationships between gonadotropins, steroid production, morphology, and oocyte maturation. *Ann. Biol. Anim. Biochim. Biophys.* **13:** 241.
Hensleigh, H.C. and R.H.F. Hunter. 1985. In vitro maturation of bovine cumulus enclosed primary oocytes and their subsequent in vitro fertilization and cleavage. *J. Dairy Sci.* **68:** 1456.
Herlands, R.L. and R.M. Schultz. 1984. Regulation of mouse oocyte growth: Probable nutritional role for intercellular communication between follicle cells and oocytes in oocyte growth. *J. Exp. Zool.* **229:** 317.
Hillensjo, T. and C.P. Channing. 1980. Gonadotropin stimulation of steroidogenesis and cellular dispersion in cultured porcine cumuli oophori. *Gamete Res.* **3:** 233.
Hillensjo, T., C. Magnusson, U. Svensson, and H. Thelander. 1981. Effect of luteinizing hormone and follicle-stimulating hormone on progesterone synthesis by cultured rat cumulus cells. *Endocrinology* **108:** 1920.
Homa, S.T. 1988. Effects of cyclic AMP on the spontaneous meiotic maturation of cumulus-free bovine oocytes cultured in chemically defined medium. *J. Exp. Zool.* **248:** 222.
Hsu, C.-J., S.D. Holmes, and J.M. Hammond. 1987. Ovarian epidermal growth factor-like activity. Concentrations in porcine follicular fluid during follicular enlargement. *Biochem. Biophys. Res. Commun.* **147:** 242.
Hsueh, A.J.W., T.H. Welsh, and P.B.C. Jones. 1981. Inhibition of ovarian and testicular steroidogenesis by epidermal growth factor. *Endocrinology* **108:** 2002.
Hubbard, C.J. 1986. Cyclic AMP changes in the component cells of Graafian follicles: Possible influences on maturation in the follicle-

enclosed oocytes of hamsters. *Dev. Biol.* **118:** 343.

Hubbard, C.J. and G.S. Greenwald. 1982. Cyclic nucleotides, DNA, and steroid levels in ovarian follicles and corpora lutea of the cyclic hamster. *Biol. Reprod.* **26:** 230.

Hulet, C.V., W.C. Foote, and D.A. Price. 1969. Ovulation rate and subsequent lamb production in the nulliparous and primiparous ewe. *J. Anim. Sci.* **28:** 512.

Hunter, R.H.F. 1966. The effect of superovulation on fertilisation and embryonic survival in the pig. *Anim. Prod.* **8:** 457.

Hunter, R.H.F. and C. Polge. 1966. Maturation of follicular oocytes in the pig after injection of human chorionic gonadotropin. *J. Reprod. Fertil.* **12:** 525.

Hyttel, P., H. Callesen, and T. Greve. 1986. Ultrastructural features of preovulatory oocyte maturation in superovulated cattle. *J. Reprod. Fertil.* **76:** 645.

Iritani, A., K. Niwa, and H. Imai. 1978. Sperm penetration *in vitro* of pig follicular oocytes matured in culture. *J. Reprod. Fertil.* **54:** 379.

Itoh, M., M. Igarashi, K. Yamada, Y. Hasegawa, M. Seki, Y. Eto, and H. Shibai. 1990. Activin A stimulates meiotic maturation of the rat oocyte in vitro. *Biochem. Biophys. Res. Commun.* **166:** 1

Iwamatsu, T. and M.C. Chang. 1972. Sperm penetration *in vitro* of mouse oocytes at various times during maturation. *J. Reprod. Fertil.* **31:** 237.

Iwamatsu, T. and R. Yanagimachi. 1975. Maturation *in vitro* of ovarian oocytes of prepubertal and adult hamsters. *J. Reprod. Fertil.* **45:** 83.

Janson, P.O., M. Brannstrom, P.V. Holmes, and J. Sogn. 1988. Studies on the mechanism of ovulation using the model of the isolated ovary. *Ann. N.Y. Acad. Sci.* **541:** 22.

Jinno, M., B.A. Sandow, and G.D. Hodgen. 1989. Enhancement of the developmental potential of mouse oocytes matured *in vitro* by gonadotropins and ethylenediaminetetraacetic acid (EDTA). *J. In Vitro Fert. Embryo Transfer* **6:** 36.

Jones, A.A., G.S. Acosta, J.E. Garcia, R.E. Bernardus, and Z. Rosenwaks. 1985. The effect of follicle-stimulating hormone without additional luteinizing hormone on follicular stimulation and oocyte development in normal ovulatory women. *Fertil. Steril.* **43:** 696.

Jones, P.B.C., T.H. Welsh, Jr., and A.J.W. Hsueh. 1982. Regulation of ovarian progestin production by epidermal growth factor in cultured granulosa cells. *J. Biol. Chem.* **257:** 11268.

Kanatani, H., H. Shirai, K. Nakanishi, and T. Kurokawa. 1969. Isolation and identification of meiosis-inducing substance in starfish *Asterias anurensis*. *Nature* **211:** 273.

Kennelly, J.J. and R.H. Foote. 1965. Superovulatory response of pre- and post-pubertal rabbits to commercially available gonadotropins. *J. Reprod. Fertil.* **9:** 177.

Komar, A. 1982. Fertilization of parthenogenetically activated mouse eggs. I. Behaviour of sperm nuclei in the cytoplasm of parthenogenetically activated eggs. *Exp. Cell. Res.* **139:** 361.

Lacroix, E., W. Eechante, and I. Leusen. 1974. The biosynthesis of estrogens by cow follicles. *Steroids* **23:** 337.

Larsen, W.J. and S.E. Wert. 1988. Roles of cell gap junctions in gametogenesis and in early embryonic development. *Tissue & Cell* **20:** 809.

Larsen, W.J., S.E. Wert, and G.D. Brunner. 1986. A dramatic loss of cumulus cell gap junctions is correlated with germinal vesicle breakdown in rat oocytes. *Dev. Biol.* **113:** 517.

Leibfried, L. and N.L. First. 1979. Characterization of bovine follicular oocytes and their ability to mature *in vitro*. *J. Anim. Sci.* **48:** 76.

———. 1980. Effect of bovine and porcine follicular fluid and granulosa cells on maturation of oocytes *in vitro*. *Biol. Reprod.* **23:** 699.

Leibfried, M.L. and B.D. Bavister. 1983. Fertilizability of *in vitro* matured oocytes from golden hamsters. *J. Exp. Zool.* **226:** 481.

Leibfried-Rutledge, M.L., E.S. Critser, and N.L. First. 1986. Effects of fetal calf serum and bovine serum albumin on *in vitro* maturation and fertilization of bovine and hamster cumulus-oocyte complexes. *Biol. Reprod.* **35:** 850.

Leibfried-Rutledge, M.L., E.S. Critser, W.H. Eyestone, D.L. Northey, and N.L. First. 1987. Development potential of bovine oocytes matured *in vitro* or *in vivo*. *Biol. Reprod.* **36:** 376.

Lindner, H.R., A. Tsafriri, M.E. Lieberman, U. Zor, Y. Koch, S. Bauminger, and A. Barnea. 1974. Gonadotropin action on cultured Graafian follicles: Induction of maturation division of the mammalian oocyte and differentiation of the luteal cell. *Recent Prog. Horm. Res.* **30:** 79.

Longenecker, D.E. and B.N. Day. 1968. Fertility level of sows superovulated at post weaning estrus. *J. Anim. Sci.* **27:** 709.

Lu, K.H., I. Gordon, M. Gallagher, and H. McGovern. 1987. Pregnancy established in cattle by transfer of embryos derived from *in vitro* fertilization of oocytes matured *in vitro*. *Vet. Rec.* **121:** 259.

Lutterbach, A., R.A. Koll, and G. Brem. 1987. *In vitro* maturation of bovine oocytes in co-culture with granulosa cells and their subsequent fertilization and development. *Zuchthygiene* **22:** 145.

Magnusson, C., T. Hillensjo, A. Tsafriri, R. Hultborn, and K. Ahren. 1977. Oxygen consumption of maturing rat oocytes. *Biol. Reprod.* **17:** 9.

Makris, A. and K.J. Ryan. 1975. Progesterone, androstenedione, testosterone, estrone and estradiol synthesis in hamster ovarian follicle cells. *Endocrinology* **96:** 694.

———. 1977. Aromatase activity of isolated and recombined hamster granulosa cells and theca. *Steroids* **29:** 65.

Mashiach, S., J. Dor, M. Goldenberg, J. Shalev, J. Blankstein, E. Rudak, Z. Shoam, Z. Finelt, L. Nebel, B. Goldman, M. and Z. Ben-

Rafael. 1988. Protocols for induction of ovulation: The concept of programmed cycles. *Ann. N.Y. Acad. Sci.* **541:** 37.

Masui, Y. and C.L. Markert. 1971. Cytoplasmic control of nuclear behavior during meiotic maturation in frog oocytes. *J. Exp. Zool.* **177:** 129.

Mattioli, M., G. Galeati, M.L. and E. Seren. 1988a. Follicular factors influence oocyte fertilizability by modulating the intercellular cooperation between cumulus cells and oocyte. *Gamete Res.* **21:** 223.

Mattioli, M., G. Galeati, M.L. Bacci, and E. Seren. 1988b. Effect of follicle somatic cells during pig oocyte maturation on egg penetrability and male pronucleus formation. *Gamete Res.* **20:** 177.

McGaughey, R.W. 1983. Regulation of oocyte maturation. *Oxford Rev. Reprod. Biol.* **5:** 107.

McGaughey, R.W. and M.C. Chang. 1969. Meiosis of mouse oocytes before and after sperm penetration. *J. Exp. Zool.* **170:** 397.

McGaughey, R.W. and C. Polge. 1971. Cytogenetic analysis of pig oocytes matured in vitro. *J. Exp. Zool.* **176:** 383.

McGaughey, R.W., C. Racowsky, V. Rider, K.V. Baldwin, A.A. DeMarais, and S.W. Webster. 1990. Ultrastructural correlates of meiotic maturation in mammalian oocytes. *J. Electron Microsc. Tech.* **16:** 257.

McLaren, A. 1967. Factors affecting the variation in response of mice to gonadotrophic hormones. *J. Endocrinol.* **37:** 147.

———. 1988. The developmental history of female germ cells in mammals. *Oxford Rev. Reprod. Biol.* **10:** 162.

Meister, A. and M.E. Anderson. 1983. Glutathione. *Annu. Rev. Biochem.* **52:** 711.

Merchant, H. and M.C. Chang. 1971. An electron microscopic study of mouse eggs matured in vivo and in vitro. *Anat. Rec.* **171:** 21.

Miller, B.G. and D.T. Armstrong. 1981a. Effects of superovulatory dose of pregnant mare serum gonadotropin on ovarian function, serum estradiol, and progesterone levels and early development in immature rats. *Biol. Reprod.* **25:** 261.

———. 1981b. Superovulatory doses of pregnant mare serum gonadotropin cause delayed implantation and infertility in immature rats. *Biol. Reprod.* **25:** 253.

Mintz, B. and E.S. Russell. 1957. Gene-induced embryological modifications of primordial germ cells in the mouse. *J. Exp. Zool.* **134:** 207.

Mondschein, J.S. and D.W. Schomberg. 1981. Growth factors modulate gonadotropin receptor induction in granulosa cell cultures. *Science* **211:** 1179.

Moor, R.M. and J.P. Heslop. 1981. Cyclic AMP in mammalian follicle cells and oocytes during maturation. *J. Exp. Zool.* **216:** 205.

Moor, R.M. and A.O. Trounson. 1977. Hormonal and follicular factors affecting maturation of sheep oocytes *in vitro* and their subsequent developmental capacity. *J. Reprod. Fertil.* **49:** 101.

Moor, R.M., C. Polge, and S.M. Willadsen. 1980a. Effect of follicular steroids on the maturation and fertilization of mammalian oocytes. *J. Embryol. Exp. Morphol.* **56:** 319.

Moor, R.M., M.W. Smith, and R.M.C. Dawson. 1980b. Measurement of intercellular coupling between oocyte and cumulus cells using intracellular markers. *Exp. Cell. Res.* **16:** 15.

Moor, R.M., J.C. Osborn, D.G. Cran, and D.E. Walters. 1981. Selective effect of gonadotrophins on cell coupling, nuclear maturation and protein synthesis in mammalian oocytes. *J. Embryol. Exp. Morphol.* **61:** 347.

Moore, P.J. and G.S. Greenwald. 1974. Seasonal variation in ovarian responsiveness of the cycling hamster to PMSG. *J. Reprod. Fertil.* **36:** 219.

Motlik, J. 1972. Cultivation of pig oocytes in vitro. *Folia Biol.* **18:** 345.

Motlik, J., N. Crozet, and J. Fulka. 1984. Meiotic competence *in vitro* of pig oocytes isolated from early antral follicles. *J. Reprod. Fertil.* **72:** 323.

Murdoch, W.J. 1988. Disruption of cellular associations within the granulosal compartment of periovulatory ovine follicles: Relationship to maturation of the oocyte and regulation by prostaglandins. *Cell Tiss. Res.* **252:** 459.

Murphy, B.D., R.J. Mapletoft, J. Manns, and W.D. Humphrey. 1984. Variability in gonadotrophin preparations as a factor in the superovulatory response. *Theriogenology* **21:** 117.

Naito, K., Y. Fukuda, and Y. Toyoda. 1988. Effects of porcine follicular fluid on male pronucleus formation in porcine oocytes matured *in vitro*. *Gamete Res.* **21:** 289.

Neal, P. and T.G. Baker. 1973. Response of mouse ovaries in vivo and in organ culture to pregnant mare's serum gonadotropin and human chorionic gonadotropin. I. Examination of critical time intervals. *J. Reprod. Fertil.* **33:** 399.

Neufeld, G., N. Ferrara, L. Schweigerer, R. Mitchell, and D. Gospodarowicz. 1987. Bovine granulosa cells produce basic fibroblast growth factor. *Endocrinology* **121:** 597.

Newcomb, R., W.B. Christie, and L.E.A. Rowson. 1978. Birth of calves after *in vivo* fertilization of oocytes removed from follicles and matured *in vitro*. *Vet. Rec.* **102:** 461.

Niwa, K. and M.C. Chang. 1975. Fertilization of rat eggs in vitro at various times before and after ovulation with special reference to fertilization of ovarian oocytes matured in culture. *J. Reprod. Fertil.* **43:** 435.

Nuti, K.M., B.N. Sridharan, and R.K. Meyer. 1975. Reproductive biology of PMSG-primed immature female rats. *Biol. Reprod.* **13:** 38.

O, W.-S., D.M. Robertson, and D.M. de Kretser. 1989. Inhibin as an oocyte meiotic inhibitor. *Mol. Cell. Endocrinol.* **62:** 307.

Odor, D.L. 1955. The temporal relation of the first maturation division of rat ova to the onset of heat. *Am. J. Anat.* **97:** 461.

Osborn, J.C. and R.M. Moor. 1983. The role of the steroid signals in the maturation of mammalian oocytes. *J. Steroid Biochem.* **19:** 133.

Papkoff, H. 1974. Chemical and biological properties of the subunits of pregnant mare serum gonadotrophin. *Biochem. Biophys. Res. Commun.* **58:** 397.

Parkes, A.S. 1943. Induction of superovulation and superfecundation in rabbits. *J. Endocrinol.* **3:** 268.

Paules, R.S., R. Buccione, R.C. Moschel, G.F. Vande Woude, and J.J. Eppig. 1989. Mouse *Mos* protooncogene product is present and functions during oogenesis. *Proc. Natl. Acad. Sci.* **86:** 5395.

Pavlok, A. 1981. Penetration of hamster and pig zona-free eggs by boar ejaculated spermatozoa preincubated *in vitro*. *Int. J. Fertil.* **26:** 101.

Perrault, S.D., R.R. Barbee, and V.L. Slott. 1988. Importance of glutathione in the acquisition and maintenance of sperm nuclear decondensing activity in maturing hamster oocytes. *Dev. Biol.* **125:** 181.

Perrault, S.D., R.A. Wolff, and B.R. Zirkin. 1984. The role of disulfide bond reduction during mammalian sperm nuclear decondensation *in vivo*. *Dev. Biol.* **101:** 160.

Peters, H. 1970. Migration of gonocytes into the mammalian gonad and their differentiation. *Philos. Trans. R. Soc. Lond. B Biol. Sci.* **259:** 91.

Pincus, G. and E.V. Enzmann. 1935. The comparative behaviour of mammalian eggs *in vivo* and *in vitro*. *J. Exp. Med.* **62:** 665.

Quirke, J.F. and I. Gurdun. 1971. Culture and fertilization of sheep ovarian oocytes. III. Evidence on fertilization in the sheep oviduct based on pronucleate and cleavage eggs. *J. Agric. Sci.* **76:** 375.

Qvist, R., L.F. Blackwell, H. Bourne, and J.B. Brown. 1990. Development of mouse ovarian follicles from primary to preovulatory stages *in vitro*. *J. Reprod. Fertil.* **89:** 169.

Racowsky, C. 1983. Androgenic modulation of cyclic adenosine monophosphate (cAMP)-dependent meiotic arrest. *Biol. Reprod.* **28:** 74.

———. 1984. Effect of forskolin on the spontaneous maturation and cyclic AMP content of rat oocyte-cumulus complexes. *J. Reprod. Fertil.* **72:** 107.

———. 1985a. Effect of forskolin on the spontaneous maturation and cyclic AMP content of hamster oocyte-cumulus complexes. *J. Exp. Zool.* **234:** 87.

———. 1985b. Antagonistic actions of estradiol and tamoxifen upon forskolin-dependent meiotic arrest, intercellular coupling and the cyclic AMP content of hamster oocyte-cumulus complexes. *J. Exp. Zool.* **234:** 251.

———. 1985c. Effect of forskolin on maintenance of meiotic arrest and stimulation of cumulus expansion, progesterone and cyclic

AMP production by pig oocyte-cumulus complexes. *J. Reprod. Fertil.* **74:** 9.

———. 1986. The releasing action of calcium upon cyclic AMP-maintained meiotic arrest in hamster oocytes. *J. Exp. Zool.* **239:** 263.

Racowsky, C. and K.V. Baldwin. 1989a. *In vitro* and *in vivo* studies reveal that hamster oocyte meiotic arrest is maintained only transiently by follicular fluid, but persistently by membrana/cumulus granulosa cell contact. *Dev. Biol.* **134:** 297.

———. 1989b. Modulation of intrafollicular oestradiol in explanted hamster follicles does not affect oocyte meiotic status. *J. Reprod. Fertil.* **87:** 409.

Racowsky, C. and R.A. Satterlie. 1985. Metabolic, fluorescent dye and electrical coupling between hamster oocytes and cumulus cells during meiotic maturation *in vivo* and *in vitro*. *Dev. Biol.* **108:** 191.

———. 1987. Decreases in heterologous metabolic and dye coupling, but not in electrical coupling, accompany meiotic resumption in hamster oocyte-cumulus complexes. *Eur. J. Cell. Biol.* **43:** 283.

Racowsky, C., R.C. Hendricks, and K.Y. Baldwin. 1989a. Direct effects of nicotine on the meiotic maturation of hamster oocytes. *Reprod. Toxicol.* **3:** 13.

Racowsky, C., K.V. Baldwin, C.A. Larabell, A.A. DeMarais, and C.J. Kazilek. 1989b. Down-regulation of membrana granulosa cell gap junctions is correlated with irreversible commitment to resume meiosis in golden Syrian hamster oocytes. *Eur. J. Cell. Biol.* **49:** 244.

Rowlands, I.W. 1944. The production of ovulation in the immature rat. *J. Endocrinol.* **3:** 384.

Roy, S.K. and G.S. Greenwald. 1985. An enzymatic method for dissociation of intact follicles from the hamster ovary: Histological and quantitative aspects. *Biol. Reprod.* **32:** 203.

Russell, J.B., M.L. Polan, and A.H. De Cherney. 1986. The use of pure follicle-stimulating hormone for ovulation induction in normal ovulatory women in an *in vitro* fertilization program. *Fertil. Steril.* **45:** 829.

Sagata, N., M. Oskarsson, S. Showalter, and G.F. Vande Woude. 1988. Function of c-*mos* proto-oncogene product in meiotic maturation in *Xenopus* oocytes. *Nature* **335:** 519.

Sagata, N., N. Watanabe, G.F. Vande Woude, and Y. Ikawa. 1989a. The c-*mos* proto-oncogene product is a cytostatic factor responsible for meiotic arrest in vertebrate eggs. *Nature* **342:** 512.

Sagata, N., I. Daar, M. Oskarsson, S. Showalter, and G.F. Vande Woude. 1989b. The product of the *mos* proto-oncogene as a candidate "initiator" for oocyte maturation. *Science* **245:** 643.

Salustri, A. and G. Siracusa. 1983. Metabolic coupling, cumulus expansion and meiotic resumption in mouse cumuli oophori cultured *in vitro* in the presence of FSH or dbcAMP, or stimulated *in vivo* by hCG. *J. Reprod. Fertil.* **68:** 335.

Sasamoto, S., K. Sato, and H. Naito. 1972. Biological active life of PMSG in mice with special reference to follicular ability to ovulate. *J. Reprod. Fertil.* **30:** 371.

Schmidt, P.M., S.L. Monfort, and D.E. Wildt. 1989. Pregnant mare's serum gonadotropin source influences fertilization and fresh and thawed embryo development, but the effect is genotype specific. *Gamete Res.* **23:** 11.

Schroeder, A.C. and J.J. Eppig. 1984. The developmental capacity of mouse oocytes that matured spontaneously *in vitro* is normal. *Dev. Biol.* **102:** 493.

———. 1989. Developmental capacity of mouse oocytes that undergo maturation in vitro: Effect of the hormonal state of the oocyte donor. *Gamete Res.* **24:** 81.

Schroeder, A.C., S.M. Downs, and J.J. Eppig. 1988. Factors affecting the developmental capacity of mouse oocytes undergoing maturation *in vitro*. *Ann. N.Y. Acad. Sci.* **541:** 197.

Schuetz, A.W. 1967. Action of hormones in germinal vesicle breakdown in frog (*Rana pipiens*) oocytes. *J. Exp. Zool.* **166:** 347.

Schultz, R.M., R.R. Montgomery, and J.R. Belanoff. 1983. Regulation of mouse oocyte meiotic maturation: Implication of a decrease in oocyte cAMP and protein dephosphorylation in commitment to resume meiosis. *Dev. Biol.* **97:** 264.

Shalgi, R., N. Dekel, and P.F. Kraicer. 1979. The effect of LH on the fertilizability and developmental capacity of rat oocytes matured *in vitro*. *J. Reprod. Fertil.* **55:** 429.

Shiigi, S.M. and R.I. Mishell. 1975. Sera and the *in vitro* induction of immune responses. I. Bacterial contamination and the generation of good fetal bovine sera. *J. Immunol.* **115:** 741.

Sirard, M.A., J.J. Parrish, C.B. Ware, M.L. Leibfried-Rutledge, and N.L. First. 1988. The culture of bovine oocytes to obtain developmentally competent embryos. *Biol. Reprod.* **39:** 546.

Skinner, M.K., D. Lobb, and J.H. Dorrington. 1987. Ovarian thecal-interstitial cells produce an epidermal growth factor-like substance. *Endocrinology* **121:** 1892.

Skoblina, M.N. 1988. Maturation of mammalian oocytes in vitro. In *Oocyte growth and maturation* (ed. T.A. Dettlaff and S.G. Vassetzky), p. 341. Consultants Bureau, Plenum Press, New York.

Smith, D.M., C.H. Conawey, and W.T. Kerber. 1978a. Influence of season and age on maturation in vitro of rhesus monkey oocytes. *J. Reprod. Fertil.* **54:** 91.

Smith, D.M., J.P.P. Tyler, and G.F. Erickson. 1978b. Effects of medium composition and progesterone on maturation *in vitro* of rabbit oocytes from Graafian follicles of different sizes. *J. Reprod. Fertil.* **54:** 393.

Smith, L.D. and R.E. Ecker. 1971. The interaction of steroids with *Rana pipiens* oocytes in the induction of maturation. *Dev. Biol.* **25:** 232.

Sorensen, R.A. and P.M. Wassarman. 1976. Relationship between oocyte growth and meiotic maturation of the mouse oocyte. *Dev. Biol.* **50:** 531.

Soupart, P. 1975. *In vitro* maturation and fertilization of human oocyte. In *La Fecondation* (ed. C. Thibault), p. 1. Masson, Paris.

St-Arnaud, R., P. Walker, P.A. Kelly, and F. Labrie. 1983. Rat ovarian epidermal growth factor receptors: Characterization and hormonal regulation. *Mol. Cell. Endocrinol.* **31:** 43.

Staigmiller, R.B. and R.M. Moor. 1984. Effects of follicle cells on the maturation and developmental competence of ovine oocytes matured outside the follicle. *Gamete Res.* **9:** 221.

Stubbings, R.B., K.J. Betteridge, and P.K. Basrur. 1988. Investigations of culture requirements for bovine oocyte maturation *in vitro*. *Theriogenology* **29:** 313. (Abstr.)

Szollosi, D. 1975. Mammalian eggs aging in the fallopian tubes. In *Aging gametes* (ed. R.J. Blandau), p. 98. Karger, Basel.

Szybek, K. 1972. *In vitro* maturation of oocytes from sexually immature mice. *J. Endocrinol.* **54:** 527.

Thadani, V.J. 1979. Injection of sperm heads into immature rat oocytes. *J. Exp. Zool.* **212:** 436.

Thibault, C. 1972. Final stages of mammalian oocyte maturation. In *Oogenesis* (ed. J.D. Biggers), p. 397. University Park Press, Baltimore.

———. 1977. Are follicular maturation and oocyte maturation independent processes? *J. Reprod. Fertil.* **51:** 1.

Thibault, C. and M. Gerard. 1970. Facteur cytoplasmique necessaire a la formation du pronucleus male dans l'ovocyte de lapine. *C.R. Acad. Sci.* **270:** 20.

Thibault, C., M. Gerard, and Y. Menezo. 1976. Nuclear and cytoplasmic aspects of mammalian oocyte maturation *in vitro* in relation to follicle size and fertilization. *Prog. Reprod. Biol.* **1:** 233.

Thibault, C., D. Szollosi, and M. Gerard. 1987. Mammalian oocyte maturation. *Reprod. Nutr. Dev.* **27:** 865.

Tsafriri, A. and C.P. Channing. 1975. Influence of follicular maturation and culture conditions on the meiosis of pig oocytes *in vitro*. *J. Reprod. Fertil.* **43:** 149.

Tsafriri, A. and P.F. Kraicer. 1972. The time sequence of ovum maturation in the rat. *J. Reprod. Fertil.* **29:** 387.

Ueno, S., T.F. Manganaro, and P.K. Donahue. 1988. Human recombinant Mullerian inhibiting substance of rat oocyte meiosis is reversed by epidermal growth factor *in vitro*. *Endocrinology* **123:** 1652.

Usui, N. and R. Yanagamachi. 1976. An inhibitory influence of granulosa cells and follicular fluid upon porcine oocyte meiosis *in vitro*. *J. Ultrastruct. Res.* **57:** 276.

Van Blerkom, J. and R.W. McGaughey. 1978. Molecular differentiation of the rabbit ovum. I. During oocyte maturation in vivo and in vitro. *Dev. Biol.* **63:** 139.

van de Sandt, J.J.M., A.C. Schroeder, and J.J. Eppig. 1990. Culture media for mouse oocyte maturation affect subsequent embryonic development. *Mol. Reprod. Dev.* **25:** 164.

Vanderhyden, B.C. and D.T. Armstrong. 1989. Role of cumulus cells and serum on the *in vitro* maturation, fertilization, and subsequent development of rat oocytes. *Biol. Reprod.* 40:720.

Walton, E.A., G. Evans, and D.T. Armstrong. 1983. Ovulation response and fertilization failure in immature rats induced to superovulate. *J. Reprod. Fertil.* **67:** 91.

Ward, M.C. 1948. The early development and implantation of the golden hamster, *Cricetus auratus*, and the associated endometrial changes. *Am. J. Anat.* **82:** 231.

Westergaard, L., A.G. Byskov, P.F.A. Van Look, R. Angell, J. Aitken, I.A. Swanston, and A.A. Templeton. 1985. Meiosis-inducing substances in human preovulatory follicular fluid related to time of follicle aspiration and to the potential of the oocyte to fertilize and cleave *in vitro*. *Fertil. Steril.* **44:** 663.

Wilson, E.D. and M.X. Zarrow. 1962. Comparison of superovulation in the immature mouse and rat. *J. Reprod. Fertil.* **3:** 148.

Xu, K.P., R. Hoier, and T. Greve. 1988. Dynamic changes of estradiol and progesterone concentrations during in vitro oocyte maturation in cattle. *Theriogenology* **30:** 245.

Yanagamachi, R. and M.C. Chang. 1961. Fertilizable life of the golden Syrian hamster ova and their morphological changes at the time of losing fertilizability. *J. Exp. Zool.* **148:** 185.

Yoshinaga, K., C. Rice, J. Krenn, and R.L. Pilot. 1979. Effects of nicotine on early pregnancy in the rat. *Biol. Reprod.* **20:** 294.

Younis, A.I., B.G. Brackett, and R.A. Fayrer-Hosken. 1989. Influence of serum and hormones on bovine oocyte maturation and fertilization *in vitro*. *Gamete Res.* **23:** 189.

Yun, Y.W., B.H. Yuen, and Y.S. Moon. 1987. Effect of superovulatory doses of pregnant mare serum gonadotropin on oocyte quality and ovulatory and steroid hormone responses in rats. *Gamete Res.* **16:** 109.

———. 1988. Effects of an antiandrogen, flutamide, on oocyte quality and embryo development in rats superovulated with pregnant mare's serum gonadotropin. *Biol. Reprod.* **39:** 279.

Yun, Y.W., F.H. Yu, B.H. Yuen, and Y.S. Moon. 1989. Effects of a superovulatory dose of pregnant mare serum gonadotropin on follicular steroid contents and oocyte maturation in rats. *Gamete Res.* **23:** 289.

Zamboni, L. 1974. Fine morphology of the follicle wall and follicle cell-oocyte association. *Biol. Reprod.* **10:** 125.

Zeilmaker, G.H. 1978. Observations on follicular lactate concentrations and the influence of granulosa cells on oocyte maturation in the rat (including data on second polar body formation). *Ann. Biol.*

Anim. Biochim. Biophys. **18:** 529.

Zhang, X. and D.T. Armstrong. 1989. Fertilization of rat oocytes is enhanced by FSH during *in vitro* maturation of oocyte-cumulus complexes. *Theriogenology* **31:** 277. (Abstr.)

Cryopreservation of the Mammalian Oocyte

J. Van Blerkom
Department of Molecular, Cellular, and Developmental Biology
University of Colorado, Boulder, Colorado 80309

INTRODUCTION

Virtually all of the current laboratory procedures utilized for human conception in vitro are derived from studies of the oocytes and embryos of experimental animals such as the mouse and rabbit and of commercially important mammals such as the cow and sheep. Indeed, the media and methods for in vitro fertilization (IVF) and support of preimplantation embryogenesis in the human species are not much different from those utilized for the mouse (Whittingham 1971), rabbit (Van Blerkom et al. 1973), and cow (Iritani 1988; Van Blerkom et al. 1990). In the same sense, methods for the cryopreservation of the human oocyte and preimplantation-stage embryo also originate from work in laboratory (Siebzehnruebl et al. 1989) and livestock species (Wilmut 1986; Iritani 1988).

The ability to cryopreserve preimplantation-stage embryos has become of central importance in the application of clinical protocols designed to maximize the potential for achieving pregnancy by in vitro methods (Mandelbaum et al. 1988b; Quigley 1990). The capacity to freeze human embryos is especially relevant when it is considered that current clinical protocols of controlled ovarian hyperstimulation frequently produce oocytes and embryos in excess of the number that reasonably can be placed in the fallopian tubes or uterus, respectively. The attendant risk of multiple pregnancies and the possibility of having to perform selective intrauterine abortion (Zaner et al. 1990) have, for many assisted reproduction programs, limited the number of oocytes and pronucleate eggs transferred to the fallopian tubes during gamete and zygote intrafallopian transfer (GIFT/ZIFT) procedures, respectively, and

 0-87969-333-9/91 $3.00 + 00

the number of cleavage-stage embryos placed in the uterus after IVF. Many assisted reproduction programs attempt to maximize the number of embryos produced on a single cycle of ovarian stimulation and use cryopreservation in the management of excess embryos. Other than intentionally not inseminating oocytes, or donating surplus oocytes to patients suffering from definitive ovarian failure such as gonadal dysgenesis (Cornet et al. 1990), embryo cryopreservation has become a viable option in assisted reproduction. Indeed, the birth of normal babies from cryopreserved human embryos demonstrates that freezing does increase the potential for pregnancy on a single IVF attempt (Mandelbaum et al. 1987; Junca et al. 1988). However, due to moral, legal (Jones 1990; Schuster 1990), and ethical problems associated with this methodology (Ethics Committee of the American Fertility Society 1990), cryopreservation of human embryos in many IVF programs either has become nonroutine or is performed only on those embryos that demonstrate the ability to develop progressively through the early cleavage stages in vitro.

At present, human embryos are frozen during the first 1–2 days after fertilization (Freedman et al. 1988; Fugger et al. 1988; Mandelbaum et al. 1988a,b; Caumus et al. 1989). This developmental period is considered the most practical because pronucleate or cleavage-stage embryos are usually transferred to the fallopian tubes or uterus for ZIFT or conventional IVF procedures, respectively. However, replacement at these early stages, especially at the pronuclear stage, does not always permit an adequate assessment of subsequent developmental potential. This is a significant issue when it is considered that the majority of human embryos, whether fertilized in vivo or in vitro, will not progress to implantation (Edwards 1986; Osborn and Moor 1988; Van Blerkom 1989a; Boklage 1990). Consequently, cryopreservation of newly fertilized eggs and early cleavage-stage embryos is necessarily associated with the potential for long-term storage of embryos with a limited capacity to develop progressively after thawing. This conclusion has led to the suggestion that excess human embryos be allowed to develop to at least the early morula stage prior to freezing (Bolton et al. 1989; Van Blerkom 1989a). Such an approach would permit the application of biochemical and cytological methods to assess developmental potential and presumably would result in the storage of only those embryos that have demonstrable developmental potential (for review, see Osborn and Moore 1988; Van Blerkom 1989a, 1990a).

Alternatively, the retrieval of a single oocyte on a natural menstrual cycle is a recent trend in clinical IVF that not only eliminates the cryopreservation option, but more importantly, appears to have the same rate of success with each attempt at initiating a pregnancy as has been reported for stimulated cycles and multiple embryos transfers (Ranoux et al. 1988). Although changes in protocols for assisted reproduction that require fertilization in vitro may eliminate the need for embryo cryopreservation by reducing the number of retrieved oocytes, the ability to retain embryos in long-term storage (Cohen et al. 1988) will probably remain an important option in the attempt to achieve pregnancy by laboratory-based technologies (Van den Abbeel et al. 1988).

Collectively, the problematic ethical and legal issues (Bonnicksen 1988) associated with attempts to define when life begins (Biggers 1990) and the rights and legal status of frozen human embryos (Ethics Committee of the American Fertility Society 1990), and concerns about the developmental potential and normality of thawed human embryos, have led to a consideration of oocyte cryopreservation as an alternative approach in the management of the multiple gametes usually retrieved after controlled ovarian hyperstimulation. Consequently, the primary focus of this paper is on the potential problems and benefits of oocyte cryopreservation in a clinical situation.

CRYOPRESERVATION OF THE OOCYTE

At present, only a handful of reports have described births from cryopreserved human oocytes after fertilization in vitro. Chen (1986, 1988) reported two births resulting from the uterine transfer of early cleavage-stage embryos obtained by in vitro fertilization of oocytes that had been frozen at metaphase II (MII). Of 50 oocytes cryopreserved at a controlled rate (slow freeze) with 1.5 M dimethylsulfoxide (DMSO) as the sole cryoprotectant, approximately 50% fertilized and divided in vitro after thawing. With a similar DMSO-based protocol, Siebzehnruebl et al. (1989) reported that after thawing and insemination in vitro, 14 of 38 MII-stage oocytes fertilized, 7 were syngamic at the uterine transfer, and a single, normal pregnancy resulted.

Although these findings are potentially encouraging for the clinical application of oocyte cryopreservation, as discussed below, several investigators have described deleterious effects of cryoprotectants, cooling, freezing, and thawing, on the organization of the cytoplasm and the integrity of the meiotic spindle. Cryomicroscopic studies of meiotically mature mouse and human oocytes (Ashwood-Smith et al. 1988) documented the following physical and physiochemical changes that may induce lethal damage during cryopreservation: (1) deformation of the oocyte, which can occur as a direct result of the initial contact with the ice front, may form nascent weak spots through which damaged cytoplasm leaks after thawing and (2) the generation of gas bubbles after thawing may disrupt intracellular structure and organization and lead to cell death or lysis. Perturbations in the structure of the cytoplasm and the organization of the metaphase spindle have been described for human oocytes (Sathananthan et al. 1987; Sathananthan and Trounson 1989) cryopreserved by either of the following four methods, each of which is currently in clinical use for embryo freezing: (1) slow cooling to low subzero temperatures (–50°C to –80°C) in 1.5 M DMSO (Trounson and Mohr 1983), (2) ultrarapid freezing in 2–4 M DMSO and 0.25 M sucrose (Trounson 1986; Wilson and Quinn 1989), (3) vitrification (Rall and Fahy 1985; Friedler et al. 1987), and (4) staged slow cooling to high subzero temperatures (–35°C) in 1.5 M DMSO or 1.5 M 1,2-propanediol, followed by direct plunge into liquid nitrogen (Testart et al. 1986). Other than complete cell lysis, subtle forms of cryoinjury were recognized by the presence of one or more of the following structural and/or cytogenetic alterations: (1) small cracks in the zona pellucida, (2) disruption of the plasma membrane, (3) extensive disorganization of the cytoplasm, (4) cortical granule exocytosis (Schalkoff et al. 1989), (5) depolymerization of spindle microtubules with the resultant scattering of chromosomes, and (6) formation of micronuclei (see also Sathananthan et al. 1987; Sathananthan and Trounson 1989; Trounson and Kirby 1989; Trounson and Sathananthan 1989).

In contrast to the comparatively high rates of survival for thawed human pronuclear eggs and early cleavage-stage embryos (Van den Abbeel et al. 1988; Siebzehnruebl et al. 1989), post-thaw survival, fertilizability, and developmental viability of cryopreserved, MII-stage human oocytes are surprisingly low (Friedler et al. 1988). Some thawed human oocytes have been fertilized in vitro and have developed progressively through gestation, indicating that the human oocyte is capable of surviving freezing without developmentally lethal insults (Van Uem et al. 1987). However, a necessary prerequisite for the routine application of oocyte cryopreservation in the human species is the identification of stage-related cellular and molecular conditions that may be associated with or have the potential to predispose an oocyte to developmentally lethal cryoinjury.

DEVELOPMENTAL EQUIVALENCY OF MATURE HUMAN OOCYTES: CYTOPLASMIC AND CYTOGENETIC FACTORS THAT MAY PREDISPOSE AN OOCYTE TO LETHAL CRYOINJURY

One obvious explanation for the apparent difference in survival rates between the mature oocyte and the pronuclear stage egg is the self-selected nature of the two groups. By definition, pronuclear eggs represent a comparatively homogeneous population with a demonstrated developmental capacity. In contrast, meiotically mature human oocytes, although presumably at the same stage of maturation at the time of freezing, cannot be assumed to be developmentally equal. This is especially relevant for those protocols of assisted reproduction in which the objective is to maximize the number of oocytes retrieved by hyperstimulating the ovaries. Perturbations in the organization and composition of the cytoplasm of meiotically mature human oocytes obtained after controlled hyperstimulation have been identified by light microscopy and have been shown to be associated with fertilization failure or postfertilization developmental arrest (Nayudu et al. 1989; Van Blerkom 1990a). Van Blerkom (1990b) estimated that approximately 10–15% of the MII-stage human oocytes obtained from hyperstimulated ovaries may exhibit at least one of the following cytoplasmic aberrations associated with reduced fertilizability of the oocyte or inability of the zygote to develop progressively after fertilization: (1) a massive accumulation of tubules of the smooth-surface endoplasmic reticulum (SER), (2) regions of intracellular degeneration or necrosis, (3) aggregation of SER vesicles and mitochondria, (4) depletion of organelles from portions of the cortical cytoplasm, (5) cytoplasmic vacuolation, (6) internalization of perivitelline fluid in pronucleus-size endocytotic vacuoles, and (7) premature and partial exocytosis of cortical granules. In contrast, human oocytes that matured in vitro to MII from the germinal vesicle (GV) stage rarely displayed these major forms of cytoplasmic alteration (Van Blerkom 1990b). In the same respect, only 1 of 42 MII-stage oocytes retrieved to date during unstimulated (spontaneous) cycles has exhibited a cytoplasmic perturbation (vacuolation) (J. Van Blerkom, unpubl.). Taken together, these observations suggest that controlled ovarian hyperstimulation may be accompanied by the generation of cohorts of meiotically mature oocytes that, by routine inspection, may appear to be morphologically equivalent but are fundamentally heterogeneous with respect to developmental potential. The occurrence and type(s) of cytoplasmic abnormality at MII not only appear to be predictive of fertilizability and subsequent developmental ability, but also may determine the capacity of the oocyte to survive the freeze/thaw process. Consequently, assumptions about the cryopreservability of the mature human oocyte are difficult to evaluate from published reports because freezing usually involves excess oocytes that are typically selected on the basis of assumed chromosomal maturity (presence of first polar body), rather than after the application of phenotypic criteria indicative of cytoplasmic normality and developmental potential.

Disruption of the meiotic metaphase spindle with an attendant increase in the frequency of chromosomal aneuploidy has often been described as consequences of oocyte cryopreservation (Pickering and Johnson 1987; Hamlett et al. 1989; Kola et al. 1988; Sathananthan et al. 1988). It is generally held that progressive cooling of oocytes prior to exposure to subzero temperatures disrupts microtubular organization (Magistrini and Szollosi 1980) such that after thawing and rehydration, a variable number of chromosomes may fail to reassociate with the reformed spindle. In this respect, recent studies have shown that transient exposure of mature oocytes to room temperature can be accompanied by irreversible spindle disruption and chromosomal scattering. Van der Elst et al. (1988) reported that nearly 90% of mouse MII oocytes exposed to either 1.5 M DMSO or 1.5 M propanediol at room temperature displayed abnormal spindles. Pickering et al. (1990) have shown that spindle disruption and chromosomal displacement occurs in mature human oocytes maintained at room temperature for as short a time as 10 minutes. These findings demonstrate the sensitivity of the meiotic spindle to reduced temperature and of particular significance, the attendant potential to introduce chromosomal perturbations in the mature oocyte by transient exposure to temperatures below 37°C.

The extent to which chromosomal scattering is an unavoidable complication of human oocyte cryopreservation is difficult to determine precisely because a significant percentage of meiotically mature human oocytes obtained from stimulated cycles may have been aneuploid prior to freezing. Estimates of the frequency of aneuploidy in newly harvested MII-stage human oocytes, which includes both hypohaploid and hyperhaploid conditions, range from 20% to 60% (for review, see Plachot et al. 1988; Van Blerkom 1989c). Indeed, Van Blerkom and Henry (1988) noted that 15% of *living* MII human oocytes obtained from hyperstimulated ovaries that appeared to be morphologically normal by routine light microscopic inspection nevertheless contained one or more non-spindle-associated chromosomes. These investigators also described a class of grossly normal-appearing human oocytes that were assumed to be at the MII stage owing to the presence of a first polar body, but were shown by DNA fluorescence and electron microscopy to contain no organized spindle and MII chromosomes individually enclosed by a nuclear membrane. The collected results of numerous cytogenetic analyses of meiotically mature human oocytes (Plachot et al. 1988) demonstrate that regardless of appearance in the light microscope, MII-stage human oocytes cannot be assumed to be chromosomally normal or equivalent.

In addition to intrafollicular conditions that may predispose an oocyte to aneuploidy (Nayudu et al. 1987), the age of the oocyte, measured in hours from the resumption of arrested meiosis, may be a principal factor associated with the generation of chromosomal abnormalities. Aging of oocytes, whether it occurs in vivo or in vitro, is often accompanied by a deterioration in the integrity of the spindle and a resultant scattering of chromosomes (Eichenlaub-Ritter et al. 1986). With current protocols of controlled ovarian stimulation, aberrant intrafollicular conditions may be associated with a spontaneous resumption of meiosis that is premature with respect to when ovulation would be expected to occur (Van Blerkom and Henry 1988; Van Blerkom 1989a). Likewise, because it cannot be assumed that all MII-stage oocytes reinitiate meiosis synchronously in vivo, the age of the multiple oocytes typically obtained at retrieval can vary significantly. Indeed, it is not uncommon to observe a mixture of mature and immature oocytes (germinal vesicle breakdown and metaphase I stages) in stimulated cycles. Further complicating the potential for age-associated cytogenetic perturbations is the practice of allowing MII-stage human oocytes to reside in vitro for a variable length of time prior to insemination. This practice is derived from the notion that although chromosomal maturation may be complete at the time of follicular aspiration, developmental changes (cytoplasmic maturation) necessary for fertilization and male pronucleus formation may require additional time in culture to complete (Trounson et al. 1982; Harrison et al. 1988). Therefore, prolonged preincubation prior to cryopreservation may, for some oocytes, lead to the generation of cytogenetic abnormalities that are not a consequence of the actual freezing process. The phenotypic and developmental heterogeneity of human oocytes associated with controlled ovarian hyperstimulation, and the potential for a deterioration in spindle stability with prolonged preincubation or exposure to temperatures below 37°C, could account for some of the cytoplasmic and cytogenetic abnormalities observed in human oocytes after thawing.

Results from the cryopreservation of mouse oocytes demonstrate that both the type of cryoreagent(s) and the specific protocol(s) of freezing can significantly influence the degree of aneuploidy, the normality of cytoplasmic organization, and post-thaw developmental ability. Johnson et al. (1988) suggested that prefreeze cooling and the addition and withdrawal of cryoprotectant(s) may be more significant factors in the survival of the oocyte than the freezing process itself. These investigators reported that cooling of MII-stage mouse oocytes to 4°C was associated with a 30% reduction in the frequency of fertilization. A significant proportion of the cooled oocytes exhibited a hardening of the zona pellucida that was similar to the physiochemical changes in zona structure (zona reaction) caused by the exocytosis and solubilization of cortical granules (cortical reaction) after contact between oolemma and fertilizing spermatozoon. Both reactions are primary mechanisms that prevent fertilization by more than a single spermatozoon. An increased resistance to chymotrypsin is one measurable indication of the hardening of the zona pellucida that occurs after cortical granule exocytosis (Boldt and Wolf 1986). Johnson et al. (1988) concluded that the reduced rate of fertilization of cooled oocytes was explicable entirely in terms of an effect on the zona pellucida. Subsequent studies by Vincent et al. (1990) indicated that cooling of meiotically mature mouse oocytes in the absence of a cryoprotectant promotes a precocious release of cortical granules and a premature zona reaction. Indeed, these investigators demonstrated that exposure of mature mouse oocytes to 1.5 M DMSO at temperatures between 20°C and 37°C was accompanied by cortical granule exocytosis, the consequence of which was a marked reduction in the fertilization rate. Similar findings were reported by Schalkoff et al. (1989), who described a premature exocytosis of 70–80% of the cortical granules in all mouse and human oocytes cultured in the presence of either 1.5 M 1,2-propanediol or 1.5 M DMSO, concentrations of cryoprotectants routinely used for embryo cryopreservation. Therefore, physiochemical modifications of the zona pellucida that ordinarily occur at fertilization can be a detrimental consequence of exposure to cryoprotectants or cooling (or both) and as a result, may render otherwise normal-appearing MII-stage oocytes impenetrable to sperm.

Numerous studies have shown that prefreeze cooling of meiotically mature mouse oocytes is accompanied by disruption of the metaphase I (MI) or II spindles (Magistrini and Szollosi 1980; Sathananthan et al. 1988; Hamlett et al. 1989). Indeed, Pickering and Johnson (1987) reported that cooling even to 25°C had a profound influence on the microtubule system of the oocyte, disrupting microtubules, microtubular organizing centers, and metaphase plate chromosomes. Many of the changes in spindle and microtubule organization induced by low temperature were reversed when oocytes were returned to 37°C. A potentially significant finding from this study was the restoration of an apparently normal spindle in many but not all oocytes. Cooling of MI- and MII-stage mouse oocytes to 4°C was shown to be accompanied by the disruption of spindle microtubules and a resultant displacement and scattering of chromosomes (Hamlett et al. 1988). Although the degree of spindle disorganization and chromosomal scattering observed for MII-stage mouse oocytes cooled to 4°C is variable, the displacement of chromosomes from the spindle has the potential to create cytogenetic abnormalities that can compromise normal development, even if spindle restoration occurs at 37°C and the oocytes are fertilizable (Van Blerkom 1989b). In support of this notion is the recent report of Pickering et al. (1990), who examined the effects that transient cooling to room temperature had on the MII spindle of the human oocyte. Cooling to room temperature for 10 minutes caused disruption of the spindle in approximately 50% of the oocytes examined. Cooling for 30 minutes caused spindle disruption in all oocytes, but there was evidence of some chromosomal dispersion in 60% of the oocytes. Of major clinical significance was the finding that returning oocytes to 37°C did not restore a normal spindle in the majority of the oocytes and for some, chromosomal dispersal was still evident.

In addition to the effects of reduced temperatures, the sensitivity of the oocyte microtubular system is demonstrated by pronounced organizational changes that occur during culture at 37°C in a solution containing 1.5 M DMSO, which is the concentration of cryoprotectant normally used for oocyte and embryo cryopreservation (Wilmut 1986; Trounson and Kirby 1989). Johnson and Pickering (1987) reported that exposure of MII-stage mouse oocytes to DMSO for 1 hour was accompanied by the loss of an identifiable spindle or the presence of a spindle remnant. In this study, DMSO was added either directly at 1.5 M or in increments of 0.25 M, with 10-minute exposures at each concentration. Both protocols of dehydration revealed that loss of a spindle was associated with a dispersal of chromosomes from the equatorial plate. Chromosomal scattering was localized to the cytocortex and more central regions of the oocyte. The removal of DMSO resulted in the restoration of the typical barrel-shape spindle in some oocytes, whereas for most others, either no spindle formed or spindle reconstitution was incomplete or abnormal (e.g., multiple poles). Chromosomal dispersal was evident in oocytes lacking a properly reconstituted spindle. These results demonstrate that both prefreeze cooling and exposure to cryoprotectants such as DMSO can have developmentally significant effects on the stability of the microtubular system of the mature mouse oocyte.

Aneuploidy that may result from spindle disruption has been reported to occur at high frequency in thawed MII-stage mouse oocytes fertilized in vitro. Kola et al. (1988) observed an incidence of chromosomally aneuploid mouse zygotes that was threefold higher than unfrozen controls. In contrast, Glenister et al. (1987) found that the incidence of aneuploidy at the first cell division in thawed mouse oocytes fertilized in vitro was not affected by cryopreservation. In addition, these authors reported an apparent doubling of the frequency of polyploid embryos, suggesting that thawed oocytes may be more susceptible to polyspermic penetration. More recently, Carroll et al. (1989) noted that the increased frequency of polyploidy in thawed mouse oocytes was not due to polyspermic penetration but rather was the result of retention of the second polar body. However, chromosomal dislocation is not a ubiquitous event in cryopreserved mouse oocytes because development to the expanded blastocyst stage (Carroll et al. 1989) and chromosomally normal offspring (Glenister et al. 1987) have been obtained from oocytes that were cryopreserved with 1.5 M DMSO and that were cooled at a controlled rate. The occurrence of aneuploid oocytes and the apparent increase risk of polyploidy observed in animal model systems such as the mouse are of significant concern with respect to the clinical application of oocyte cryopreservation.

In addition to chromosomally abnormal zygotes that develop from oocytes fertilized after exposure to cryoreagents, teratogenic effects on fetal development have also been observed. Kola et al. (1988) reported that vitrification, or exposure of mature mouse oocytes to vitrification reagents (Rall and Fahy 1985), was equally associated with an unusually high incidence of developmental malformations, of which central nervous system (CNS)-type defects such as exencephaly and hydrocephaly were the most frequent. These investigators suggested that chemicals used for vitrification may have an effect on oocyte DNA, perhaps by inducing mutations. Alternatively, these chemicals may accumulate in the oocyte cytoplasm and persist to later stages of embryogenesis, where their presence may act as endogenous mutagens.

It is doubtful that potentially mutagenic agents persist in sufficient quantities during early embryogenesis to have a teratogenic effect during organogenesis. It is of interest to note that in the study of Kola et al. (1988), cytogenetic analyses demonstrated that all of the malformed fetuses were chromosomally normal. In contrast, Nakagata (1989) observed no abnormalities in the live young that developed from mouse oocytes vitrified by the method of Rall and Fahy (1985). In this investigation, approximately 88% of the oocytes survived thawing, of which 82% fertilized in vitro and 78% progressed to the two-cell stage at the time of oviductal transfer. Approximately 46% of these embryos developed to birth, a figure that is comparable to the frequency of successful gestation observed after oviductal transfer of nonfrozen mouse embryos. However, it would have been of significant value had this study determined the extent, if any, to which loss of 64% of the transferred embryos was the result of chromosomal or developmental abnormalities such as those described by Kola et al. (1988).

CRYOPRESERVATION OF THE IMMATURE OOCYTE

Clearly, cryopreservation of meiotically mature oocytes has the potential to induce chromosomal and possibly congenital abnormalities of developmental significance. These findings led the Ethics Committee of the American Fertility Society (1990) to conclude that "without good evidence of the potential safety of the procedure in animals or in humans, the cryopreservation of eggs for clinical use is of grave concern." However, the reservations expressed by this committee appear to be directed primarily at the cryopreservation of the meiotically mature oocyte.

For the reasons described below, freezing of the fully grown, meiotically immature oocyte (GV stage) may be an alternative approach to the cryopreservation of the female gamete. The GV-stage oocytes of laboratory and commercial species such as the cow (Lu et al. 1987; Xu et al. 1987; Van Blerkom et al. 1990), pig (McGaughey and Van Blerkom 1977; Cheng et al. 1986; Mattioli et al. 1989), sheep (Cheng et al. 1986; Crozet et al. 1987), rabbit (Van Blerkom and McGaughey 1978), and mouse (Stern and Wassarman 1974; Van Blerkom and Runner 1984) not only mature in vitro at high frequency (>80%), but, after fertilization in vitro, are frequently capable of development through the preimplantation stages. Normal gestation and births have been reported to occur after uterine transfer of these embryos (Schroeder and Eppig 1984; Crozet et al. 1987; Xu et al. 1987; for review, see Racowsky, this volume). A similar high frequency of spontaneous maturation to MII has been described for cultured GV-stage human oocytes (Jagiello et al. 1976), and such oocytes have been shown to be fertilizable (Veeck et al. 1983).

One potential advantage of cryopreservation at the GV stage is the fact that freezing is performed prior to chromosomal condensation and as a consequence, aneuploidy that may result from cooling-induced disruptions of the MII spindle would not occur. Furthermore, because fully grown GV-stage oocytes can be obtained from small follicles that have not been subjected to elevated levels of circulating gonadotropins such as follicle-stimulating hormone and luteinizing hormone (FSH and LH), a greater degree of developmental synchrony and normality may occur than if obtained from antral follicles after controlled hyperstimulation (Van Blerkom et al. 1990). Indeed, our observations of human oocytes support this notion, since GV-stage oocytes retrieved from *unstimulated* ovaries and matured in vitro rarely display the types of cytoplasmic pathologies observed at MII after controlled *hyperstimulation* (Van Blerkom 1990b). In contrast, GV-stage oocytes that fail to resume meiosis after the ovulation induction in controlled hyperstimulated cycles display a variable degree of vesiculation, vacuolation, and organelle clustering indicative of a premorbid or degenerative state (Nayudu et al. 1989). Although most GV-stage oocytes obtained from hyperstimulated ovaries fail to resume meiosis or arrest maturation prior to MII, those that do mature to MII generally undergo fertilization and normal preimplantation development (Nayudu et al. 1989; J. Van Blerkom, unpubl.).

A second advantage of freezing human oocytes at the GV stage is that the frequency of aneuploidy after the spontaneous resumption of meiosis in vitro is quite low. Although the reported frequency of aneuploidy after controlled ovarian hyperstimulation ranges from 20% to 60% (Plachot et al. 1988; Van Blerkom 1990b), Jagiello et al. (1976) found that less than 1% of the approximately 600 human GV-stage oocytes that matured to MII in vitro were chromosomally abnormal. A similar low frequency of aneuploidy (<1%) was observed by Van Blerkom (1989a, 1990a) at MII for human oocytes harvested at the GV stage from unstimulated ovaries. A third benefit of cryopreservation at the GV stage is the ability to monitor the progression of nuclear (chromosomal) maturation and to assess the normality of cytoplasmic maturation at the light microscopic level. In this respect, maturation in vitro apparently precludes the exposure of the oocyte to intrafollicular conditions that may be inconsistent with a normal program of nuclear and cytoplasmic differentiation (Nayudu et al. 1989).

At present, attempts to obtain developmentally viable oocytes after freezing at the GV stage have had limited success. Two protocols of cryopreservation that have been successfully applied to the freezing of human pronuclear stage eggs and early cleavage-stage embryos, were utilized by Junca et al. (1988) for immature hamster and human oocytes. Either 1.5 M propanediol and 0.1 M sucrose (Mandelbaum et al. 1987) or 1.5 M DMSO (Trounson 1986) are used as cryoprotectants. In both protocols, oocytes are cooled from −1°C to −40°C at 0.3°C/min. Junca et al. (1988) reported that although 77% (56/73) of the hamster oocytes resumed meiosis after thawing, only 22% (16/73) abstricted the first polar body. GV-stage human oocytes demonstrated a low survival rate after thawing and a poor maturational frequency, with only a few percentage of the oocytes competent to resume meiosis. The contrast between these findings and the successful cryopreservation of pronuclear and early cleavage-stage embryos suggests that fertilization may be accompanied by changes in the structure or organization of the plasma membrane and cytoplasm that fundamentally alter the ability of the cell to retain developmental viability after thawing. However, the slow-cooling protocol used for the immature oocyte would be expected to be accompanied by the same types of physical deformations and physiochemical perturbations described by Ashwood-Smith et al. (1988) for mature oocytes cooled in such a progressive manner.

CRYOPRESERVATION BY VITRIFICATION

A significant advantage of the vitrification method for oocyte cryopreservation is that cooling occurs so rapidly that intracellular water transforms directly from a liquid to a glass-like state. Consequently, cellular damage that may ordinarily occur with the formation of extra- and intracellular ice crystals during slow or controlled-rate freezing would not be expected (Rall and Fahy 1985). To date, the most detailed study of the cellular and developmental characteristics of cryopreserved GV-stage mouse oocytes has been described by Van Blerkom (1989b), who used the vitrification protocol of Rall and Fahy (1985). In this study, 96% (719/749) of the mouse oocytes vitrified at the GV stage resumed meiosis after devitrification and 83% progressed to MII. A fertilization rate of approximately 32% was observed after insemination of MII-stage mouse oocytes that had been vitrified at the GV stage. More recently, very similar results have been obtained after vitrification in a solution containing 3.5 M DMSO and 0.5 M sucrose (J. Van Blerkom and P. Davis, in prep.). However, this relatively high rate of post-thaw survival and retained meiotic competence occurred against a background of profound nuclear and cytoplasmic alteration (Van Blerkom 1989b).

The cellular organization observed in the dehydrated GV-stage oocyte just prior to placement in liquid nitrogen is as fol-

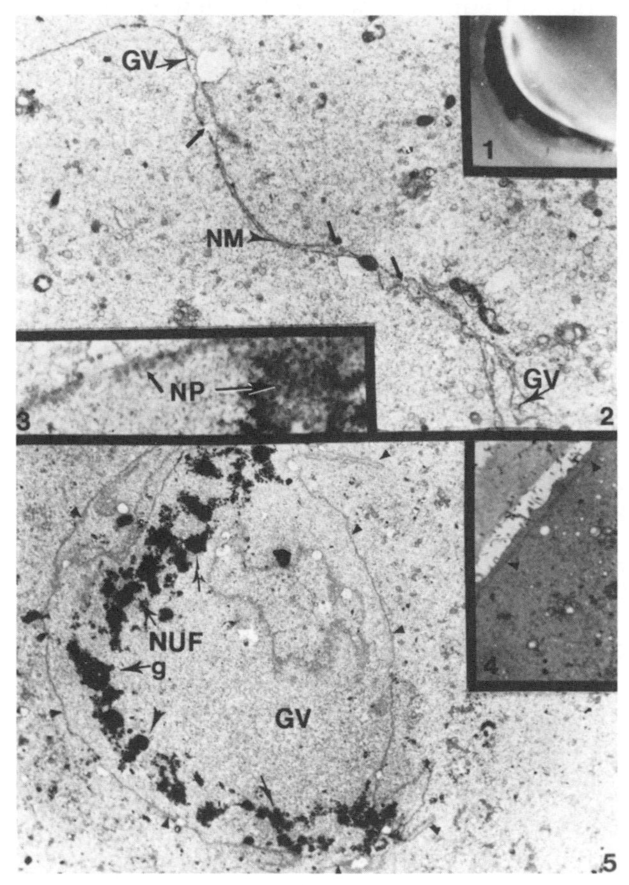

FIGURE 1 Light (frame *1*) and transmission electron microscopic views (frames *2–5*) of a mouse oocyte just prior to placement in liquid nitrogen. During initial stages of dehydration, the oocyte becomes crescent-shaped and the normally round germinal vesicle (GV) collapses into a highly compacted sac in which opposite regions of the nuclear membrane (NM) become juxtaposed. The nuclear membrane is highly porous (NP, panel *3*) and frequently discontinuous (arrows, frame *2*). The single nucleolus, or nucleoli, disassociates into numerous fragments (NUF, frame *5*), some of which have a granular texture (arrow *g*). Apparent displacement of subplasmalemmal microfilaments into the cortical cytoplasm is clearly evident in the final vitrification solution (arrowheads, frame *4*). Magnifications: (*1*) 300x; (*2*) 4000x; (*3*) 11,000x; (*4*) 2500x; (*5*) 4500x. (Frames *2, 3, 4, 5* were reprinted, with permission of Oxford University Press, from Van Blerkom 1989b.)

lows: (1) The oocyte contracts into a thin, crescent-shaped cell (Fig. 1, frame 1) that contains a collapsed and highly compact germinal vesicle (Fig. 1, frame 2); (2) the nuclear envelope is highly interdigitated (Fig. 1, frame 5), porous (Fig. 1, frame 3), and in some locations, discontinuous (arrows, Fig. 1, frame 2); (3) during the initial stages of dehydration, the nucleolus dissociates into scores of small spherical fragments (Fig. 1, frame 5, and Fig. 2, frame 8 [protocol of Rall and Fahy 1985]; Fig. 2, frame 9 [3.5 M DMSO/0.5 M sucrose]). At the cytoplas-

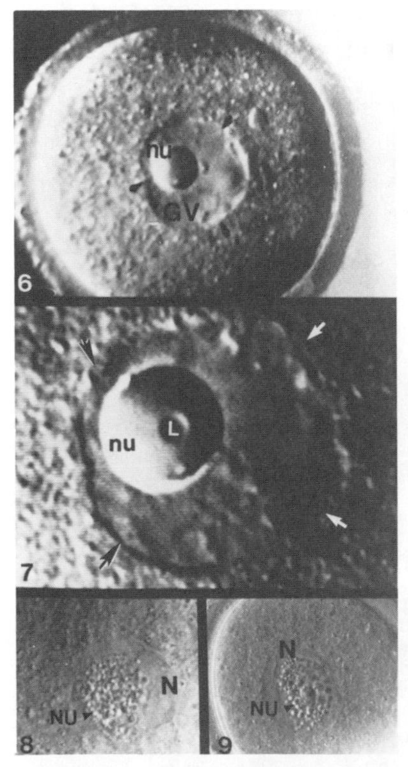

FIGURE 2 Video images of a germinal vesicle (GV)-stage mouse oocyte (frame *6*) and its nucleolus (nu) (frame *7*) immediately preceding vitrification in a solution of 3.5 M DMSO and 0.5 M sucrose. A small nucleolar lacuna (L, frame 7) has developed just prior to vitrification. Arrows outline the oocyte nucleus. Exposure to the first vitrification solution is accompanied by dissociation of the nucleolus (nu) into numerous fragments (frames *8* and *9*). Magnifications: (*6*) 550x; (*7*) 1100x; (*8*) 320x; (*9*) 320x.

mic level, vitrification is accompanied by the displacement of subplasmalemmal microfilaments into the pericortical cytoplasm (Fig. 1, frame 4), and by the apparent breakdown of the fibrillar arrays (lattices) that densely populate the mouse ooplasm. In contrast, the integrity and distribution of other cytoplasmic components such as Golgi complexes and mitochondria remain unchanged during vitrification (see Fig. 7, frames 34 and 36).

The profound effects of vitrification on nuclear, nucleolar, and cytoplasmic structure and organization are progressively reversed after devitrification. Within seconds, oocytes returned to a spherical state, and within minutes, the GV had reexpanded fully. Nucleolar reformation was the most striking

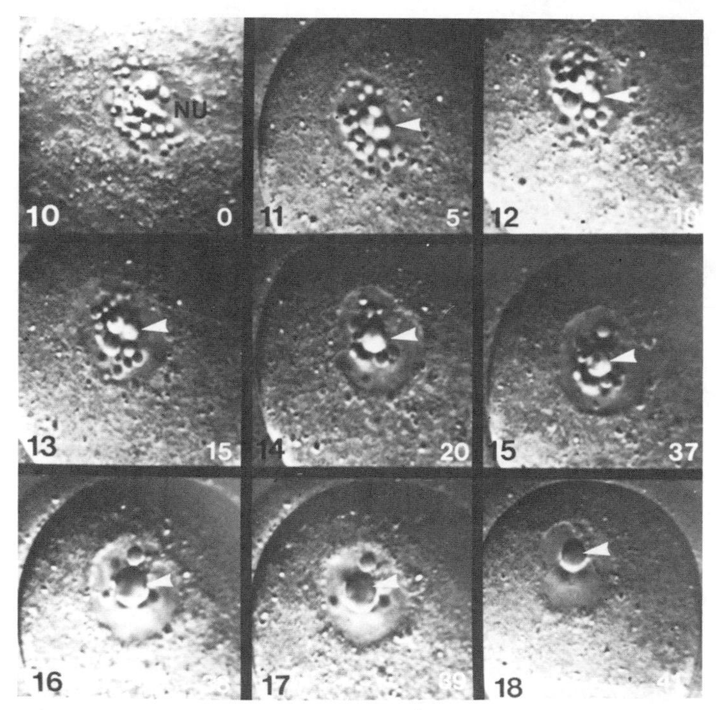

FIGURE 3 Time-lapse video images of the progressive stages in nucleolar reformation of the oocyte shown in Fig. 2. Images were taken after thawing and devitrification (DMSO and sucrose protocol). Time (in minutes) after placement in normal culture medium is indicated in the lower right-hand corner of the photomicrographs. Magnifications: (*10–18*) 420x.

morphodynamic process to occur after thawing and culturing of the oocytes. With both types of vitrification solutions, the first 30–40 minutes after placement of devitrified oocytes in normal culture medium was accompanied by the aggregation of nucleolar fragments to re-form a nucleolus of the same size, appearance, and position as that which existed prior to vitrification. This observation also held for oocytes that contained multiple nucleoli (Van Blerkom 1989b). The video images shown in Figure 2, frames 6 and 7, demonstrate the characteristic appearance of a fully grown, immature mouse oocyte and at higher magnification, the oocyte nucleolus just prior to placement in the first vitrification solution (0.875 M DMSO and 0.125 M sucrose in phosphate-buffered saline). Time-lapse video images of nucleolar disaggregation and assembly in this oocyte are presented in Figure 3, frames 10–18. After thawing, devitrification, and placement in normal culture medium (Fig. 3, frame 10; time 0), the numerous nucleolar fragments that appeared during the dehydration phase were still clearly evident.

Figure 4, frames 19 and 20, demonstrates an identical process of nucleolar disaggregation observed by light (frame 19) and electron microscopy (frame 20) in an oocyte vitrified in the solution of Rall and Fahy (1985; see also Van Blerkom 1989b). During the next 40 minutes of culture, these fragments aggregated to re-form a single nucleolus (or nucleoli, if multiple nucleoli existed prior to freezing; Van Blerkom 1989b; J. Van Blerkom and P. Davis, in prep.). Approximately 10 to 20 minutes after nucleolar re-formation was complete, the rapid (1–2 min) dissolution of the nucleolus that normally precedes germinal vesicie breakdown was observed (GVBD, see Van Blerkom 1989d). After GVBD, both chromosomal maturation and stage-related changes in cytoplasmic organization that normally occur during the maturation of the mouse oocyte in vitro were observed.

As described in detail by Van Blerkom and Runner (1984), such stage-related changes included a perinuclear accumulation of mitochondria during MI (Fig. 5, frame 23) and a progressive redistribution of these organelles during the subsequent stages of meiosis (Fig. 6, frames 24, 25, 27, 28, 30, and 31). The normality of metaphase spindle organization was as-

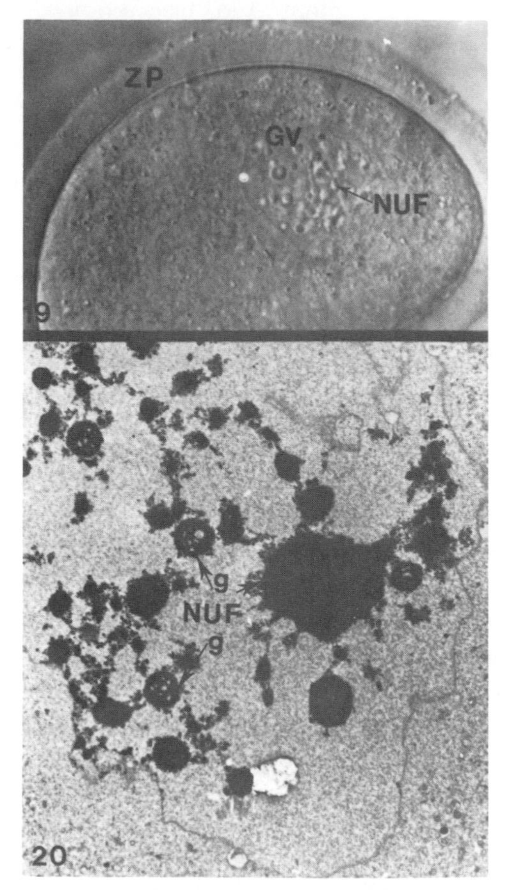

FIGURE 4 Light (frame *19*) and electron microscopic views (frame *20*) of the germinal vesicle (GV) of an immature mouse oocyte 3 min after thawing and devitrification. Numerous nucleolar fragments (NUF) are generated during vitrification, some of which have a granular composition (arrows, *g*). (ZP) Zona pellucida. Magnifications: (*19*) 650x; (*20*) 4500x. (Frame *20* was reprinted, with permission of Oxford University Press, from Van Blerkom 1989b.)

sessed by anti-tubulin immunostaining of MI-stage (Fig. 5, frame 21) through MII-stage oocytes (Fig. 6, frames 24, 26, 27, 29, 30, and 32). In combination with 4′,6-diamino-2-phenylindole (DAPI) (chromosomal) fluorescence (Fig. 5, frames 22 and 22b, Fig. 6, frame 29), these studies indicated that vitrified GV-stage oocytes typically formed MI and MII spindles that were normal in organization and chromosomal constitution. Cytogenetic analysis at MII indicated a frequency of aneuploidy of less than 3% (15/538) for oocytes vitrified at the GV stage in either cryoprotectant mixture (J. Van Blerkom and P. Davis, in prep.).

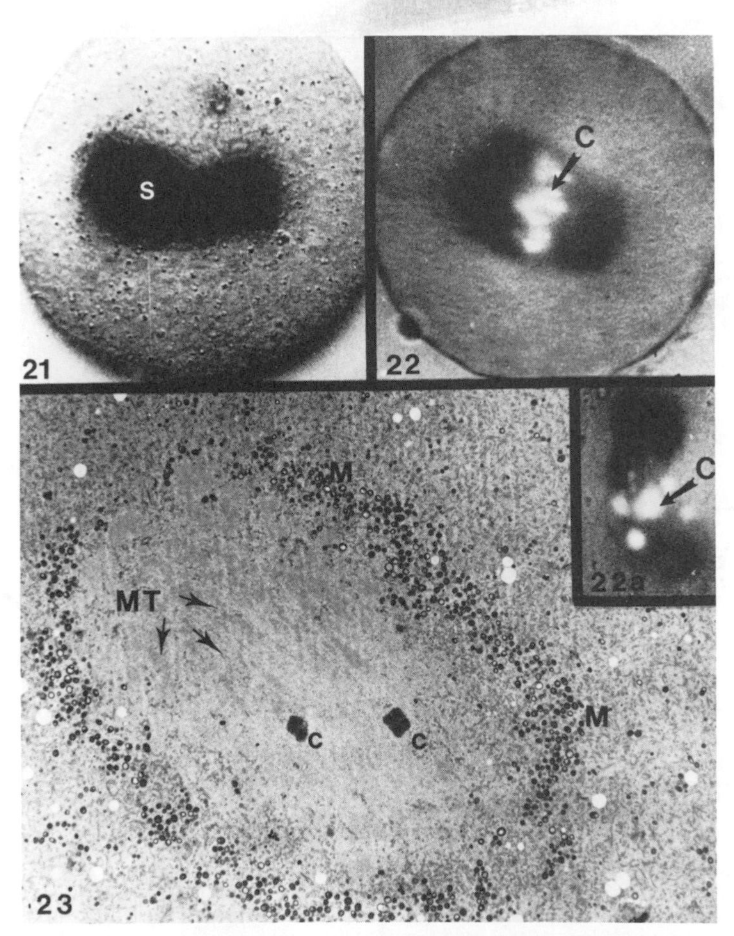

FIGURE 5 Anti-tubulin immunostaining of MI-spindle (S, frame *21*) and associated chromosomes (C, frames *22* and *22a*; DAPI fluorescence) in mouse oocytes that had been vitrified at the GV stage and allowed to mature in vitro after thawing. Normality of cytoplasmic maturation in vitro is indicated by the presence of perinuclear mitochondria (M) and normally arranged microtubules (MT) and bivalent chromosomes (C). Magnifications: (*21* and *27*) 700x; (*23*) 1300x.

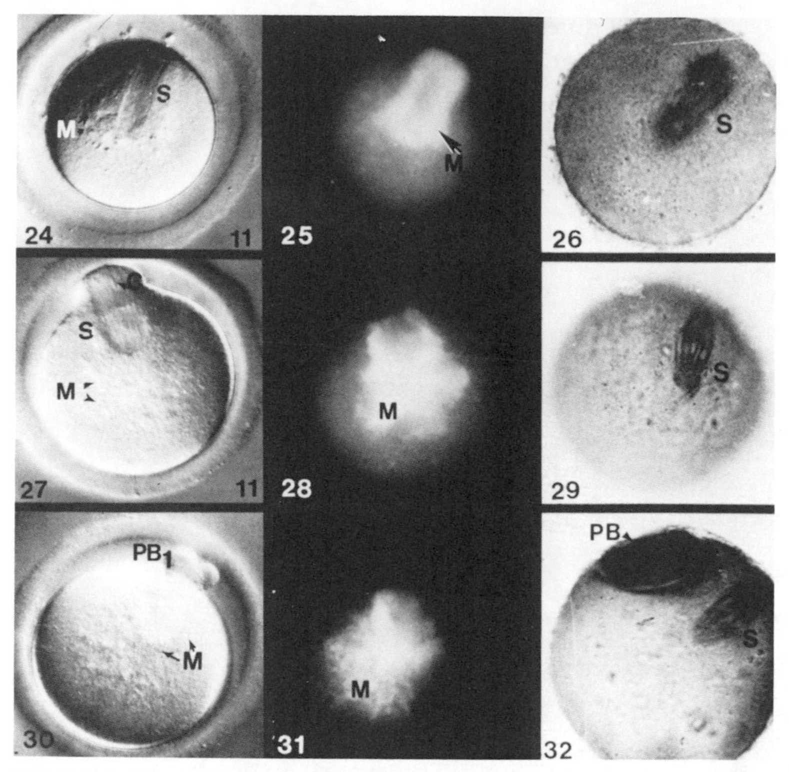

FIGURE 6 Normality of chromosomal and cytoplasmic maturation during final stages of meiotic maturation in vitro of mouse oocytes vitrified at the GV stage in a solution of DMSO and sucrose is indicated by light microscopic images (differential interference contrast optics, frames *24, 27, 30*), by stage-related changes in the pattern of mitochondrial distribution (M) as detected by rhodamine-123 fluorescence (frames *25, 28, 31*) and by the organization and position of the spindle (S) as resolved by anti-tubulin immunostaining (frames *26, 29, 32*). (PB1) Polar body 1. Magnifications: (*24, 26, 27, 29, 30,* and *32*) 413x.

For the vast majority of GV-stage oocytes, devitrification and culture were accompanied by a restoration of normal cytoplasmic structure and organization (for details, see Van Blerkom 1989b). However, the temporal pattern of cellular restoration differed for each affected cytoplasmic component. Although a return to the previtrification state of cytocortical microfilament organization occurred within a few minutes of culture, fibrillar arrays (lattices) were not apparent until 30–45 minutes and returned to a normal density only after 3–4 hours of culture (Van Blerkom 1989b). The organization and distribution of other major cytoplasmic organelles, such as pericortical Golgi complexes (Fig. 7, frame 36) and cortical granules, do not appear to be perturbed by vitrification.

After thawing and maturation in vitro, no degenerative or pathological changes in cytoplasmic organization or structure were evident at the light and electron microscopic levels. However, an apparent exchange of nuclear and cytoplasmic components was observed to have occurred during vitrification in approximately 15% of the GV-stage oocytes (Van Blerkom 1989b). Direct evidence for such an exchange was demonstrated at the fine structural level by the presence of mitochondria (Fig. 7, frame 33) and fibrillar lattices (Fig. 7, frame 35) within the re-expanded germinal vesicle and by nucleolar fragments in the cytoplasm (Fig. 7, frame 34). Alterations in the organization and geometry of the GV during the vitrification process suggest a structural basis for the observed interchange between nucleoplasm and cytoplasm. The extensive interdigitation of a highly porous nuclear envelope that occurs during dehydration appears to be associated with the generation of breaks or discontinuities in the nuclear envelope (Fig. 7, frame 35). At present, it is assumed that such breaks may present a conduit through which small nucleolar fragments exit the nucleus and perinuclear lattices and mitochondria enter the nucleus (Van Blerkom 1989b). Remarkably, this interchange did not prevent the re-formation of the nucleolus, the resumption of meiosis, and the progression of chromosomal maturation to MII. Cytoplasmic maturation of the mouse oocyte, which includes a perinuclear translocation of mitochondria (Van Blerkom and Runner 1984), was also not affected by this apparent exchange of nuclear and cytoplasmic components, as nucleolar fragments were still detectable in the cytoplasm at MII. These findings indicate that despite profound changes in nuclear, nucleolar, and cytoplasmic organization associated with vitrification of the GV-stage mouse oocyte, cytoplasmic and nuclear alterations are reversible and meiotic competence, as demonstrated by the high frequency of

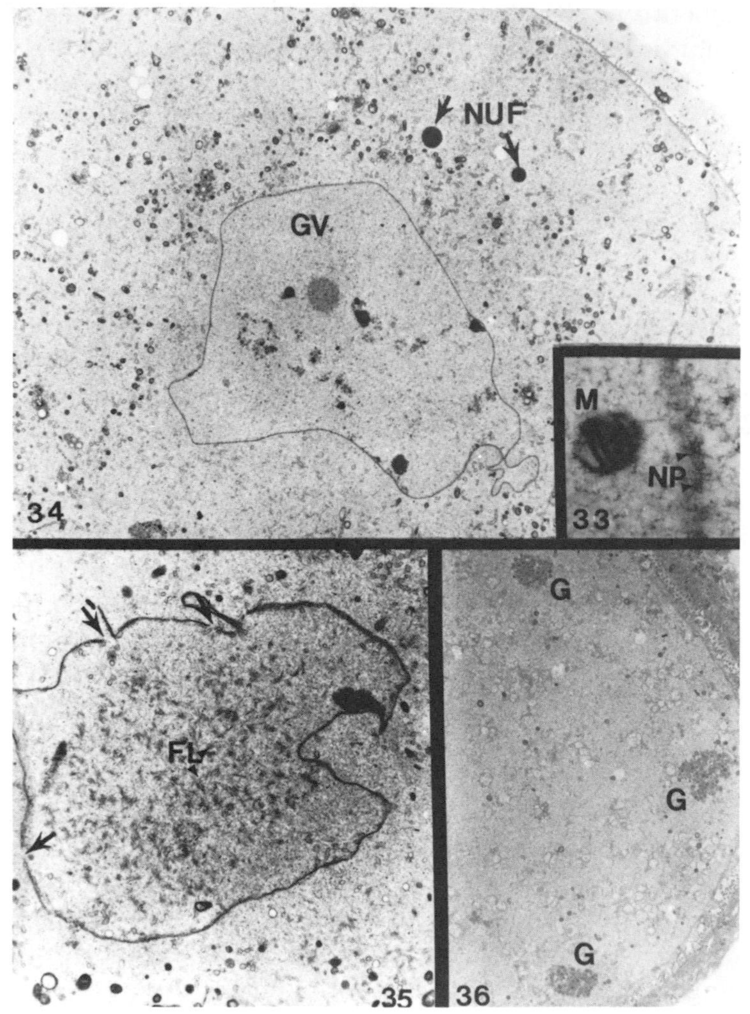

FIGURE 7 Apparent exchange of nuclear and cytoplasmic components evident in ~15% of GV-stage oocytes after devitrification. This phenomenon is indicated by the presence of nucleolar fragments in the ooplasm (NUF, frame *34*), and by the occurrence of mitochondria (M, frame *33*) and fibrillar lattices (FL, frame *35*) within the re-expanded germinal vesicle. The most likely means of exchange is through breaks in the nuclear envelope (arrows, frame *35*). In contrast, other major cytoplasmic elements such as cortical Golgi complexes (G, frame *36*) remain unchanged in location and organization during the vitrification/devitrification process. Magnifications: (*33*) 16,900x; (*34*) 1300x; (*35*) 1300x; (*36*) 975x. (Frames *33* and *34* were reprinted, with permission of Oxford University Press, from Van Blerkom 1989b.)

maturation to MII, is retained. The retention of developmental potential was further demonstrated by the fertilizability of oocytes that matured in vitro after thawing (Van Blerkom 1989b) and by the capacity of these embryos to continue gestation after transfer to surrogate uteri (J. Van Blerkom and P. Davis, in prep.).

As described previously, Kola et al. (1988) observed that preimplantation-stage embryos and day-16 mouse fetuses derived from oocytes vitrified at MII displayed a high incidence of aneuploidy and developmental aberration, respectively. In contrast, vitrification of GV-stage mouse oocytes in either 3.5 M DMSO and 0.5 M sucrose, or the solution described by Rall and Fahy (1985), was associated with a low frequency of aneuploidy at MII or at cleavage stages after in vitro fertilization (~2%; J. Van Blerkom and P. Davis, in prep.). In contrast to the report of Kola et al. (1988), our current findings demonstrate that none of the 12 fetuses examined on day 16 of gestation, nor any of the 27 neonates produced to date from oocytes vitrified at the GV stage (3.5 M DMSO/0.5 M sucrose solution), have been aneuploid or exhibited obvious developmental abnormalities. These results suggest that despite profound structural reorganization, vitrification at the GV stage is not accompanied by an increased rate of chromosomal anomalies that can be detected by karyotypic analysis of oocytes or preimplantation-stage embryos.

The above results suggest that vitrification at the GV stage may have clinical applications. However, a finding of potential concern is that after thawing, approximately 5% of the oocytes vitrified by the method of Rall and Fahy (1985) were observed by DNA fluorescence microscopy to contain small fragments of chromatin on the cytoplasmic side of the nuclear envelope (Fig. 8, frame 37) (Van Blerkom 1989b). DNA fluorescence (Fig. 8, frame 37) and electron microscopic analyses (Fig. 8, frames 38 and 39) of oocytes during vitrification, devitrification, and subsequent culture suggested the following mechanism(s) by which chromatin fragments could be externalized: (1) Changes in the physical state of DNA during vitrification may facilitate escape of chromatin fragments if a means of exit from the GV existed, (2) extensive interdigitation coupled with discontinuities in the nuclear envelope could result in an inversion of the

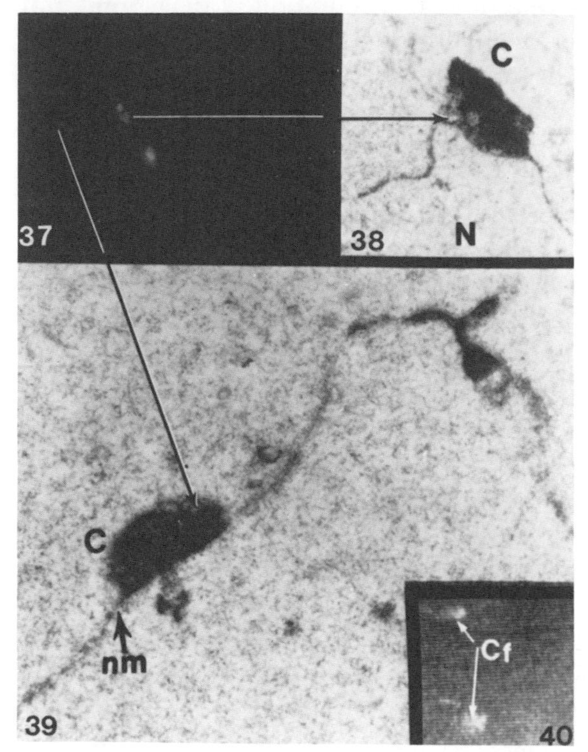

FIGURE 8 Observation of small fragments of chromatin (C) on the cytoplasmic side of the reexpanded nuclear membrane (nm) in vitrified GV-stage oocytes. Fragments are detected by DAPI fluorescence (frame *37*) and confirmed by electron microscopy (frames *38* and *39*). With image amplification, processing, and noise reduction, small fragments of excluded chromatin (Cf) were observed to persist in the cytoplasm through MII (frame *40*). Magnifications: (*38*) 5000x; (*39*) 13,000x. (Frames *37*, *38*, and *39* were reprinted, with permission of Oxford University Press, from Van Blerkom 1989b.)

nuclear membrane and associated chromatin, and (3) breaks in the nuclear envelope may provide an exit point for chromatin in proximity to the inner nuclear membrane during the collapse of the GV. At present, we have followed the fate of 27 oocytes in which extranuclear chromatin fragments were identified by DNA fluorescence in thawed oocytes; 24 of these oocytes underwent an apparently normal process of bivalent formation and chromosomal maturation to MII (Van Blerkom 1989b). After electronic enhancement and signal amplification (Van Blerkom and Henry 1988), video images of DNA fluorescence demonstrated that some of the larger chromatin fragments were still detectable at MII (Fig. 8, frame 40) and after fertilization. Therefore, fertilized eggs can arise from vitrified oocytes in which small fragments of chromatin were completely excluded from the process of chromosomal condensation.

Collectively, the above results support the notion that aneuploidy resulting from microtubule disruption or perturbation may not be a factor when freezing is accomplished at the GV stage. However, because approximately 5% of the vitrified oocytes contained chromosomal fragments on the exterior surface of the re-expanded GV, there is a very real possibility of generating fertilizable oocytes that carry subtle chromosomal deletions. At present, it is unknown whether this phenomenon is stage-related and therefore unavoidable at the GV stage or whether it is specifically associated with the vitrification method of cryopreservation.

Kola et al. (1988) reported that exposure of MII-stage oocytes (without freezing) to the mixture of vitrification reagents described by Rall and Fahy (1985) was accompanied by a high incidence of malformed fetuses with normal karyotypes. This observation is of central concern in the clinical application of this particular protocol of vitrification to the mature oocyte. Fetal malformations of the types described by these investigators may originate from small deletions in MII chromosomes that result from intracellular reactions involving a particular combination of vitrification reagents (Kola et al. 1988). Whether such deletions can result in fetal malformations or, for that matter, can even be detected by high-resolution cytogenetics (e.g., chromosomal banding) remains to be determined.

Although the absence of aneuploidy and developmental anomalies in fetal and neonatal mice produced to date in our studies of GV-stage oocytes vitrified in a solution containing 3.5 M DMSO and 0.5 M sucrose is encouraging, any potential clinical application will, as discussed below, require an unambiguous indication that chromosomal deletions do not occur in human oocytes vitrified at the GV stage.

POTENTIAL FOR OOCYTE CRYOPRESERVATION IN ASSISTED REPRODUCTION

From the preceding discussion, it is apparent that cryopreservation of the mature oocyte may be associated with an increased frequency of aneuploidy, whereas in the immature oocyte, it may induce subtle and perhaps undetectable chromosomal deletions. These factors represent significant concerns when oocyte cryopreservation is considered in a clinical situation. For the immature oocyte, modifications of the current vitrification protocol(s), as well as nonvitrification-based methods (e.g., controlled rate freezing), may avoid the types of cellular and nuclear alterations that have been observed (for review of methodologies, see Kuzan and Quinn 1988). Consequently, it is of particular relevance to determine whether the collapse of the nucleus and the coincident fragmentation of the nucleolus are obligatory for the GV-stage oocyte. Methods of dehydration and rehydration that minimize or control the rate of GV collapse, or do not alter the structure of the nuclear envelope, may prevent the artifactual generation of chromosomal abnormalities that may compromise subsequent development. The potential teratogenic effects associated with exposure of immature oocytes to cryoreagents will have to be unambiguously excluded in animals systems before clinical application can be contemplated.

At present, cryopreservation of immature oocytes has more direct and widespread application in commercially important species (such as the bovine) where ovaries are routinely available (for review, see Van Blerkom et al. 1990). The ability to mature oocytes at high frequency in vitro and to achieve pregnancies after fertilization and preimplantation development in culture demonstrates the potential benefits of oocyte cryopreservation in the laboratory-based production of embryos. Cryopreservation of the immature oocytes of rare and endangered species is another important application of this technology. Cryopreservation of the GV-stage human oocyte would be of direct clinical relevance in the creation of a pool of donor gametes. The use of such oocytes would be indicated in cases of premature ovarian failure, maternal age, ovarian dysfunction, or when known genetic conditions strongly dictate against the use of maternal oocytes. In preliminary studies from our laboratory, 62% (30/49) of fully grown GV-stage oocytes obtained from normal, unstimulated human ovaries matured to MII after vitrification in 3.5 M DMSO and 0.5 M sucrose. None of the mature oocytes were aneuploid, nor were there any of the cytoplasmic phenotypes (Nayudu et al. 1989; Van Blerkom 1990b;) that correlate with reduced fertilizability or developmental potential. Occasionally, GV-stage oocytes are obtained in assisted reproduction cycles (e.g., IVF or GIFT) that involve controlled ovarian hyperstimulation and ovulation induction. To date, only 8% (7/84) of GV-stage oocytes obtained from hyperstimulated ovaries have matured to MII after vitrification. Although none of the mature oocytes were aneuploid at MII, one oocyte exhibited a central clustering of organelles and two were vacuolated (Van Blerkom 1990b). With respect to post-thaw viability and developmental potential, our preliminary findings suggest that small antral follicles from unstimulated human ovaries may be an appropriate source of immature oocytes. In contrast, GV-stage oocytes that occur after ovulation induction cannot be assumed to be meiotically competent (Nayudu et al. 1989). If they are competent, they might not undergo a normal program of cytoplasmic maturation (see also Racowsky, this volume).

The two major consequences of cryopreservation for MI- or MII-stage oocytes that can have deleterious developmental consequences are (1) spindle disruption with an increased risk of aneuploidy and (2) premature exocytosis of cortical granules that can result in a fertilization-inhibiting hardening of the zona pellucida. However, because cryopreserved MII-stage mouse (Johnson 1989) and human oocytes (Van Uem et al. 1987) can be fertilized and have a demonstrated capacity to develop progressively, albeit at a low frequency (Chen 1988; Kola et al. 1988), developmental perturbation or lethality are not an obligate aspect of freezing. In contrast, the following factors, which are intrinsic to the oocyte and are therefore independent of the cryopreservation process, may influence post-thaw developmental ability: (1) Alterations or pathological changes in cytoplasmic organization and composition associated with controlled hyperstimulation clearly influence fertilizability and developmental outcome and may exist prior to cryopreservation (Nayudu et al. 1989; Van Blerkom 1990a,b), (2) the age of an oocyte, as measured in hours from the resumption of meiosis, may influence the stability of the spindle after a freeze/thawing cycle, and (3) the precise contribution of freezing to the incidence of aneuploidy may be obscured by a significant background frequency of aneuploidy.

When it is considered that meiotically mature human oocytes harvested after controlled ovarian hyperstimulation and ovulation induction are developmentally heterogeneous, the ability to select only those oocytes with high developmental potential for use in fertilization and cryopreservation becomes critical. As discussed previously, high-resolution light microscopy can identify some of the cytoplasmic alterations or degenerative changes that preclude fertilization or are associated with developmental arrest during the early preimplantation stages (see Nayudu et al. 1989; Van Blerkom 1990a,b). In contrast, the chromosomal normality of an oocyte cannot be readily assessed by light microscopic examination. Although DNA fluorescence microscopy is capable of detecting aneuploidy in living MII-stage human oocytes (Van Blerkom and Henry 1988), there are developmental consequences of using fluorescent probes to visualize and quantify chromatin (Ebert et al. 1985), thus rendering such probes clinically unacceptable.

Biochemical indications of developmental potential can be determined by measurements of specific follicular components and levels of oocyte metabolism (for review, see Van Blerkom 1990a). However, such assays may be difficult to perform on a routine basis, and the results may not be available in a timely fashion. Consequently, detailed high-resolution light microscopy is currently the most direct and rapid approach to the selection of oocytes for cryopreservation. To determine whether specific oocytes may be predisposed to freezing-induced disorders, we have initiated studies that characterize cytoplasmic organization and ploidy of living MII-stage human oocytes (Van Blerkom and Henry 1988) before and after cryopreservation by controlled-rate freezing as described by Chen (1988) or by vitrification with 3.5 M DMSO and 0.5 M sucrose. It is anticipated that this approach will establish not only whether an increased frequency of aneuploidy or cytoplasmic disorganization is associated with cryopreservation, but also the extent to which these perturbations are related to the cytoplasmic and cytogenetic state of the oocyte that exists at retrieval. Collectively, these studies could assist in the assessment of the developmental risks associated with the cryopreservation of both immature and mature oocytes. Parallel studies that utilize animal oocytes should provide a clear indication of whether or not oocyte cryopreservation should have a central role in treatments of human infertility that require laboratory-based methods of conception.

ACKNOWLEDGMENTS

The assistance of Mr. Patrick Davis in these studies is gratefully acknowledged. I also thank Dr. Cathy Van Blerkom for her critical comments during the preparation of the manuscript. This work was supported by grants from the National Institutes of Health (HD-21582) and National Science Foundation (BBS-8601231) and by funds provided by Reproductive Genetics In Vitro, Denver, Colorado.

REFERENCES

Ashwood-Smith, N.J., G.W. Morris, R. Fowler, T.C. Appleton, and R. Ashorn. 1988. Physical factors are involved in the destruction of embryos and oocytes during freezing and thawing procedures. *Human Reprod.* **3:** 795.

Biggers, J.D. 1990. Arbitrary partitions of prenatal life. *Human Reprod.* **5:** 1.

Boklage, C.E. 1990. Survival probability of human conceptions from fertilization to term. *Int. J. Fertil.* **35:** 75.

Boldt, J. and D.P. Wolf. 1986. An improved method for the isolation of fertile zona-free mouse eggs. *Gamete Res.* **13:** 213.

Bolton, V., S. Hawes, C. Taylor, and J. Parsons. 1989. Development of spare human preimplantation embryos in vitro: An analysis of the correlations among gross morphology cleavage rates and development to the blastocyst. *J. In Vitro Fertil. Embryo Transfer* **6:** 30.

Bonnicksen, A.L. 1988. Embryo freezing: Ethical issues in the clinical setting. *Hastings Cent. Rep.* **18:** 26.

Carroll, J., G.M. Warnes, and C.D. Matthews. 1989. Increase in digyny explains polyploidy after in vitro fertilization of frozen-thawed mouse oocytes. *J. Reprod. Fertil.* **85:** 489.

Caumus, M., E. Van den Abbeel, L. Van Waesberghe, A. Wisanto, P. Devroey, and A.C. Van Steirteghem. 1989. Human embryo viability after freezing with dimethylsulfoxide as a cryoprotectant. *Fertil. Steril.* **51:** 460.

Chen, C. 1986. Pregnancy after human oocyte cryopreservation. *Lancet* **I:** 884.

———. 1988. Pregnancies after human oocyte cryopreservation. *Ann. N.Y. Acad. Sci.* **541:** 541.

Cheng, W.K.T., R. Moor, and C. Polge. 1986. In vitro fertilization of pig and sheep oocytes matured in vivo and in vitro. *Theriogenology* **25:** 146.

Cohen, J., G.W. DeVane, C.W. Elsner, C.B. Fehilly, H.I. Kort, J.B. Massey, and T.G. Turner. 1988. Cryopreservation of zygotes and early cleaved human embryos. *Fertil. Steril.* **49:** 283.

Cornet, D., S. Alvarez, J.M. Antoine, C. Tibi, J. Mandelbaum, M. Plachot, and J. Salat-Baroux. 1990. Pregnancies following ovum donation in gonadal dysgenesis. *Human Reprod.* **5:** 291.

Crozet N., D. Huneau, V. Desmedet, M.C. Theron, D. Szollosi, S. Torres, and C. Sevellec. 1987. In vitro fertilization with normal development in the sheep. *Gamete Res.* **16:** 159.

Ebert, K.M., E. Hammer, and V.E. Papaioannou. 1985. A simple method for counting nuclei in the preimplantation mouse embryo. *Experientia* **41:** 1207.

Edwards, R.G. 1986. Causes of pregnancy loss. *Human Reprod.* **1:** 185.

Eichenlaub-Ritter, U., A. Stahl, and J. Luciani. 1986. The microtubular cytoskeleton and chromosomes of unfertilized human oocytes aged in vitro. *Human Genet.* **80:** 259.

Ethics Committee of the American Fertility Society. 1990. Ethical considerations of new reproductive technologies. *Fertil. Steril.* (suppl.) **2:** 53s.

Freedman, M., M. Farber, L. Farmer, S.P. Leibo, S. Heyner, and W.F. Rall. 1988. Pregnancy resulting from cryopreserved human embryos using a one-step in situ dilution procedure. *Obstet. Gynecol.* **72:** 502.

Friedler, S., L.C. Giudice, and E.J. Lamb. 1988. Cryopreservation of embryos and ova. *Fertil. Steril.* **49:** 743.

Friedler, S., E. Shen, and E.J. Lamb. 1987. Cryopreservation of mouse 2-cell embryos and ova by vitrification: Methodological studies. *Fertil. Steril.* **48:** 306.

Fugger, E.F., M. Bustillo, L.P. Katz, A.D. Dorfmann, S.D. Bender, and J. Schulman. 1988. Embryonic development and pregnancy from fresh and cryopreserved sibling pronucleate human zygotes. *Fertil. Steril.* **50:** 273.

Glenister, P.H., M.J. Wood, C. Kirby, and D.G. Wittingham. 1987. Incidence of chromosomal anomalies in first-cleavage mouse embryos obtained from frozen-thawed oocytes fertilized in vitro. *Gamete Res.* **16:** 205.

Hamlett, D.K., D.R. Franken, H.S. Cronje, and H. Luus. 1989. Murine oocyte cryopreservation: Comparison between fertilization success rates of fresh and frozen metaphase I and II oocytes. *Arch. Androl.* **23:** 27.

Harrison, K.L., L. Wilson, T. Reen, A.K. Pope, J.M. Cummins, and J.F. Hennessey. 1988. Fertilization of human oocytes in relation to varying delay before insemination. *Fertil. Steril.* **50:** 294.

Iritani, A. 1988. Current status of biotechnological studies in mammalian reproduction. *Fertil. Steril.* **50:** 543.

Jagiello, G., M. Ducayen, J. Fang, and J. Graffeo. 1976. Cytogenetic observations in mammalian oocytes. In *Chromosomes today* (ed. P. Pearson and K. Lewis), vol. 5, p. 43. Wiley & Sons, New York.

Johnson, M.H. 1989. The effects on fertilization of exposure of mouse oocytes to dimethylsulfoxide: An optimal protocol. *J. In Vitro Fert. Embryo Transfer* **6:** 168.

Johnson, M.H. and S.J. Pickering. 1987. The effects of dimethyl sulfoxide on the microotubular system of mouse oocyte. *Development* **100:** 313.

Johnson, M.H., S.J. Pickering, and M.A. George. 1988. The influence of cooling on the properties of the zona pellucida of the mouse oocyte. *Hum. Reprod.* **3:** 383.

Jones, H. 1990. Cryopreservation and its problems. *Fertil. Steril.* **53:** 780.

Junca, A.M., J. Mandelbaum, M.O. Alnot, M. Plachot, J. Cohen, and J. Salat-Baroux. 1988. Factors involved in the success of human embryo freezing. Does cryopreservation really improve the IVF results? *Ann. N.Y. Acad. Sci.* **541:** 575.

Kola, I., C. Kirby, J. Shaw, A. Davey, and A.O. Trounson. 1988. Vitrification of mouse oocytes results in aneuploid zygotes and malformed fetuses. *Teratology* **38:** 467.

Kuzan, F.B. and P. Quinn. 1988. Cryopreservation of mammalian embryos. In *In vitro fertilization and embryo transfer* (ed. D.P. Wolf), p. 301. Plenum Press, New York.

Lu, K.H., I. Gordon, M. Gallangher, and H. McGovern. 1987. Pregnancy established in cattle by transfer of embryos derived from in vitro fertilization of oocytes matured in vitro. *Vet. Rec.* **121:** 259.

Magistrini, M. and D. Szollosi. 1980. Effects of cold and isopropyl-*N*-phenylcarbamate on the second meiotic spindle of mouse oocytes. *Eur. J. Cell Biol.* **22:** 699.

Mandelbaum, J., A.M. Junca, M. Plachot, and M.D. Alnot. 1987. Cryopreservation of human embryos and oocytes. *Human Reprod.* **3:** 117.

Mandelbaum, J., A.M. Junca, C. Tibi, M. Plachot, M.D. Alnot, H. Rim, J. Salat-Barouxr, and J. Cohen. 1988a. Cryopreservation of immature and mature hamster and human oocytes. *Ann. N.Y. Acad. Sci.*

541: 550.

Mandelbaum, J., A.M. Junca, M. Plachot, M. Alnot, J. Salat-Baroux, S. Alvarez, C. Tibi, J. Cohen, C. Debache, and L. Tesquier. 1988b. Cryopreservation of human embryos and oocytes. *Human Reprod.* **3:** 117.

Mattioli, M., M.L. Bacci, G. Galeati, and E. Seren. 1989. Developmental competence of pig oocytes matured and fertilized in vitro. *Theriogenology* **31:** 1201.

McGaughey, R.W. and J. Van Blerkom. 1977. Patterns of polypeptide synthesis of porcine oocytes during maturation in vitro. *Dev. Biol.* **56:** 241.

Nakagata, N. 1989. High survival rates of unfertilized mouse oocytes after vitrification. *J. Reprod. Fertil.* **87:** 479.

Nayudu, P.L., D.A. Gook, A. Lopata, S.J. Sheather, C.W. Lloyd-Smith, P. Cadusch, and W. Johnson. 1987. Follicular characteristics associated with viable pregnancy after in vitro fertilization in humans. *Gamete Res.* **18:** 37.

Nayudu, P., A. Lopata, G. Jones, D. Gook, H. Bourne, S. Sheather, T. Brown, and W. Johnson. 1989. An analysis of human oocytes and follicles from stimulated cycles: Oocyte morphology and associated follicular fluid characteristics. *Human Reprod.* **4:** 558.

Osborn, J. and R. Moor. 1988. An assessment of the factors causing embryonic loss after fertilization in vitro. *J. Reprod. Fertil.* (suppl.) **36:** 59.

Pickering, S.J. and M.H. Johnson. 1987. The influence of cooling on the organization of the meiotic spindle of the mouse oocyte. *Human Reprod.* **2:** 207.

Pickering, S.J., P.R. Braude, M.H. Johnson, R.M. Cant, and R.M. Currie. 1990. Transient cooling to room temperature can cause irreversible disruption of the meiotic spindle, in the human oocyte. *Fertil. Steril.* **54:** 102.

Plachot, M., A. Veiga, J. Montagut, J. de Grouchy, G. Calderone, S. Lepretre, A.-M. Junca, J. Santalo, E. Carles, J. Mandelbaum, P. Barri, J. Degoy, J. Cohen, J. Egozcue, J.C. Sabatier, and J. Salat-Baroux. 1988. Are clinical and biological parameters correlated with chromosomal disorders in early life: A multicentric study. *Human Reprod.* **3:** 627.

Quigley, M.M. 1990. In vitro fertilization-embryo transfer in the United States 1988 results from the IVF-ET Registery. *Fertil. Steril.* **53:** 13.

Rall, W.F. and G.M. Fahy. 1985. Ice-free cryopreservation of mouse embryos at −196°C by vitrification. *Nature* **313:** 573.

Ranoux, C., H. Foulot, J. Dubuisson, D. Rambaud, F. Aubriot, C. Poirot, and V. Cardone. 1988. Returning to spontaneous in vitro fertilization. *J. In Vitro Fertil. Embryo Transfer* **5:** 304.

Sathananthan, A.H. and A.O. Trounson. 1989. Effects of culture and cryopreservation on human oocyte and embryo ultrastructure and function. In *Ultrastructure of human gametogenesis and early embryogenesis* (ed. J. Van Blerkom and P. Notta), p. 181. Kluwer Academic Publishers, Boston.

Sathananthan, A.H., A.O. Trounson, and L. Freeman. 1987. Morphology and fertilizability of frozen human oocytes. *Gamete Res.* **16:** 343.

Sathananthan, A.H., S.C. Ng, A.O. Trounson, A. Bongso, S.S. Ratnam, A. Ho, H. Mok, and M.N. Lee. 1988. The effects of ultrarapid freezing on meiotic and mitotic spindles of mouse oocytes and embryos. *Gamete Res.* **21:** 385.

Schalkoff, M.E., S.P. Oskowitz, and R.D. Powers. 1989. Ultrastructural observations of human and mouse oocytes treated with cryopreservatives. *Biol. Reprod.* **40:** 379.

Schroeder, A.C. and J.J. Eppig. 1984. The developmental capacity of mouse oocytes that matured spontaneously in vitro is normal. *Dev. Biol.* **102:** 493.

Schuster, E. 1990. Seven embryos in search of legitimacy. *Fertil. Steril.* **53:** 975.

Siebzehnruebl, E.R., S. Todorow, J. Van Uem, R. Koch, L. Wildt, and N. Lang. 1989. Cryopreservation of human and rabbit oocytes and one-cell embryos: A comparison of DMSO and propanediol. *Human Reprod.* **4:** 312.

Stern, S. and P. Wassarman. 1974. Meiotic maturation of the mammalian oocyte in vitro: Effects of dibutryl-cyclic AMP on protein synthesis. *J. Exp. Zool.* **189:** 275.

Testart, J., B. Lasalle, J. Belaisch-Allart, A. Hazout, R. Forman, J.D. Rainhorn, and R. Frydman. 1986. High pregnancy rate after early human embryo freezing. *Fertil. Steril.* **46:** 268.

Trounson, A.O. 1986. Preservation of human eggs and embryos. *Fertil. Steril.* **46:** 1.

Trounson, A.O and C. Kirby. 1989. Problems in the cryopreservation of unfertilized eggs by slow cooling in dimethyl sulfoxide. *Fertil. Steril.* **52:** 778.

Trounson A.O. and L. Mohr. 1983. Human pregnancy following cryopreservation, thawing and transfer of an 8-cell embryo. *Nature* **305:** 707.

Trounson, A.O. and H. Sathananthan. 1989. Human oocyte and embryo freezing. *Prog. Clin. Biol. Res.* **296:** 355.

Trounson, A.O., L.R. Mohr, C. Wood, and J.F. Leeton. 1982. Effects of delayed insemination on in vitro fertilization, culture and transfer of human embryos. *J. Reprod. Fertil.* **64:** 285.

Van Blerkom, J. 1989a. Developmental failure in human reproduction associated with preovulatory oogenesis and preimplantation embryogenesis. In *Ultrastructure of human gametogenesis and early embryogenesis* (ed. J. Van Blerkom and P. Motta), p. 125. Kluwer Academic Publishers, Boston.

———. 1989b. Maturation at high frequency of germinal vesicle-stage mouse oocytes after cryopreservation: Alterations in cytoplasmic, nuclear, nucleolar and chromosomal structure and organization associated with vitrification. *Human Reprod.* **4:** 883.

———. 1989c. The origin and detection of chromosomal abnormalities in meiotically mature human oocytes obtained from stimulated follicles and after failed fertilization in vitro. *Prog. Clin. Biol. Res.* **296:** 299.

———. 1989d. Morphodynamics of nuclear and cytoplasmic reorganization during the resumption of arrested meiosis in the mouse oocyte. *Prog. Clin. Biol. Res.* **294:** 33.

———. 1990a. Extrinsic and intrinsic influences on human oocyte and early embryo developmental potential. In *Elements of fertilization* (ed. P. Wassarman), p. 82. CRC Press, Boca Raton, Florida.

———. 1990b. Occurrence and developmental consequences of aberrant cellular organization in meiotically mature human oocytes after exogenous ovarian hyperstimulation. *J. Electron Microsc. Tech.* **16:** 324.

Van Blerkom, J. and G. Henry. 1988. Cytogenetic analysis of living human oocytes: Cellular basis and developmental consequences of perturbations in chromosomal organization and complement. *Human Reprod.* **3:** 777.

Van Blerkom, J. and R.W. McGaughey. 1978. Molecular differentiation of the rabbit ovum. I. During the in vivo and in vitro maturation of the oocyte. *Dev. Biol.* **63:** 139.

Van Blerkom, J. and M. Runner. 1984. Mitochondrial reorganization during resumption of arrested meiosis in the mouse. *Am. J. Anat.* **171:** 335.

Van Blerkom, J., H. Bell, and D. Weipz. 1990. Cellular and developmental biological aspects of bovine meiotic maturation, fertilization and preimplantation embryogenesis in vitro. *J. Electron. Microsc. Tech.* **16:** 298.

Van Blerkom, J., C. Manes, and J.C. Daniel. 1973. Development of preimplantation rabbit embryos in vivo and in vitro. I. An ultrastructural comparison. *Dev. Biol.* **35:** 262.

Van den Abbeel, E., J. Van der Elst, L. Van Waesberghe, M. Camus, P. Devroey, I. Kahn, J. Smitz, C. Straessen, A. Wisanto, and A. Van Steriteghem. 1988. Hyperstimulation: The need for cryopreservation of embryos. *Human Reprod.* (suppl.) **3:** 53.

Van der Elst, J., E. Van den Abbeel, R. Jacobs, E. Wisse, and A. Van Steirtegham. 1988. Effect of 1,2-propanediol and dimethylsulfoxide on the meiotic spindle of the mouse oocyte. *Human Reprod.* **3:** 960.

Van Uem, J.F.H.M., E.R. Siebzehnrubl, B. Schuh, R. Koch. S. Trotnov, and N. Lang. 1987. Birth after cryopreservation of unfertilized oocytes. *Lancet* **III:** 752.

Veeck, L., J.W. Wortham, J. Witmyer, B.A. Sandowm, A. Acosta, G.S. Garcia, G. Jones, and H. Jones. 1983. Maturation and fertilization of morphologically immature human oocytes in a program of in vitro fertilization. *Fertil. Steril.* **39:** 594.

Vincent, C., S.J. Pickering, and M.H. Johnson, 1990. The hardening effect of dimethylsulphoxide on the mouse zona pellucida requires the presence of an oocyte and is associated with a reduction in the number of cortical granules present. *J. Reprod. Fertil.* **89:** 253.

Whittingham, D.G. 1971. Culture of mouse ova. *J. Reprod. Fertil.* (suppl.) **14:** 7.

Wilmut, I. 1986. Cryopreservation of mammalian eggs and embryos. In *Manipulation of mammalian development* (ed. R.L. Gwatkin), p. 217. Plenum Press, New York.

Wilson, L. and P. Quinn. 1989. Development of mouse embryos cryopreserved by an ultra-rapid method of freezing. *Human Reprod.* **4:** 86.

Xu, K.P., T. Greve, H. Callensen, and P. Hytell. 1987. Pregnancy resulting from in vitro fertilization of bovine oocytes matured in vitro. *J. Reprod. Fertil.* **81:** 501.

Zaner, R.M., F.H. Roehm, and G.A. Hill. 1990. Selective termination in multiple pregnancies: Ethical considerations. *Fertil. Steril.* **54:** 203.

Ions and Preimplantation Development

J.D. Biggers,[1] J.M. Baltz,[1] and C. Lechene[2]
[1]Laboratory of Human Reproduction and Reproductive Biology
and Department of Cellular and Molecular Physiology
Harvard Medical School, Boston, Massachusetts 02115
[2]Department of Medicine, Harvard Medical School and Brigham and
Women's Hospital, Boston, Massachusetts 02115

INTRODUCTION

A major characteristic of life is the creation and maintenance of a large difference in ionic composition between extracellular fluid, which is rich in sodium and poor in potassium, and intracellular fluid, which is poor in sodium and rich in potassium. Constancy of the composition of this "*milieu interieur*" (Bernard 1878) achieved by the homeostatic functions of the cell (Cannon 1932) is essential for mammalian life, with relatively small alterations leading to death. It is now also clear that the difference in ionic composition between the cell and its microenvironment is a universal source of cellular potential energy (Lechene 1988). Transmembrane ionic gradients—particularly that of sodium—are the driving force used to accomplish both specialized organ-specific cellular functions (e.g., action potentials in nerve, reabsorption and secretion by kidney, absorption by intestine, and cerebrospinal fluid formation) and general cellular functions essential for cell growth and division (e.g., entry of nutrients, amino acids, glucose, and phosphate, regulation of intracellular pH and free calcium, and maintenance of cell volume). Both constancy and controlled variation of the intracellular ionic composition are essential. For example, general ion constancy is necessary for optimal enzyme function (Somero 1985), and hydrogen ion concentration changes play a critical role in cellular growth and division, specifically by an alkalinization in response to growth factors necessary for initiation of DNA synthesis (Pouyssegur

Animal Applications of Research in Mammalian Development
 0-87969-333-9/91 $3.00 + 00

et al. 1985; Moolenaar et al. 1988). Such fundamental roles of ions in cellular functions extend to intracellular organelles as well. For example, the proton electromotive force created by the respiratory chain is used by a mitochondrial proton pump functioning in reverse to generate ATP, the energy currency of the cell (Mitchell 1967). In addition, a proton pump and perm-selectivity of its membrane allow a lysosome to maintain the acid pH necessary for proper functioning of its hydrolases.

In recent years, however, ions and the mechanisms controlling them have not generally been regarded as having much developmental importance. They are often dismissed by developmental biologists as being involved solely in merely homeostatic or so-called "housekeeping" activities. This view seems to have been first codified by Raff and Kaufman (1983), who proposed that the genome be conceptually divided into two parts: one part consisting of those genes directly involved in determination, differentiation, and morphogenesis, and the other part consisting of those genes that regulate metabolic functions. Unfortunately, Raff and Kaufman (1983) dismiss the latter as having no developmental interest. This myopic view of the control of development is now being challenged, as in the review by Nijhout (1990), in which it is pointed out that ions and small molecules operating in complex networks are likely to be involved in the sequential control of developmentally important genes. Furthermore, even apparently purely homeostatic functions should not be ignored by developmental biologists. As pointed out elsewhere (Biggers 1980), the apparently static set points may change with development, and what appears to be housekeeping on one time scale can be a signal or driving force for change on a longer time scale; ionic signaling occurs when an ovum is activated at fertilization, and ion transport systems become active at specific times to produce fluid-filled body cavities in the embryo such as the blastocoel and the amnion (Biggers 1972).

In this paper, we first review the current information on ions in the preimplantation mammalian embryo. We then discuss, in more detail, two specific physiological systems in which ions are involved in developmentally important phenomena in the early embryo. One is the regulation of intracellular pH (pH_i), which emphasizes the fact that some transport systems change with development according to the embryo's needs at each stage of the life cycle. The second involves the regionalization of Na^+/K^+-ATPase in the embryo, which results in the formation of the blastocyst through cavity formation, thereby establishing the earliest patterned structure in mammalian development.

GENERAL CONCEPTS

A major concept in physiology is the dynamic state of the body constituents, first explicitly recognized by Schoenheimer (1940). Theoretical models that display the participant transport systems are particularly useful in analyzing such dynamic systems. Biggers et al. (1977) proposed two such models to show the major changes in transport systems in the preimplantation embryo brought about by compaction (Fig. 1). Prior to compaction, the blastomeres can be considered as an assembly of loosely connected cells whose transport systems can be represented by a two-compartment model (intracellular and extracellular). The transport of ions across the membranes of these cells regulates important functions such as pH_i. Compaction results in the formation of an outer epithelium called the trophectoderm, which allows the formation of extracellular fluid by the embryo itself. Thus, the embryo becomes an organized multicellular organism that can be represented by a four-compartment model (intracellular: trophoblast cell, inner cell mass cell; extracellular: uterine, blastocoel).

The ion fluxes that participate in early embryo development can be described according to current physiological thinking about the ion fluxes generally found in most cells, as summarized schematically in Figure 2. In general, the intracellular concentrations of Na^+ and K^+ are remarkably constant. However, they are maintained at concentrations very different from those in the immediate environment of the cells. The intracellular concentration of Na^+ is low and its extracellular concentration is high. In contrast, the intracellular concentration of K^+ is high and its extracellular concentration is low. Consequently, a large electrochemical gradient due to

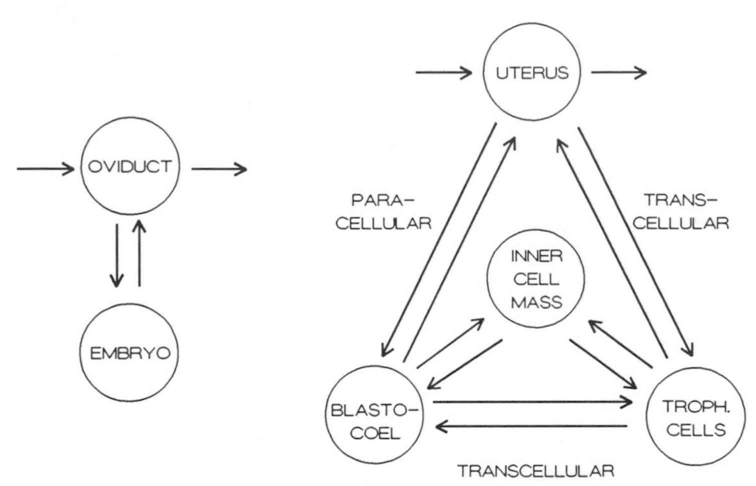

FIGURE 1 Theoretical models showing changes in exchange and transport pathways in mammalian preimplantation embryos as a result of compaction. The two stages represented are a precompacted stage and a cavitating morula or blastocyst. The precompacted embryo is represented by a two-compartment model (*left*). The two compartments are the oviduct and the embryo cells. Substances are transported into and out of the oviduct from the bloodstream and between the oviduct and the embryo. The cavitating morula or blastocyst is represented by a four-compartment model (*right*). The compartments are the uterus, trophoblast, inner cell mass, and blastocoel. Substances are transported into and out of the uterus from the bloodstream and then between the three compartments of the embryo via the pathways shown. (Adapted, with permission, from Biggers et al. 1977.)

these two ions exists across the cell membrane. Despite these constancies, both of the ions are in a state of flux. At a given time, their flux is the result of the rate at which the ion is lost by leaking down its electrochemical gradient and the rate at which it is retrieved by pumps spending chemical energy. Na^+ and K^+ fluxes are maintained by (sodium and potassium) pumps that are located within the cell membrane. The pumps are molecules of the enzyme Na^+/K^+-ATPase, which uses the energy derived from the hydrolysis of ATP to transport Na^+ and K^+ up their electrochemical gradients and maintain gradients of these species across the plasma membrane. The energy, in the form of these ion gradients (mainly of Na^+), is then used for other energy-requiring transports, such as the co- and count-

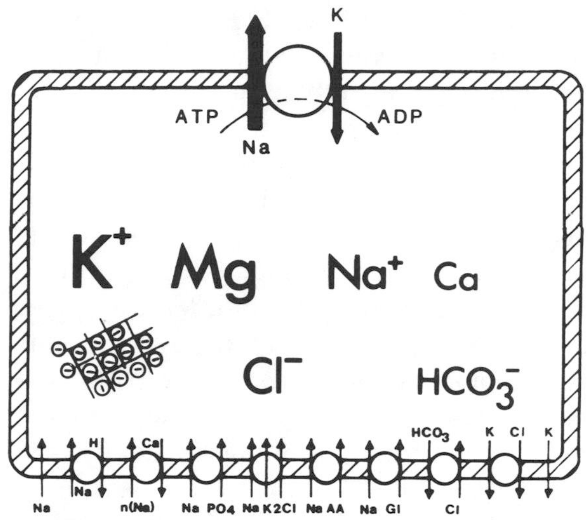

FIGURE 2 Cells may be visualized as performing their functions through the dissipation of ionic gradients using a variety of specialized leak pathways. The Na^+,K^+ pump creates and maintains the potential energy of the cells, adjusting its activity in response to modulations in the intensity of the leaks. Chemical energy from ATP is converted to potential energy of ion gradients, which, in turn, drive numerous transmembrane transport processes. These processes, along with the fixed negative charges within the cell (lattice at the lower left) maintain an ionic environment different from the extracellular fluid. (AA) Amino acid; (Gl) glucose. (Reprinted from Lechene, *Prog. Clin. Biol. Res.* copyright 1988 by permission of Wiley-Liss, a division of John Wiley and Sons, Inc.)

ertransport of protons, chloride, phosphate, glucose, and amino acids, some of which are in turn involved in transporting still other substances. If the intracellular Na^+ concentration rises because Na^+ moves down its gradient while performing such functions, the pump responds quickly by increasing the Na^+/K^+-ATPase activity and soon thereafter by producing more pumps. For a further discussion of these dynamic homeostatic systems, see Lechene (1988) and Cohen and Lechene (1989).

Although these systems are important throughout development, their participation in the formation of the trophectoderm has attracted special interest, since it is the first differentiated tissue to appear in the mammalian embryo. The initial compaction step results in a radical change in the transport properties of the outer cells of the embryo (Fig. 1). The transport systems are redistributed to create a dynamic system that allows the newly formed epithelial structure to transport substances vectorially between the apical and basal surfaces. Compaction is frequently discussed in terms of the cell polarity that is necessary for a simple epithelium to develop (Ziomek and Johnson 1980; Wiley et al. 1990). However, it should also be emphasized that the developed epithelium is a closed sheet that produces fluid which collects internally. As a result, the trophectoderm needs to develop as a compliant structure that allows expansion as the fluid collects. This system requires the close coordination of transport mechanisms and cell multiplication while maintaining the permeability seal of the trophectoderm (Biggers et al. 1988).

There are considerable differences between the blastocysts of mammalian species, particularly in the extent to which they expand after compaction. Although the species can be arranged as a continuum, it has been useful to speak of minimally expanding blastocysts and maximally expanding blastocysts (Biggers 1972). The former is represented by the mouse blastocyst, which increases in volume only about five times, and the latter is represented by the rabbit blastocyst, which increases in volume about 1500 times (see Cruz and Pedersen, this volume). The mechanisms involved in the development of cell polarity and compaction have been almost exclusively studied in the minimally expanding mouse blastocyst. Only recently has the subject been studied in the rabbit (Ziomek et al. 1990). In contrast, the transport mechanisms involved in the expansion of the blastocyst have been studied mainly in the rabbit; this is the animal model of choice for physiological studies of expansion because of the large degree of blastocyst swelling that occurs.

IONS AND ION TRANSPORT IN PREIMPLANTATION EMBRYOS

Intracellular Ion Concentrations and Membrane Permeabilities

The intracellular concentrations of four inorganic ions—Na^+, K^+, Cl^-, and H^+—have been determined for the cells of preimplantation mammalian embryos. Using radioactive isotopes, Powers and Tupper (1977) measured the intracellular contents of K^+, Na^+, and Cl^- in the two-cell stage mouse embryo. These were found to be 130 mM K^+, 151 mM Na^+, and 52 mM Cl^-. Flame photometry was also used to confirm their determinations of K^+ and Na^+ concentrations; these measurements yielded 142 and 143 mM, respectively. The extremely high Na^+ concentration relative to that usually found in healthy cells may indicate that these cells were somehow damaged and no longer able to maintain normal ion gradients actively (see below).

More recently, Lee (1987) used ion-sensitive microelectrodes to measure the intracellular activities of K^+ and Na^+ in the mouse embryo at various stages of development. He found K^+ concentrations of 105, 100, and 90 mM in the one-cell, two-cell, and eight-cell stage, respectively. The Na^+ concentrations were found to be 15 and 25 mM at the one-cell and two-cell stages, respectively. He also reported preliminary values of K^+ concentration for cells of the blastocyst: 80 mM in the trophectoderm, 50 mM in the polar trophectoderm, and 100–140 mM in the inner cell mass.

It is apparent that the measurements from these two studies are not in agreement. At the two-cell stage, Powers and Tupper found an Na^+ concentration that is sixfold greater than that obtained by Lee. Since somatic cells normally have much higher K^+ levels than Na^+ in their cytoplasm, the low K^+/Na^+ ratio may indicate that the embryos used in the Powers and Tupper study were no longer healthy. However, Powers and Tupper measured permeabilities (see below) in the normal range and were able to detect significant ouabain-sensitive components of both the Na^+ and K^+ fluxes (implying Na^+/K^+-ATPase activity), which argues conversely that these cells *were* functioning normally. Lee does not cite the Powers and Tupper work, so there is no comment on this discrepancy in his paper. Because all of the cells whose plasma membranes are permeable to water maintain iso-osmolarity with their environment, these large variations in ionic concentration must reflect either impermeability of the plasma membranes to water (which is not true for embryos), experimental error, or the existence of osmolytes that were not detected in these studies.

The only measurements of the permeability (P) of inorganic ions through the plasma membranes of mammalian embryos are (1) those by Powers and Tupper (1977) for Na^+ and K^+ permeabilities in the two-cell mouse embryo, where P_{Na} was found to be 1.6 x 10^{-7} cm/sec and P_K was found to be 2.1 x 10^{-7} cm/sec and (2) our measurement of the proton permeability of the two-cell mouse embryo plasma membrane, which was 0.23 cm/sec (Baltz et al. 1990). The values for Na^+ and K^+ are in the normal range for other cells (Jain and Wagner 1980). The permeability of two-cell mouse embryos to protons is very high, being several orders of magnitude larger than normal for other cell types (Deamer 1984; Gutknecht 1987; Baltz et al. 1990). The physiological basis for this high proton permeability in early mouse embryos is not known.

Membrane Potential

The membrane potential of a cell is determined by charge separation across the plasma membrane resulting from the distribution of charged ions and their permeabilities (Goldman 1943). A number of measurements of the membrane potential of preimplantation mammalian embryos have been reported (Table 1). Measurements of the same stages by different investigators do not precisely agree, probably reflecting differences of medium composition, temperature, and embryo handling. However, the membrane potential of mouse embryos measured with microelectrodes appears to be from approximately –20 to –40 mV, whereas that of the hamster is somewhat higher. The membrane potential of mouse embryos seems to decrease (depolarize) as development progresses (Lee 1987), whereas that of the hamster increases (hyperpolarizes; Mitani 1985).

Ion Channels

Eusebi et al. (1983) first showed that the excitable Ca^{++} channel that had been identified in oocytes (Okamoto et al. 1977) persisted after fertilization in the mouse. Eusebi et al. found that these Ca^{++} channels were present at least up to the four-cell stage but that membrane excitability decreased throughout early cleavage. They did not examine embryos past the four-cell stage. Subsequent work has established that Ca^{++} channels disappear completely at the 8–16-cell stage in the mouse embryo (Mitani 1985; Yoshida 1985; Lee 1987). Ca^{++}

TABLE 1 *MEMBRANE POTENTIAL MEASUREMENTS ON PREIMPLANTATION EMBRYOS*

Species	*Stage*				*Study*
	1-cell	*2-cell*	*4-cell*	*8-cell*	
Mouse					
	–	-19[a]	–	–	Powers and Tupper (1977)
	-19	–	–	–	Eusebi et al. (1983)
	-7 to -15[b]	-7 to -15[b]	-7 to -15[b]	-7 to -15[b]	Yoshida (1985)
	-30	-22	–	–	Mitani (1985)
	-40	-35	-25	-25	Lee (1987)
Hamster					
	-16	-34	-49	-56	Mitani (1985)

[a]Membrane potentials are given in millivolts.
[b]This range was given for oocytes and embryos up to the 16-cell stage.

channels have also been found in the hamster embryo, also persisting to the eight-cell stage (Mitani 1985). These channels are voltage-sensitive Ca^{++} channels, affected by pharmacological agents known to inhibit Ca^{++} channels in other tissues (Yoshida 1985; Lee 1987). The threshold membrane potential for activation of the channels is –50 mV, with the maximal current at approximately –20 mV (Mitani 1985).

Voltage-sensitive Na^+ channels are apparently lacking in the cleavage-stage mouse embryo. In Ca^{++}-containing medium, action potentials are independent of Na^+ concentration (Eusebi et al. 1983). In Ca^{++}-free, high-Na^+ medium, action potentials can be elicited that are mediated by Na^+ (Yoshida, 1985). However, these are insensitive to tetrodotoxin, a specific voltage-sensitive Na^+ channel inhibitor, and sensitive to Ca^{++} channel blockers. This indicates that under these extreme conditions, the Ca^{++} channel of the mouse embryo will conduct Na^+.

In the mouse blastocyst, there is some evidence for the existence of amiloride-sensitive Na^+ channels. Manejwala et al. (1989) have found that benzamil, an amiloride-derivative Na^+ channel blocker, inhibits the influx of Na^+ into the blastocoel under conditions of low extracellular Na^+.

A Ca^{++}-activated K^+ conductance has also been reported in the hamster embryo (one-cell stage immediately following fertilization; Miyazaki and Igusa 1982; Igusa and Miyazaki 1983). This conductance apparently underlies the recurring hyperpolarizations that follow fertilization in the hamster. It is not known whether this conductance persists throughout development. A much smaller K^+ conductance has been reported in mouse embryos upon fertilization (Igusa et al. 1983).

Amino Acid Transport

The area of amino acid transport in embryos has recently been reviewed by Van Winkle (1988) and will be only briefly summarized. Amino acid transport is of interest because of those systems that use the driving force of electrochemical gradients of inorganic ions to transport amino acids. Although there are a number of Na^+-dependent amino acid transport systems, there is one that is Na^+- and Cl^--dependent. This is the Gly system, which transports glycine and its analog, sarcosine.

In precompaction mouse embryos, the only inorganic ion-dependent amino acid flux is due to this Gly system, and glycine transport is affected by Na^+ and Cl^- fluxes at these stages. By the time the embryo reaches the blastocyst stage, the Gly system has disappeared. Indeed, it seems that all other amino acid transporters that were present during cleavage either are no longer active or are obscured by two novel, broad-spectrum, embryo-specific transport systems designated $b^{0,+}$ and $B^{0,+}$, of which $B^{0,+}$ is Na^+-dependent, and possibly Cl^--dependent (Van Winkle et al. 1988). Although only glycine was transported by a Na^+/Cl^--dependent transporter during cleavage, at later stages of development, most amino acids are transported, at least in part, by a Na^+/Cl^--dependent process. Benos (1981) examined the transport of methionine into the rabbit blastocyst. He found that methionine uptake is Na^+-dependent in day-5 and day-6 blastocysts, but Na^+-independent in day-7 blastocysts.

Metabolite Transport

Powers and Tupper (1977) found no Na^+-dependent components of glucose or pyruvate uptake at the two-cell stage of

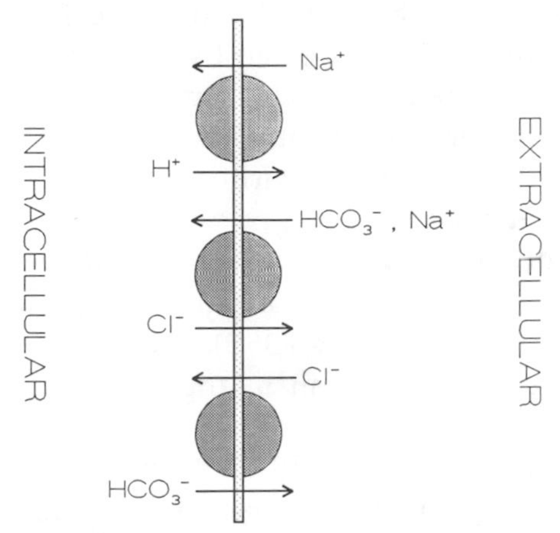

FIGURE 3 Common intracellular pH regulatory systems in vertebrate cells. The Na^+/H^+ antiport and the Na^+-dependent HCO_3^-/Cl^- exchanger are active in relieving acid loads. Both depend on the Na^+ gradient (and the presence of external Na^+) for energy. Amiloride and its derivatives inhibit the Na^+/H^+ antiport, and the Na^+-dependent HCO_3^-/Cl^- exchanger is inhibited by stilbene derivatives such as DIDS. The Na^+-independent HCO_3^-/Cl^- exchanger is active in relieving alkaline loads. It depends on the Cl^- gradient (and the presence of external Cl^-) for its energy. It is also inhibited by stilbene derivatives such as DIDS, but is unaffected by the absence of Na^+.

the mouse embryo and likewise, Gardner and Leese (1988) found that mouse blastocysts similarly exhibited entirely Na^+-independent glucose transport. In rabbit blastocysts (days 5–7), glucose transport is also not coupled to Na^+ (Benos 1981). A recent study by Robinson et al. (1990) has shown that a Na^+-independent glucose transport system is present in both the apical and basolateral membranes of 6- and 7-day-old rabbit trophoblast cells, similar to that found in the mouse.

REGULATION OF INTRACELLULAR pH

Intracellular pH (pH_i) is regulated by specific plasma membrane systems in almost all cell types (Fig. 3) (for review, see Roos and Boron 1981; Boron 1987). The most common systems for alleviating acid loads are (1) the Na^+/H^+ antiporter, a system that removes excess protons from the cell by exchanging them for Na^+, which runs down its gradient, and (2) the Na^+-dependent HCO_3^-/Cl^- exchanger, a system that also uses the energy of the Na^+ gradient to import bicarbonate (along with Na^+) into the cell in exchange for Cl^-.

The Na^+/H^+ antiport is nearly ubiquitous in vertebrate cells, being absent from only a very few specialized cell types (Gillespie and Greenwell 1988; Bidani et al. 1989). We examined the role of this system in acid extrusion in the two-cell-stage mouse embryo. The embryos were acid-loaded by the ammonium chloride–pulse method (Boron and DeWeer 1976; Roos and Boron 1981) in nominally bicarbonate-free medium to prevent any HCO_3^-/Cl^- exchange from obscuring the results. The acid-loaded embryos recovered from the acidosis with a time constant on the order of 5 minutes (Baltz et al. 1990). Unexpectedly, however, we could not detect a component of this recovery that was due to Na^+/H^+ antiport activity. The recovery rate was insensitive to removal of extracellular Na^+ and also to the Na^+/H^+ antiport inhibitor amiloride and its more specific analog ethylisopropylamiloride (EIPA) (Fig. 4A). Thus, mouse embryos (at least at the two-cell stage) are unique in lacking the otherwise ubiquitous Na^+/H^+ antiport.

Clearly, the Na^+/H^+ antiport must appear at some point, because it is present in the adult mouse. We have not yet done a developmental study to determine at which stage pH_i regulation by Na^+/H^+ antiport begins. However, evidence indicates that the antiport may mediate part of the Na^+ flux into the mouse blastocoel; Manejwala et al. (1989) have found that this Na^+ flux and blastocoel expansion can be inhibited by EIPA, which specifically inhibits the antiport. pH_i regulation in the blastocyst has not yet been studied, nor has Na^+ flux dependence on amiloride at the two-cell stage. It is possible that the Na^+/H^+ antiport is present at all stages and affects Na^+ transport, but at a lower level as compared to the apparently passive proton flux. If this were true, then its effect on proton flux would be masked during acid load recovery.

Even though we found that pH_i regulation is not mediated by Na^+/H^+ antiport activity, it is still possible that acid loads

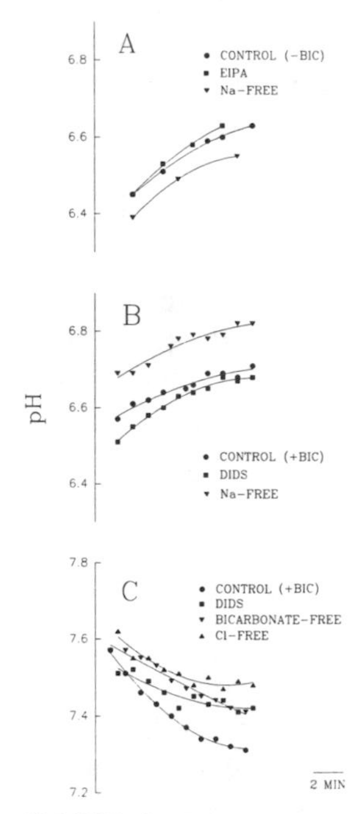

FIGURE 4 Experiments to detect the presence of pH regulatory systems in the two-cell stage mouse embryo. Each trace is one representative experiment (the mean pH_i of the 10 embryos in the experiment is shown) of at least five experiments. Data are shown beginning just after the cells have been acid- or alkaline-loaded (i.e., at the start of recovery). Only the first 10 min of each recovery are shown (the curves eventually level off). Inhibition of recovery is manifest as a reduction in the slope of the curve compared to the control. (*A*) Two-cell embryos were acid-loaded by the ammonia-pulse technique and recovery followed in bicarbonate-free medium (–BIC). The recovery was not slowed by either the presence of the Na^+/H^+ antiport inhibitor EIPA or the absence of external Na^+ (Na-FREE), indicating a lack of participation by the antiport in acid-load recovery. (*B*) Two-cell embryos, acid-loaded as in *A*, were allowed to recover in a bicarbonate/ CO_2-buffered medium (+BIC). Recovery was not slowed by either the anion exchange inhibitor, DIDS, or the absence of external Na^+ (Na-FREE). Thus, the Na^+-dependent HCO_3^-/Cl^- exchanger does not participate in recovery from acid-loading. (*C*) Two-cell mouse embryos were alkaline-loaded by exposure to ammonium salt. The subsequent recovery was very significantly slowed by either the presence of DIDS, the absence of HCO_3 (BICARBONATE-FREE), or the absence of Cl^- (Cl-FREE). Note the much faster fall in pH for the control curve versus the three curves, where HCO_3^-/Cl^- exchange is inhibited. This indicates that a great proportion of the recovery from alkaline loads is mediated by the HCO_3^-/Cl^- exchanger in the two-cell mouse embryo. (Data from Baltz et al. 1990, 1991.)

could be specifically relieved in the presence of bicarbonate by the Na^+-dependent HCO_3^-/Cl^- exchanger. We have done acid-loading experiments in a bicarbonate/CO_2-buffered medium to test for any bicarbonate-sensitive component of acid-load recovery. Again, the recovery seems not to be mediated by a specific system, as indicated by the following observations: (1) The recovery is the same rate in the presence or absence of bicarbonate, (2) the anion exchange inhibitor 4,4′-diisothiocyanostilbene disulfonic acid (DIDS) has no effect on the recovery, and (3) the recovery is insensitive to a lack of extracellular Na^+ (Fig. 4B) (Baltz et al. 1991). Thus, it seems that, at least in culture media, there is no specific relief of acid loads in the two-cell mouse embryo. Instead, protons appear to move passively across the plasma membrane. Since the passive equilibrium distribution of charged species is determined by the membrane potential, baseline pH_i should be sensitive to extracellular K^+ concentration if protons are in passive equilibrium. We found that by raising extracellular K^+ to the intracellular level (~100 mM), the pH_i could rapidly be made to take on the value of the pH of the medium. Conversely, removing all extracellular K^+ caused the pH_i to fall somewhat (Baltz et al. 1990). Recovery from acid-loading under these conditions takes place by means of proton flux until equilibrium is reestablished through the apparently high proton permeability of the plasma membrane.

Alkaline loads can be relieved by another system—the Na^+-independent HCO_3^-/Cl^- exchanger (Fig. 3). This system uses the inwardly directed Cl^- gradient to remove bicarbonate from the cell. In contrast to the lack of systems for the relief of alkaline loads, two-cell-stage mouse embryos appear to possess this specific system for the relief of alkaline loads. When the medium surrounding two-cell mouse embryos is changed from normal medium to a Cl^--free medium, a rapid intracellular alkalinization occurs (Baltz et al. 1991) consistent with the reversed Cl^- gradient, causing intracellular Cl^- to be exchanged for extracellular HCO_3^-. This alkalinization is absent in bicarbonate-free medium or if DIDS is present. Additionally, the alkalinization does not require extracellular Na^+ (results not shown). To assess alkaline load reduction directly, cells were alkalinized with ammonium chloride. The subsequent recovery is inhibited either by DIDS, by the lack of bicarbonate or by the lack of extracellular Cl^- (Fig. 4C) (J.M. Baltz et al., unpubl.), demonstrating that alkaline-load recovery is specifically mediated by the exchanger.

The mouse embryo is very different from differentiated cells in its pH-regulating abilities. At the two-cell stage, it does not possess any of the specific systems for acid-load relief that are ubiquitous in adult cell types. It does, however, have a means to relieve alkaline loads rapidly. It is possible that this reflects the unique environment of the preimplantation embryo. The oviductal fluid has both high pH and high K^+ concentration (Leese 1988), both of which tend to alkalinize the cells of the embryo. Therefore, an acid-load relieving mechanism may not be necessary in this milieu. The developmental regulation of the appearance of these systems after the two-cell stage must be carefully coordinated with the needs of the embryo to control the various ion gradients involved. We are currently investigating when these systems appear and what determines when they become necessary to the embryo. We can already speculate that pH regulation will be an example of an apparent housekeeping function that will be shown to participate intimately in development.

REGIONALIZATION OF NA^+/K^+-ATPASE

The potential energy used for co- and countertransport in cells is ultimately derived from the activity of Na^+/K^+-ATPase. The enzyme plays such a key role in homeostasis that it occurs in all cells, being localized throughout the cell membrane. When cells are aggregated into epithelial sheets bound together by tight junctions, the Na^+/K^+-ATPase becomes asymmetrically distributed and will be found exclusively in the basolateral regions of the cell membrane, where it allows the cells to perform vectorial transport of salts and water. In recent years, rapid advances have been made in the molecular biology of Na^+/K^+-ATPase. The following summary is taken from four recent reviews by Fambrough (1988), Glynn (1988), Broude et al. (1989), and McDonough et al. (1990).

Na^+/K^+-ATPase is a member of the E1E2 type of transport ATPases that form phospho derivatives of aspartic acid as an

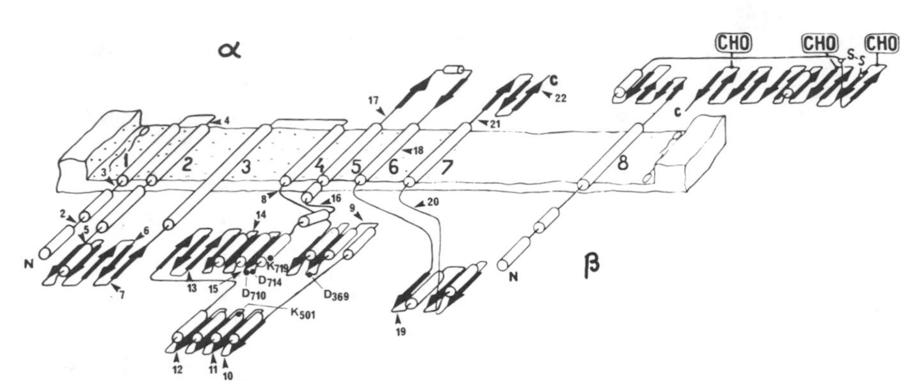

FIGURE 5 Model for membrane arrangement of the Na^+/K^+-ATPase catalytic subunit. Cylinders and arrows indicate α-helical and β-sheet regions, respectively. (Reprinted, with permission, from Broude et al. 1989)

intermediate. However, the molecule is unusual in that it is the only member of the E1E2 type that is made up of two subunits, the α and β subunits (Fig. 5). The α-subunit (110 kD) is the catalytic subunit and contains all the functional sites of the enzyme. The β-subunit (35 kD) is a glycosylated peptide and has no specifically identified function. Nevertheless, both subunits are required for biological activity. There is also evidence that the β-subunit regulates, through the assembly of αβ heterodimers, the number of sodium pumps in the plasma membrane (McDonough et al. 1990). The α-subunit spans the cell membrane, with binding sites for ATP and Na^+ and the phosphorylation site residing in the cytoplasmic region, and binding sites for K^+ and the enzyme inhibitor ouabain exposed on the external side of the cell membrane. The β-subunit spans the cell membrane as well, with most of the molecule located external to the lipid bilayer. The glycosylated region is also located in this external position. Three isoforms of the α-subunit are known (α1, α2, α3). These isoforms have distinct sodium affinities, sensitivity to ouabain, and tissue distribution (Sweadner 1989). Although two isoforms of the β-subunit have been proposed (β1 and β2), their separate existence is still uncertain (McDonough et al. 1990).

So far, investigators have not been able to demonstrate Na^+/K^+-ATPase in the precompacted embryo by either histochemical methods (Vorbrodt et al. 1977; M.K. Kim and J.D. Biggers [unpubl.], quoted by Borland 1977) or fluorescence immunochemical methods (Watson and Kidder 1988). Nevertheless, indirect evidence suggests that the enzyme is present. Studies on the intracellular ion concentrations show clear evidence that large K^+ gradients are maintained and partial evidence that Na^+ gradients are maintained (Powers and Tupper 1977; Lee 1987). Furthermore, using isotope flux measurements, Powers and Tupper (1977) suggested that the enzyme is present in the mouse unfertilized ovum and the two-cell embryo. The physiological roles of Na^+ in the very early stages of preimplantation development need to be investigated because at least some of the Na^+-dependent transport systems that relieve acid loads are absent or nonfunctional in the two-cell mouse embryo.

Extensive work exposing maximally expanding rabbit blastocysts to ouabain, a specific inhibitor of Na^+/K^+-ATPase, has provided strong evidence that this enzyme mediates the formation of blastocoel fluid (Smith 1970; Gamow and Daniel 1970; Biggers et al. 1978). Similar observations have also been made on the minimally expanding mouse blastocyst (DiZio and Tasca 1977; Wiley 1984; Manejwala et al. 1989). By comparing the action of ouabain applied externally to the rabbit blastocyst with its action when microinjected into the blastocoel, Biggers et al. (1977) were able to demonstrate that the Na^+/K^+-ATPase is located primarily on the basolateral (blastocoel-facing) domains of the trophoblast cells. By actively exporting Na^+ from the cell across the basolateral membrane, the enzyme allows the passive movement of water into the blastocoel cavity. The Na^+ is derived from exogenous sources and must initially enter the trophoblast cell through the permeability pathways in the apical membrane as the blastocyst expands.

Several transport systems that may contribute to Na^+ permeability have been demonstrated in the apical membrane of the rabbit trophectoderm (Fig. 6). These systems are (1) a sodium channel; (2) an amiloride-sensitive component; (3) an Na^+/Cl^- symport, a system that cotransports Na^+ and Cl^-; and (4) an Na^+-dependent amino acid transport system. Evidence indicates that the presence of these transport systems may vary

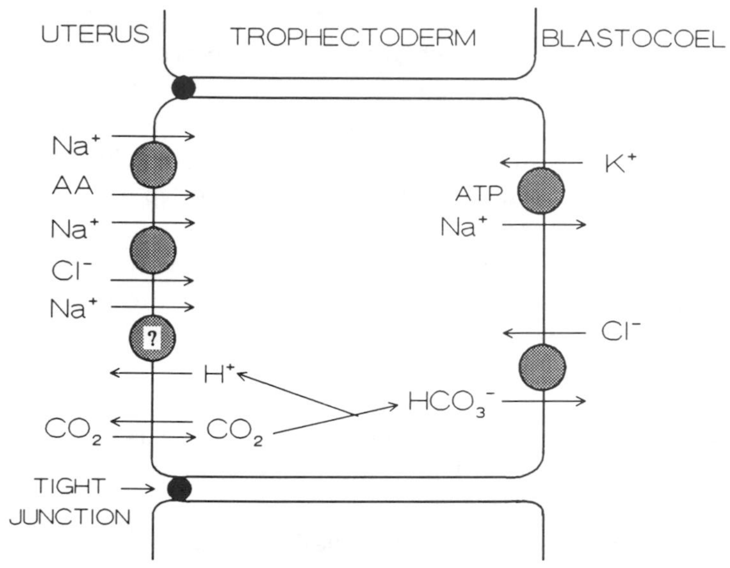

FIGURE 6 Transport systems in the rabbit blastocyst trophoblast cell. A large Na^+ flux is driven by a high concentration of Na^+/K^+-ATPase at the basolateral membrane (facing blastocoel). Na^+ influx at the apical (uterine) membrane occurs via Na^+ and Cl^- cotransport, Na^+ and amino acid (AA) cotransport and an amiloride-sensitive component(s), that may be Na^+ channels and/or an Na^+/H^+ antiport (question mark). Bicarbonate and pH equilibration can be effected by CO_2 diffusion (as the hydrated form H_2CO_3) through the membrane. CO_2, as H_2CO_3, also can dissociate into HCO_3^- and H^+, with HCO_3^- transported by the HCO_3^-/Cl^- exchanger.

with the age of the blastocyst. For example, in the rabbit, the furosemide-sensitive Na^+/Cl^- symport appears only after the fifth day, when the blastocoel cavity is beginning its rapid expansion (Benos and Biggers 1983). For a more detailed discussion of these mechanisms, see Biggers et al. (1988). No experimental evidence has demonstrated the presence of an Na^+-dependent transport system for glucose in the apical membrane of the trophoblast cell of the mouse (Dabich and Acey 1982; Gardner and Leese 1988) and the rabbit (Robinson et al. 1990). In view of these results, the report by Wiley et al. (1991) that an antibody to purified renal Na^+/glucose cotransporter localizes in isolated mouse blastomeres needs critical evaluation.

A topic of major developmental interest is how the blastocoel cavity begins to develop. The phenomenon has been the subject of recent reviews (Wiley 1987; Biggers ct al. 1988) and is only considered briefly here. The discovery of junctional complexes in the early blastocyst by Enders and Schlafke (1965) led to the recognition that the trophectoderm is the first tissue to develop in the mammalian embryo. Enders (1971) determined that it is an epithelium that creates a permeability seal around the embryo, which confers the ability to control the composition of its extracellular fluid, and Borland et al. (1977) demonstrated that the trophectoderm is able to create a unique internal fluid in the mouse. We now know that the final step of compaction is the formation of tight junctions (Ziomek and Johnson 1980). Recent work on the mouse demonstrated that the formation of these junctions involves a complex system including the appearance of the ZO-1 protein and uvomorulin (Fleming et al. 1989; Watson et al. 1990a). These mechanisms lead to a partitioning of transport systems between the apical and basolateral membranes of the trophoblast cell.

It is important to recognize that the formation of tight junctions does not immediately seal the blastocyst wall. Several lines of evidence show that the seal gradually tightens after compaction. The first indication that the paracellular permeability of the blastocyst may change with development was the finding by McLaren and Smith (1977) that early mouse blastocysts could not exclude antisera used in the technique of immunosurgery for the isolation of the inner cell mass, whereas older blastocysts could. Physiological confirmation that the trophectoderm changes from a relatively leaky epithelium to a relatively tight epithelium was provided by Benos (1981) using the rabbit. He showed that the electrical resistance increased from 35 Mohms cm^2 on day 4 to 1758 Mohms cm^2 on day 6. Furthermore, he found that the transtrophectodermal flux of radioactively labeled lanthanum was 0.46 nmole $cm^{-2}h^{-1}$ on day 4 and undetectable on day 5. This evidence suggests that the trophectoderm of the rabbit has developed its maximum degree of tightness by the fifth day postcoitum. This conclusion is consistent with the work of Borland and Biggers (unpubl.) showing that the hydraulic conductivity of the rabbit trophectoderm does not change between the fifth and sixth days of development, being 70 mm $sec^{-1}atm^{-1}$ on day 6 and 62 mm $sec^{-1}atm^{-1}$ on day 7. The gradual closure of the tight junctions between the cells over the entire surface of the morula, together with the large increase in Na^+/K^+-ATPase in the basolateral membrane, must ultimately create local conditions for the establishment of osmotic gradients necessary for the accumulation of water. For reasons discussed elsewhere (Biggers et al. 1988), the evidence for the local priming of this system by metabolically produced water as proposed by Wiley (1987) is speculative and seems unnecessary.

Several studies have recently been made on the levels of mRNA for the α and β subunits of Na^+/K^+-ATPase in the preimplantation embryo of the rabbit (Gardiner et al. 1990a) and the mouse (Gardiner et al. 1990b; Watson et al. 1990b). The subject can be discussed in two parts. The first concerns the mRNAs of the two subunits present in the already formed, expanding blastocyst, and the second concerns the role that the mRNA plays in initiating cavitation. Gardiner et al. (1990a) estimated the mRNA levels of the Na^+/K^+-ATPase α-subunit in rabbit blastocysts 4–7 days after fertilization, using a cDNA for the α-subunit from sheep kidney. The content of the α-subunit mRNA increased 35-fold over this period, corresponding to the same order of magnitude of increase in the surface area of the rabbit blastocyst. Using the same analytical techniques on mouse preimplantation embryos developing in vitro from the two-cell stage to the expanded blastocyst, it was demonstrated that the amount of the α-subunit mRNA increased approximately 45 times (Gardiner et al. 1990b). Most of the increase occurred after compaction; furthermore, the increase is the same order of magnitude as the increase in cell number expected over this period.

The increase in α-subunit mRNA per embryo in both species probably reflects the increase in number of cells in the trophectoderm and not an increase in the levels of mRNA per cell. This implies that the turnover of the α-subunit mRNA is approximately constant. Watson et al. (1990b) reported that in the mouse α and β subunits, mRNA does not increase until after compaction. They also found that the appearance of the β-subunit mRNA lags behind that of the α-subunit mRNA.

This result raises the question as to whether the production of the β-subunit mRNA limits the amount of functional αβ heterodimer (Na^+/K^+-ATPase) produced. The increase in the amount of mRNA per embryo, however, may not correlate with the increase in the amount of subunits per embryo that are produced. Gardiner et al. (1990a) found that in the rabbit blastocyst, the amount of α-subunit increased approximately 22 times between 4 and 6 days of development, and Overström et al. (1989) found that the rate of incorporation of [^{35}S]methionine increased approximately 90 times. In contrast, Gardiner et al. (1990b) found that the amount of α-subunit in the preimplantation mouse embryo remained constant from about the time of compaction until the expanded blastocyst stage, whereas the amount of the β-subunit increased nine times. It must be remembered that the net amount of subunit will be the resultant between subunit synthesis and subunit degradation. The loss of subunit may be considerable, since the half-life of Na^+/K^+-ATPase has been estimated to be relatively short, about 18 hours (Karin and Cook 1986).

The large increases in the amounts of α- and β-subunit mRNAs per embryo after compaction may suggest that the production of functional Na^+/K^+-ATPase is critical in the initiation of cavitation. A complete understanding of the molecular processes that result in cavitation, however, requires the investigation of the kinetics of the production of the α and β subunits on a per cell basis. It is possible that the amount of enzyme produced per cell is constant and that as the tight junctions develop, the α and β subunits are redirected from their site of synthesis in the endoplasmic reticulum (McDonough et al. 1990) to the newly produced basolateral membrane. Furthermore, a full understanding of the mechanisms that initiate cavitation requires that the kinetics of the molecular mechanisms controlling tight junction formation be correlated with the kinetics of Na^+/K^+-ATPase formation.

CONCLUSIONS

Techniques are now available that allow the quantitative study of ions and their transport throughout preimplantation development. Several techniques are so sensitive that they can be applied to single blastomeres. The results reviewed here show that the function of ions changes throughout preimplantation development. A critical juncture occurs upon compaction, which confers the ability to regulate the composition of an extracellular fluid and produces an integrated multicellular stage from a loose assembly of single cells. An important area for future research is the study of the regulation of the synthesis and location of the transporters involved in early development and the factors that determine when they become functional.

ACKNOWLEDGMENTS

This work was part of the National Cooperative Program on Non-Human In Vitro Fertilization and Preimplantation Development, grant HD-21988, and was also sponsored by NICHD (NIH) grants HD-21581 and 5P44 1RR02064.

REFERENCES

Baltz, J.M., J.D. Biggers, and C. Lechene. 1990. Apparent absence of Na^+M^+ antiport activity in the two-cell mouse embryo. *Dev. Biol.* **138:** 421.

———. 1991. Two-cell-stage mouse embryos appear to lack mechanisms for alleviating intracellular acid loads. *J. Biol. Chem.* (in press).

Benos, D.J. 1981. Developmental changes in epithelial transport characteristics of preimplantation rabbit blastocysts. *J. Physiol.* **316:** 191.

Benos, D.J. and J.D. Biggers. 1983. Sodium and chloride cotransport by preimplantation rabbit blastocysts. *J. Physiol.* **342:** 23.

Bernard, Claude. 1878. *Lecons sur les phenomenes de la vie communs aux animaux et aux vegetaux.* Balliere, Paris.

Bidani, A., S.E.S. Brown, T.A. Heming, R. Gurich, and T.D. Dubose. 1989. Cytoplasmic pH in pulmonary macrophages: Recovery from acid load is Na^+ independent and NEM sensitive. *Am. J. Physiol.* **257:** C65.

Biggers, J.D. 1972. Mammalian blastocyst and amnion formation. In *The water metabolism of the fetus* (ed. A.C. Barnes and A.E. Seeds), p. 3. Charles C. Thomas, Springfield, Illinois.

———. 1980. Fetal and neonatal physiology. In *Medical physiology* (ed. V.B. Mountcastle), vol. 2, p. 1947. Mosby, St. Louis.

Biggers, J.D., J.E. Bell, and D.J. Benos. 1988. Mammalian blastocyst: Transport functions in a developing epithelium. *Am. J. Physiol.* **255:** C419.

Biggers, J.D., R.M. Borland, and C.P. Lechene. 1978. Ouabain-sensitive fluid accumulation and ion transport by rabbit blastocysts. *J. Physiol.* **280:** 319.

Biggers, J.D., R.M. Borland, and R.D. Powers. 1977. Transport mechanisms in the preimplantarton mammalian embryo. *Ciba Found. Symp.* **52:** 129.

Borland, R.M. 1977. Transport processes in the mammalian blastocyst. In *Development in mammals* (ed. M.H. Johnson), vol. 1, p. 31. Elsevier-North Holland Biomedical Press, The Netherlands.

Borland, R.M., J.D. Biggers, and C.P. Lechene. 1977. Studies on the composition of mouse blastocoele fluid using electron probe microanalysis. *Dev. Biol.* **55:** 1.

Boron, W.F. 1987. Intracellular pH regulation. In *Membrane transport processes in organized systems* (ed. T.E. Andreoli et al.), p. 39. Plenum Press, New York.

Boron, W.F. and P. DeWeer. 1976. Intracellular pH transients in squid giant axons caused by CO_2, NH_3 and metabolic inhibitors. *J. Gen. Physiol.* **67:** 91.

Broude, N.E., N.N. Modyanov, G.S. Monastyrskaya, and E.D. Sverdlov. 1989. Advances in Na^+/K^+-ATPase studies: From protein to gene and back to protein. *FEBS Lett.* **257:** 1.

Cannon, W.B. 1932. *Wisdom of the body*. W.W. Norton, New York.

Cohen, B.J. and C. Lechene. 1989. (Na,K)-pump: Cellular role and regulation in nonexcitable cells. *Biol. Cell* **66:** 191.

Dabich, D. and R.A. Acey. 1982. Transport of glucosamine (aldhexoses) by preimplantation mouse blastocysts. *Biochim. Biophys. Acta* **684:** 146.

Deamer, D.W. 1984. Proton flux across model and biological membranes. In *Membrane processes—Molecular biology and medical applications* (ed. G. Benga et al.), p. 111. Springer-Verlag, New York.

DiZio, S.M. and R.J. Tasca. 1977. Sodium-dependent amino acid transport in preimplantation mouse embryos. III. Na^+/K^+-ATPase-linked mechanisms in blastocysts. *Dev. Biol.* **59:** 198.

Enders, A.C. 1971. The fine structure of the blastocyst. In *The biology of the blastocyst* (ed. R.J. Blandau), p. 71. University of Chicago Press, Illinois.

Enders, A.C. and S.J. Schlafke. 1965. The fine structure of the blastocyst: Some comparative studies. In *Preimplantation stages of pregnancy* (ed. G.E.W. Wolstenholme and M. O'Connor), p. 29. Churchill, London.

Eusebi, F., R. Colonna, and F. Mangia. 1983. Development of membrane excitability in mammalian oocytes and early embryos. *Gamete Res.* **7:** 39.

Fambrough, D.M. 1988. The sodium pump becomes a family. *Trends Neurosci.* **11:** 325.

Fleming, T.P., J. McConnell, M.H. Johnson, and B.R. Stevenson. 1989. Development of tight junctions de novo in the mouse early embryo: Control of assembly of the tight junction-specific protein, ZO-1. *J. Cell Biol.* **108:** 1407.

Gamow, G. and J.C. Daniel. 1970. Fluid transport in the rabbit blastocyst. *Wilhelm Roux' Arch. Entwicklungssmech. Org.* **164:** 261.

Gardiner, C.S., M.A. Grobner, and A.R. Menino. 1990a. Sodium/potassium adenosine triphosphatase α-subunit and α-subunit mRNA levels in early rabbit embryos. *Biol. Reprod.* **42:** 539.

Gardiner, C.S., J.S. Williams, and A.R. Menino, Jr. 1990b. Sodium/potassium adenosine triphosphate α- and β-subunit and α-subunit mRNA levels during mouse embryo development in vitro. *Biol. Reprod.* **43:** 788.

Gardner, D.K. and H.J. Leese. 1988. The role of glucose and pyruvate transport in regulating nutrient utilization by preimplantation mouse embryos. *Development* **104:** 423.

Gillespie, J.I. and J.R. Greenwell. 1988. Changes in intracellular pH and pH regulating mechanisms in somitic cells of the early chick embryos: A study using fluorescent pH-sensitive dye. *J. Physiol.* **405:** 385.

Glynn, I.M. 1988. How does the sodium pump pump? In *Cell physiology of blood* (ed. R.B. Gunn and C. Parker), p. 2. Rockefeller University Press, New York.

Goldman, D.E. 1943. Potential, impedance, and rectification in membranes. *J. Gen. Physiol.* **27:** 37.

Gutknecht, J. 1987. Proton conductance through phospholipid bilayers: Water wires or weak acids? *J. Bioeng. Biomembr.* **19:** 427.

Igusa, Y. and S. Miyazaki. 1983. Effects of altered extracellular and intracellular calcium concentration on hyperpolarizing responses of the hamster egg. *J. Physiol.* **340:** 611.

Igusa, Y., S. Miyazaki, and N. Yamashita. 1983. Periodic hyperpolarizing responses in hamster and mouse eggs fertilized with mouse sperm. *J. Physiol.* **340:** 633.

Jain, M.K and R.C. Wagner. 1980. *Introduction to biological membrane*, p. 134. John Wiley, New York.

Karin, N.J. and J.S. Cook. 1986. Turnover of the catalytic subunit of Na^+/K^+-ATPase in HTC cells. *J. Biol. Chem.* **261:** 10422.

Lechene, C. 1988. Physiological role of the NA-K pump. *Prog. Clin. Biol. Res.* **2688:** 171.

Lee, S. 1987. Membrane properties in preimplantation mouse embryos. *J. In Vitro Fertil. Embryo Transfer* **4:** 331.

Leese, H.J. 1988. The formation and function of oviduct fluid. *J. Reprod. Fertil.* **82:** 843.

Manejwala, F.M., E.J. Cragoe, and R.M. Schultz. 1989. Blastocoel expansion in the preimplantation mouse embryo: Role of extracellular sodium and chloride and possible apical routes of their entry. *Dev. Biol.* **133:** 210.

McDonough, A.A., K. Geering, and R.A. Farley. 1990. The sodium pump needs its β-subunit. *Fed. Am. Soc. Exp. Biol. J.* **4:** 1598.

McLaren, A. and R. Smith. 1977. Functional test of tight junctions in the mouse blastocyst. *Nature* **267:** 351.

Mitani, S. 1985. The reduction of calcium current associated with early differentiation of the murine embryo. *J. Physiol.* **363:** 71.

Mitchell, P. 1967. Proton-translocation phosphorylation in mitochondria, chloroplasts and bacteria: Natural fuel cells and solar cells. *Fed. Proc.* **26:** 1370.

Miyazaki, S. and Y. Igusa. 1982. Ca-mediated activation of a K current at fertilization of golden hamster eggs. *Proc. Natl. Acad. Sci.* **79:** 931.

Moolenaar, W.H., A.J. Bierman, and S.W. de Laat. 1988. Effects of growth factors on Na^+/H^+ exchange. In *Na^+/H^+ exchange* (ed. S. Grinstein), p. 227. CRC Press, Boca Raton, Florida.

Nijhout, H.F. 1990. Metaphors and the role of genes in development. *BioEssays* **12:** 441.

Okamoto, H., K. Takahashi, and N. Yamashita. 1977. Ionic currents through the membrane of the mammalian oocyte and their comparison with those in the tunicate and sea urchin. *J. Physiol.* **267:** 465.

Overström, E.W., D.J. Benos, and J.D. Biggers. 1989. Synthesis of Na^+/K^+-ATPase by the preimplantation rabbit blastocyst. *J. Reprod. Fertil.* **85:** 283.

Pouyssegur, J., A. Franchi, G. Lallemain, and S. Paris. 1985. Cytoplasmic pH, a key determinant of growth factor-induced DNA synthesis in quiescent fibroblasts. *FEBS Lett.* **190:** 115.

Powers, R.D. and J.T. Tupper. 1977. Developmental changes in membrane transport and permeability in the early mouse embryo. *Dev. Biol.* **56:** 306.

Raff, R.A. and T.C. Kaufman. 1983. *Embryos, genes, and evolution*, p. 201. Macmillan, New York.

Robinson, D.H., P.R. Smith, and D.J. Benos. 1990. Hexose transport in preimplantation rabbit blastocysts. *J. Reprod. Fertil.* **89:** 1.

Roos, A. and W. F. Boron. 1981. Intracellular pH. *Physiol. Rev.* **61:** 296.

Schoenheimer, R. 1940. *The dynamic state of body constituents*. Harvard University Press, Cambridge, Massachusetts.

Somero, G.N. 1985. Intracellular pH, buffering substances and proteins: Imidazole protonation and the conservation of protein structure and function. In *Transport processes, iono- and osmoregulation* (ed. R. Gilles and M. Gilles-Baillien), p. 454. Springer-Verlag, Berlin.

Smith, M.W. 1970. Active transport in the rabbit blastocyst. *Experientia* **26:** 736.

Sweadner, K.J. 1989. Isozymes of the Na^+/K^+-ATPase. *Biochim. Biophys. Acta* **988:** 185.

Van Winkle, L.J. 1988. Amino acid transport in developing animal oocytes and early conceptuses. *Biochim. Biophys. Acta* **947:** 173.

Van Winkle, L.J., N. Haghighat, A.L. Campione, and J.M. Gorman. 1988. Glycine transport in mouse eggs and preimplantation conceptuses. *Biochim. Biophys. Acta* **941:** 241.

Vorbrodt, A., M. Konwinski, D. Solter, and H. Koprowski. 1977. Ultrastructural cytochemistry of membrane bound phosphatases in preimplantation mouse embryos. *Dev. Biol.* **55:** 117.

Watson, A.J. and G.M. Kidder. 1988. Immunofluorescence assessment of the timing of appearance and cellular distribution of Na^+/K^+-ATPase during mouse embryogenesis. *Dev. Biol.* **126:** 80.

Watson, A.J., C.H. Damsky, and G.M. Kidder. 1990a. Differentiation of an epithelium: Factors affecting the polarized distribution of Na^+/K^+-ATPase in mouse trophectoderm. *Dev. Biol.* **141:** 104.

Watson, A.J., C. Pape, J.R. Emanuel, R. Levenson, and G.M. Kidder. 1990b. Expression of Na^+/K^+-ATPase α and β subunit genes during preimplantation development of the mouse. *Dev. Genet.* **11:** 41.

Wiley, L.M. 1984. Cavitation in the mouse preimplantation embryo: Na^+/K^+-ATPase and the origin of nascent blastocoele fluid. *Dev. Biol.* **105:** 330.

———. 1987. Development of the blastocyst: Role of cell polarity in cavitation and cell differentiation. In *The mammalian preimplantation embryo: Requirements of growth and differentiation in vitro* (ed. B.D. Bavister), p. 65. Plenum Press, New York.

Wiley, L.M., G.M. Kidder, and A.J. Watson. 1990. Cell polarity and development of the first epithelium. *BioEssays* **12:** 67.

Wiley, L.M., J.E. Lever, C. Pape, and G.M. Kidder. 1991. Antibodies to a renal Na+/glucose cotransport system localize to the apical plasma membrane domain of polar mouse embryo blastomeres. *Dev. Biol.* **143:** 149.

Yoshida, S. 1985. Action potentials dependent on monovalent cations in developing mouse embryos. *Dev. Biol.* **110:** 200.

Ziomek, C.A. and M.H. Johnson. 1980. Cell surface interaction induces polarization of mouse 8-cell blastomeres at compaction. *Cell* **21:** 935.

Ziomek, C.A., C.L. Chatot, and C. Manes. 1990. Polarization of blastomeres in the cleaving rabbit embryo. *J. Exp. Zool.* **256:** 84.

Origin of Embryonic and Extraembryonic Cell Lineages in Mammalian Embryos

Y.P. Cruz[1] *and R.A. Pedersen*[2]
[1]*Department of Biology*
Oberlin College, Oberlin, Ohio 44074
[2]*Laboratory of Radiobiology and Environmental Health*
and Department of Anatomy
University of California, San Francisco, California 94143

INTRODUCTION

The principal morphogenetic process that occurs during preimplantation development of eutherian (frequently called placental) and metatherian (marsupial) mammals is the differentiation of an outer epithelium. This cell layer is the trophectoderm in eutherians or the protoderm in metatherians. The subsequent proliferation of trophectoderm cells in eutherian embryos generates the trophoblast lineage, which differentiates into outer, trophoblast cells separating the conceptus from the maternal decidual tissues and also forms the diploid chorionic component of the chorioallantoic placenta. Concomitantly with trophectoderm differentiation in most eutherian mammals, a group of pluripotent cells is internalized during late cleavage and is sequestered at one pole of the nascent blastocyst, thus constituting the inner cell mass (ICM). The ICM cells facing the blastocyst cavity subsequently differentiate into the primitive endoderm layer (also known as the hypoblast), which ultimately forms the extraembryonic endoderm of the visceral and parietal yolk sacs and the allantois. The remaining ICM cells form the primitive ectoderm (also designated as epiblast), which is the source of the three primary germ layers (embryonic ectoderm, endoderm, and mesoderm), as well as the extraembryonic mesoderm lineage,

 0-87969-333-9/91 $3.00 + 00

that arise during gastrulation. Early development of the metatherian embryo differs primarily in that progenitors of both the embryonic and extraembryonic lineages are combined in the protoderm. The epiblast forms from outer cells at one pole of the protoderm, where the primary endoderm cells are internalized and differentiate as hypoblast. The remaining protoderm cells develop into chorionic ectoderm separating the embryo from the maternal tissues, as in eutherians, but generally without invading the uterus. Thus, the metatherian early embryo does not sequester its pluripotent cells in the form of an ICM, in contrast to eutherians.

Although the formation and relationship of these early embryonic and extraembryonic tissue lineages have been studied extensively in mouse embryos, there is little information about their formation and fate in domestic species. Therefore, we have compared early embryogenesis in representative domestic species to examine the basis for any generalizations about basic mechanisms of lineage formation and fate among eutherian mammals. There are substantial similarities in cleavage, compaction, and blastocyst formation between most laboratory mammals and domestic eutherian species; moreover, analysis of cell lineages during gastrulation reveals substantial similarity between the mouse gastrula fate map and that of other nonmammalian vertebrates that have been studied. These observations led to the view that mechanisms of establishment of the extraembryonic lineages and the allocation of epiblast cells to the primary germ layers have been highly conserved during mammalian evolution, despite substantial divergence in the extracellular coats, in the topographical form of the germinal disc, and in the morphology of the extraembryonic membranes.

OVERVIEW OF EARLY MOUSE DEVELOPMENT

Although classical studies of early development of eutherian mammals focused on the rabbit embryo as a model system (Rauber 1875; Koelliker 1880; Assheton 1895), most mammalian experimental embryology during the past three decades has been carried out on the mouse (for review, see Biggers

1987). The combination of its small size, short life cycle, genetic resources, and ease of obtaining and culturing early embryos have made the house mouse, *Mus musculus*, the principal laboratory model for the study of mammalian development. Despite the rapid progress that these practical advantages have fostered in our understanding of early eutherian development, we should also acknowledge the limitations of the rodent as an exemplar for domestic species and, more generally, the differences between eutherian and metatherian embryos at their early stages. In this section, we review briefly the main morphogenetic events of mouse pre- and early postimplantation development and point out the peculiarities of rodent embryo topography that are relevant to our subsequent discussion of representative domestic species.

Cleavage and Potency

Cleavage of mouse embryos begins with prolonged first and second cell cycles (~24 hr and 20 hr, respectively), which then shorten to 10–12-hour intervals in the third and subsequent cycles (for review, see Pedersen 1986). The G_1 phase of the cell cycle is short in the second and third cell cycles, and a typical cell cycle becomes established only in the fifth cycle, with a variable 2-hour G_1 phase, an 8–9-hour S phase, and 2-hour G_2/M phase (Chisholm 1988). During cleavage, mitosis appears to be regulated by cycling factors that govern nuclear envelope breakdown and mitotic chromosome condensation, including maturation-promoting, or M-phase-inducing, substances (Balakier 1978; Sorensen et al. 1985; Pratt and Muggleton-Harris 1988; McConnell and Lee 1989; for review, see Pratt 1989). Thus, the mouse embryo shares its general features of early cleavage with other systems, such as *Drosophila* and *Xenopus*, where truncated cell cycles give way to typical cycles as cleavage proceeds (for review, see Pratt 1989). The mouse embryo is different from vertebrate and invertebrate species that initiate development with synchronous cleavages, however, because it has asynchronous cleavages from the earliest possible stage (second cleavage). Moreover, the transition from maternal to zygotic gene expression that occurs at the two-cell stage in the mouse (for review, see

Schulz 1986) precedes the restoration of normal cell cycles, whereas in *Xenopus*, the two change simultaneously at the midblastula transition (Newport and Kirschner 1982). This transition occurs at later cleavage stages in all other eutherian embryos that have been studied (for details, see Prather and Robl, this volume).

Mammalian embryos are characterized by an extended period of totipotency during early cleavage stages. This has been demonstrated in mouse embryos at the 2-, 4-, 8-, and 16-cell (morula) stages by either isolating or disaggregating blastomeres to obtain individual cells. Their potency was assessed by their capacity to develop as separate blastomeres or to contribute to the various embryonic and extraembryonic lineages in aggregation chimeras (for review, see Pedersen 1986). These studies showed that single isolated mouse embryo blastomeres can develop into intact, but small, blastocysts in some cases, but not in others. The resulting embryos were either morphologically normal blastocysts, with trophectoderm and ICM (isolated two-cell blastomeres), or structures lacking an ICM and consisting only of trophectoderm (most four-cell stages and all later stages). These observations led to the hypothesis that differentiation of mouse blastomeres depended on their position in the morula, with outer blastomeres forming the trophectoderm and inner blastomeres forming the ICM (Tarkowski and Wroblewska 1967). When sufficient cytoplasmic mass was present to enclose some blastomeres at the time of compaction and development into a morula, an ICM was formed; otherwise, all blastomeres remained outside and differentiated into trophectoderm. However, by aggregating the isolated blastomere with other genetically distinct blastomeres at the same stage, it was possible to examine blastomere potency at the later cleavage stages by placing the donor cell in either an inside position or an outside position in the aggregate. These studies showed that mouse blastomeres remain totipotent throughout cleavage, retaining the ability to differentiate into either the trophectoderm or the ICM lineage (for review, see Pedersen 1986). Even inner cells of morulae and early blastocysts are totipotent, remaining capable of differentiation into either cell lineage. The totipotency of these inner cells is finally lost at the expanded blastocyst stage, corresponding approximately to the division from the 32- to 64-cell stage, when isolated ICMs differentiate into structures composed of primitive endoderm and ectoderm, rather than trophectoderm and ICM (for review, see Rossant 1986). The capacity of individual blastomeres for development in isolation is somewhat greater for embryos of domestic species than it is for the mouse, as discussed subsequently, and the embryonic nuclei of the domestic species apparently also have greater capacity to support preimplantation development after nuclear transfer than that of the mouse, as discussed by Prather and Robl (this volume).

Compaction and Blastocyst Formation

Compaction of the mouse embryo refers to the gradual flattening of the spherical blastomeres that typify early cleavage stages, so that the irregularly shaped early-cleavage embryo appears to coalesce into a smoothly contoured morula (Lewis and Wright 1935; Ducibella and Anderson 1975). Compaction begins as early as the four-cell stage and continues through the eight-cell stage (Sutherland et al. 1990). At the same time, a localized redistribution of cell surface microvilli and cytocortical components to the outer blastomere surfaces takes place (Handyside 1980; Sutherland and Calarco-Gillam 1983). This morphogenetic phenomenon, referred to as polarization, occurs as a result of contact between apposed surfaces of competent blastomeres (Johnson and Ziomek 1981; Ziomek and Johnson 1981). Polarization involves cell surface adhesion (Shirayoshi et al. 1983; Vestweber et al. 1987), leading to a reorganization of cytoskeletal components (Sobel 1983; Johnson and Maro 1984, 1985; Houliston et al. 1987; Sobel et al. 1988; Maro et al. 1990), orientation of endocytosis to the outer, apical surface (Reeve 1981; Fleming and Pickering 1985; Fleming et al. 1986), and establishment of the intercellular junctions that characterize the differentiating trophectoderm (Fleming et al. 1989) (for review, see Johnson and Maro 1986; Fleming and Johnson 1988; Pratt 1989; Sobel 1990). Polarization not only marks the beginning of epithelium formation by the incipient trophectoderm cells that differentiate on the outer surface of the embryo, but also constitutes a critical event in establishing the fate of cells inheriting the outer surface (for review, see Pedersen 1986, 1988).

Blastocyst formation in mouse embryos is thus the culmination of the process of epithelium formation that begins with compaction. Nascent blastocysts acquire a blastocyst cavity on completion of the fifth cleavage division, at approximately the 32-cell stage. Cell lineage analysis of blastomere fate shows that these outer cells have descendants only in the trophectoderm lineage, and therefore remain completely distinct from the inner cell lineage (Copp 1979; Cruz and Pedersen 1985). Thus, it is interesting to review the process in which blastomeres become allocated to either of the two lineages. All blastomeres are external at early cleavage stages, including the eight-cell stage, when extensive compaction has occurred. The first cleavage-stage blastomeres to become internalized arise in the division from the 8- to the 16-cell stage (Pedersen et al. 1986), and thereafter have only inner descendants in the early blastocyst (Fleming 1987). Clearly, blastomeres that are initially external (early cleavage stages) are capable of forming both internal and external descendants in intact embryos, as might be expected from their totipotency in the blastomere isolation and aggregation studies. This is confirmed by the fate of two- and eight-cell blastomeres in intact embryos, where they generally produce descendants in both ICM and trophectoderm (Balakier and Pedersen 1982). Outer blastomeres of 16-cell embryos also occasionally generate inner descendants in the fifth cleavage division (to the 32-cell stage), which again is consistent with their totipotency in aggregation chimera studies (Pedersen et al. 1986). Interestingly, the number of cells recruited into the inner population in the fifth cleavage division is complementary to the number recruited in the fourth cleavage division, thus making the number of inner cells a relatively constant fraction of total cells at the early blastocyst stage in different embryos (Fleming 1987). The tendency of some, but not all, outer blastomeres to produce inner descendants during cleavage may arise from their plane of cleavage, which may be either periclinal (forming outer-inner pairs), anticlinal (forming outer-outer pairs), or oblique (forming mainly outer-outer pairs); the frequency of periclinal divisions, plus most of the oblique divisions, equals the frequency of inner cell generation in the division to the 16-cell stage (Sutherland et al. 1989). Although the mechanisms for cell allocation in mouse embryos remain unclear, the cell adhesion molecule uvomorulin (also known as gp120, cell CAM120/80, L-CAM, and E-cadherin) and other associated cytoskeletal components could play an important role in this process through their effects on cell polarization, as in other epithelial cell systems (McNeill et al. 1990). Because the outer blastomeres of mouse embryos invariably have at least one outer descendant, which arises from the polarized portion of the progenitor cell (Ziomek and Johnson 1980; Johnson and Ziomek 1981), polarization appears to be a significant determinant in the allocation of cells to the trophectoderm and ICM fates in mouse embryos.

Fate of Trophectoderm and ICM

During late preimplantation and early postimplantation development of the mouse embryo, the trophoblast surrounding the blastocyst (mural trophectoderm) expands as fluid accumulates in the blastocyst cavity. Cell lineage studies have revealed that trophectoderm cells overlying the ICM (polar trophectoderm) contribute descendants to the mural trophectoderm during blastocyst development (Copp 1979; Cruz and Pedersen 1985). Thus, despite the cessation of cell division in trophectoderm cells as they transform into trophoblast giant cells (Dickson 1966), there is continued growth of the mural trophectoderm population by accretion from the polar trophectoderm. After implantation occurs, polar trophectoderm cells no longer remain organized as a unilaminar epithelium, but accumulate as a compact mass at the embryonic pole. The outermost cells become invasive, forming the ectoplacental cone, and the innermost cells become organized into the extraembryonic ectoderm, initially occupying the extraembryonic portion of the egg cylinder and subsequently forming the ectodermal component of the chorion. On the basis of the tendency of cultured ectoplacental cone cells to differentiate rapidly into trophoblast giant cells, in contrast with the slower giant-cell transformation of cultured extraembryonic ectoderm, Rossant and Tamura-Lis (1981) proposed a model for the lineage relationships of trophectoderm. This model envisions polar trophectoderm as the progenitor population for all diploid cells of the trophoblast lineage, forming the extraembryonic ectoderm and ectoplacental cone; the extraembryonic ectoderm is considered to be the diploid stem cell population for the trophoblast lineage after implantation and gives rise to the ectoplacental cone, which in turn gives rise to trophoblast giant cells. Although the morphology of the trophoblast lineage supports this view (giant cells are adjacent to the ectoplacental cone, which is contiguous with extraembryonic ectoderm), confirmation of actual lineage relationships by cell-marking experiments remains to be done.

About the time of implantation, the ICM of the mouse embryo differentiates into a blastocoelic surface layer of primitive endoderm, or hypoblast, and a group of enclosed cells, the primitive ectoderm, or epiblast. Studies of inner cell fate by cell transplantation into blastocysts or trophoblast vesicles show that the entire fetus and its embryonic membranes originate from the ICM; the hypoblast forms the endoderm of the visceral yolk sac and allantois and the parietal endoderm, whereas the epiblast forms all of the primary germ layers of the fetus, including endoderm, mesoderm, and ectoderm, plus the extraembryonic mesoderm of the amnion, yolk sac, and chorion (for review, see Beddington 1986; Rossant 1986). Thus, the long-term fate of the ICM is clear. On the other hand, there is relatively little information about the development of the ICM during the peri-implantation period. With few exceptions, ICMs transferred to trophectodermal vesicles (reconstituted blastocysts) contribute to the embryonic and extraembryonic tissues described above but not to the trophectoderm lineage (Gardner and Papaioannou 1975; Papaioannou 1982; for review, see Rossant 1986). Individual cells derived from expanded blastocysts at 3.5 days of gestation (d.g.) or late blastocysts (4.5 d.g.) contribute either to hypoblast or to epiblast, but not to both cell types (Gardner and Papaioannou 1975; Gardner and Rossant 1979), implying that such inner cells are committed to either a hypoblast or epiblast cell fate. This finding contrasts with the fate of the inner cells of the early blastocyst stage (also 3.5 d.g.), when the cells are still totipotent, as discussed previously. Moreover, when inner cells of early blastocysts were used in blastocyst reconstitution experiments, they were found in several cases to contribute not only to the fetus and yolk sac, but also to the trophectoderm lineage (Rossant and Croy 1985). An analysis of inner cell fate using microinjected rhodamine and horseradish peroxidase (HRP) also showed labeled descendants of inner cells in the trophectoderm following culture for 24 or 48 hours (Winkel and Pedersen 1988). Moreover, all regions of the ICM were capable of contributing to primitive endoderm-like cells, as well as core cells of the ICM, implying that the progenitor inner cells were not determined at the time of injection.

One possible explanation for the difference between this result and earlier chimera studies is that the pluripotent progenitors in the HRP-labeling study may have all been at the early blastocyst stage. Another approach to this issue has been to use fluorescent beads to label polar trophectoderm and thereby detect cells originating from the ICM (Dyce et al. 1987). Their study found only limited contribution from ICM to trophectoderm, either in frequency or in extent. It is possible that either the HRP-labeling or the fluorescent bead-labeling generates artifactual results that confound the analysis. Although these experiments do not resolve the short-term fate of the mouse ICM, it is nevertheless clear that inner cells do become determined at about the time of implantation as either hypoblast or epiblast progenitors, and thereafter have quite divergent fates.

Fate of Hypoblast

The differentiation of the hypoblast has been studied extensively in aggregation chimeras and by other approaches to determine cell fate. All of these approaches concur in the conclusion that the hypoblast of the mouse embryo, like that of the chick (Vakaet 1984), contributes solely to the extraembryonic endoderm and does not provide descendants to the embryonic gut endoderm. In initial studies, primitive endoderm cells from late blastocysts (4.5 d.g.) were injected into 3.5-d.g. expanded blastocysts that were transferred to foster mothers for development to midgestation stages. Analysis of fetal and yolk sac tissues by glucose phosphate isomerase

(GPI) electrophoresis showed that primitive endoderm contribution was solely to the visceral extraembryonic endoderm of the yolk sac (Rossant et al. 1978; Gardner and Rossant 1979). Further detailed analysis using GPI as a marker revealed that primitive endoderm cells from 4.5-d.g. blastocysts also contributed to the parietal endoderm cell population that lines the mural trophectoderm after implantation (Gardner 1982). Indeed, the contribution of hypoblast donor cells from either embryonic or extraembryonic regions of 4.5-, 5.5-, and 6.5-d.g. embryos was strongly biased toward parietal endoderm descendants; when parietal endoderm donor cells were studied, however, they contributed only to parietal, and not visceral, endoderm descendants (Gardner 1982). These observations support the view that both visceral extraembryonic endoderm and parietal endoderm are derived from a common progenitor pool of primitive endoderm cells (Gardner and Papaioannou 1975). Moreover, the observation that visceral endoderm could produce parietal endoderm descendants is consistent with the findings by Hogan and Tilly (1981) from experiments with cultured 6.5-d.g. embryos that visceral endoderm cells (or a subpopulation of cells intercalated among them) are capable of differentiating into parietal endoderm.

A model for cell differentiation in the primitive endoderm lineage that includes these observations (Hogan and Tilly 1981; Gardner 1982) proposes that visceral endoderm contains the stem cells for both the visceral and parietal endoderm populations and that their differentiation depends on environmental stimuli; specifically, proximity to trophoblast giant cells (or distance from other ICM cell lineages) leads to differentiation into parietal endoderm. Because visceral endoderm cells can form parietal endoderm, but not vice versa, parietal cells appear to be the terminal cell type in this lineage. In this model, cells derived from primitive endoderm are envisioned to move from the tip of the egg cylinder (embryonic region) through the visceral extraembryonic endoderm region and into the parietal endoderm region during egg cylinder growth (Hogan and Tilly 1981). The observation that primitive endoderm cells and their descendants actually move from the embryonic region to the extraembryonic region of the egg cylinder during early gastrulation is consistent with this proposal (Lawson et al. 1986; Lawson and Pedersen 1987). Although it is clear that visceral endoderm cells injected into blastocysts can give rise to parietal endoderm (Gardner 1982), whether visceral endoderm actually contributes to the parietal endoderm population during egg cylinder growth, remains to be determined. This would require a means for marking endoderm cells in the intact conceptus, and techniques for manipulating postimplantation embryo exo utero, combined with retroviral tracing methods, might provide such an approach (cf. Copp and Cockroft 1990). Alternatively, activation of reporter transgenes mediated by the yeast FLP recombinase (O'Gorman et al. 1991) might permit such lineage-specific tracing studies. For the moment, however, this model for the primitive endoderm lineage, together with that for the trophectoderm lineage (Rossant and Tamura-Lis 1981), provides a dynamic view of the relationships between extraembryonic cell types during peri-implantation differentiation. In this view, descendants from the stem cells of these two extraembryonic lineages move jointly away from the embryonic region during egg cylinder growth, contributing to the parietal endoderm cells and trophoblast giant cells, respectively, as the result of inductive interactions between them as they develop.

Fate of Epiblast during Gastrulation

The development of the embryonic region of the mouse embryo during postimplantation stages is essentially the development of the epiblast layer. As revealed by the studies described above, primitive endoderm contributes no descendants to the embryonic or fetal cell types; rather, these are derived exclusively from the epiblast layer during gastrulation (Gardner and Papaioannou 1975; Gardner and Rossant 1979; Rossant et al. 1983). Thus, cells originating from the primitive streak replace or displace the hypoblast cells adjacent to the epiblast layer in the embryonic region so that, between 6.5 and 7.5 d.g., the endoderm becomes a mixed population of primitive endoderm-derived and epiblast-derived cells (Lawson et al. 1986; Lawson and Pedersen 1987). By 8.5 d.g., the descendants of primitive endoderm have assumed positions in the yolk sac endoderm, and all endoderm descendants in the embryonic region are derived from epiblast. The pattern of migration of epiblast descendants into embryonic endoderm and mesoderm occurs consistently between embryos, so that a fate map of the initial epiblast locations of these primary germ layers can be drawn. The fate map also includes the progenitors of embryonic ectoderm (surface ectoderm and neuroectoderm), consisting of the epiblast progenitors and their descendants that do not emerge from the primitive streak, but remain in the epiblast/ectoderm layer during subsequent development (K.A. Lawson et al., unpubl.). In the epiblast fate map of embryos cultured from 6.5 to 7.5 d.g., the initial locations of progenitors of the primary germ layers are strikingly similar to those of the chick embryo at similar stages of gastrulation (Vakaet 1984). Epiblast zones close to the anterior tip of the primitive streak give rise to the endoderm and head process, whose progenitors enter the primitive streak early in gastrulation, zones close to the middle of the primitive streak give rise to embryonic mesoderm, and zones at the posterior of the streak give rise to extraembyronic mesoderm. Cells in epiblast zones distant from the streak (anterior and anterolateral) do not enter the streak and generally give rise to embryonic ectoderm.

These observations (K.A. Lawson et al., unpubl.) substantiate the conclusion from blastocyst-injection chimeras made with epiblast cells that epiblast alone contributes descendants to the fetus, and they reveal additional information about the time and site of primary germ layer formation. However, they do not address the potency of epiblast cells or the role of cell determination in the origin of the primary germ layers. An analysis of the fate of epiblast explants grafted to ectopic sites (Beddington 1981) indicates that small clumps of approximately 20 cells remain pluripotent even at the mid- to late-gastrula stages (7.5 d.g.) and are capable of generating descendants of all three germ layers, regardless of their site of origin within the epiblast layer. However, epiblast cells labeled with a retroviral vector at late gastrula/headfold stages (7.5–7.8 d.g.) are evidently allocated to individual tissue layers (F.C. Carey et al., unpubl.). Thus, epiblast cell determination does not appear to be the major factor guiding the fate of epiblast cells during gastrulation. How cells are channeled into the specific germ layers of endoderm, embyronic mesoderm, extraembryonic mesoderm, and embryonic ectoderm during gastrulation remains to be determined.

Genomic Imprinting

A major advance in understanding the development of embryonic and extraembryonic lineages of mammals has emerged from studies on the fate of parthenogenetic (or gynogenetic) embryos and their androgenetic counterparts. Diploid parthenogenotes produced by experimentally activating mouse oocytes can develop to midgestation stages, but then they die with a characteristic phenotype: The most advanced embryos have extensive development of the axial embryonic structures (brain and neural tube, somites), and other embryonic organs, but they have only rudimentary development of the trophoblast lineage (Surani et al. 1984; K.S. Sturm et al., unpubl.). Diploid androgenotes (i.e., embryos with only paternally derived chromosomes) manufactured by nuclear transfer (McGrath and Solter 1983) also die at midgestation stages, with grossly normal trophoblast but with retarded development of the embryo proper (Surani et al. 1984, 1986). The conclusion drawn from these and other studies (for review, see Surani 1986; Solter 1988; Surani et al. 1990) is that maternal and paternal gametes make distinct and complementary contributions to the developing conceptus, so that normal mouse development requires both maternal and paternal haploid genomes (Barton et al. 1984; McGrath and Solter 1984; Surani et al. 1984). This phenomenon, referred to as genomic imprinting, thus appears to have profound functional consequences, but only in eutherian mammals (metatheria and prototheria have not been studied), because examples of viable parthenogenesis are known among fish, amphibia, reptiles, and birds (Beatty 1967). The analysis of genomic imprinting in mouse development has focused on the effects of imprinting on the embryonic and extraembryonic cell lineages, on defining the genetic basis of imprinting, and on its molecular mechanism.

The fate of parthenogenetic or androgenetic embryos has been studied in aggregation chimeras, where the isoparental embryo is combined with a normal diploid embryo at early cleavage stages and then examined at the blastocyst stage or returned to the uterus of a foster mother for further development. Parthenogenetic and androgenetic embryos contribute to both the ICM and trophectoderm cell populations of such chimeras examined at the blastocyst stage, but they are selectively eliminated at later stages (Clarke et al. 1988; Thomson and Solter 1989). Parthenogenetic cells are eliminated from the trophectoderm and primitive endoderm lineages by midgestation stages but persist in most lineages of the embryo proper, whereas androgenetic cells are eliminated from embryonic lineages but persist in the extraembryonic lineages, the trophectoderm, and primitive endoderm (Clarke et al. 1988; Thomson and Solter 1988). With further development to term and to adult stages, parthenogenetically derived cells persist in some tissue lineages (e.g., brain, heart, kidney, spleen, and female germ cells) but are systematically eliminated from other tissues (e.g., skeletal muscle, liver, and pancreas) (Fundele et al. 1989; Nagy et al. 1989). These observations indicate that parthenogenetic cells are at a strong selective disadvantage in extraembryonic lineages and a lesser disadvantage in the embryonic lineages, where they may contribute extensively to normal tissues. However, adult mice with extensive parthenogenetic contribution have reduced viability and retarded postnatal development (Paldi et al. 1989).

Androgenetic cells are eliminated from the embryonic lineages by midgestation stages, but persist in the trophoblast and primitive endoderm lineages of chimeric embryos (Clarke et al. 1988; Thomson and Solter 1988). The contribution of androgenetic cells to term placenta has not been studied. However, embryonic stem (ES) cells produced from androgenetic mouse embryos were found to contribute to fetal tissues in chimeras (Mann et al. 1990). Many of these chimeras died at early postnatal stages with skeletal abnormalities. Subcutaneous tumors produced by injecting these androgenetic ES cells into normal mice were composed mostly of striated muscle. The contrast of this androgenetic ES cell phenotype (extensive muscle contribution) and the parthenogenetic phenotype (skeletal muscle deficiency) is interesting, because it echoes the complementarity of the isoparental effects on development of the embryonic and extraembryonic lineages.

These morphological observations may provide some clues about the genetic and molecular basis of imprinting in early mouse development. At least some of the parthenogenetic and androgenetic deficiencies are cell autonomous, because they are not corrected in chimeras by the cell-cell associations between them and the cells derived from the normally fertilized embryo. These deficiencies may be in factors that regulate growth and differentiation of the various cell types, as suggested by Surani and co-workers (1988). Indeed, recent work has shown that the expression of insulin-like growth factor II (IGF-II) is regulated by imprinting (DeChiara et al. 1990, 1991). This gene is transcribed only when it is inherited from the father; the maternal allele is not expressed. Expression of the IGF-II/mannose-6-phosphate (type 2) receptor, is also affected by imprinting and is transcribed only when it is maternally inherited (Barlow et al. 1991). Thus, the simplest conceptual model for the developmental effects of genomic imprinting is that endogenous imprinted genes are transcribed when they are inherited from one parent but not when they are inherited from the other parent, so that the normally fertilized individual is functionally hemizygous for such genes. Accordingly, the perturbations arising from imprinting in the development of isoparental embryos could arise from either underexpression or overexpression of the imprinted genes.

The molecular mechanism of genomic imprinting has not been elucidated. The concordance among DNA methylation levels, parental origin, and transcription observed for certain imprinted transgenes (Swain et al. 1987) suggests a role of DNA methylation in imprinting (for review, see Solter 1988; Sapienza et al. 1989). Indeed, the extraembryonic cell lineages that are strongly affected by imprinting in parthenogenetic embryos are also characterized by distinctly low levels of DNA methylation (Monk et al. 1987; Sanford et al. 1987; for review, see Sanford et al. 1985). These same cell lineages (trophectoderm and primitive endoderm) show preferential (i.e., nonrandom) inactivation of the paternal X chromosome in female embryos (for review, see Grant and Chapman 1988). Thus, lineage-specific patterns of DNA methylation and X chromosome inactivation both correlate with the early developmental consequences of imprinting, albeit in paradoxical ways. If X chromosome inactivation is taken as a paradigm for imprinting effects on autosomal genes, the role of DNA methylation in imprinting may be addressed indirectly. Studies of DNA methylation during inactivation of the X-linked hypoxanthine-guanine phosphoribosyl transferase gene indicate that methylation of the sites involved in regulating the transcription of this gene occurs after its expression has ceased (Lock et al. 1987). Studies of X-inactivation in a marsupial, the North American opossum, showed that there were low levels of DNA methylation in the DNase-hypersensitive 5′sites of the X-linked glucose-6-phosphate dehydrogenase gene, regardless of whether it was inactive or active (Kaslow and Migeon 1987). These observations indicate that inactivation of X-linked loci during mammalian development occurs independently of the state of DNA methylation at sites involved in regulating transcription, thus suggesting that methylation has a secondary role in imprinting; this role might be to stabilize the imprinting pattern present at early developmental stages so that it persists into adulthood. Whether DNA methylation plays a similar role in the imprinting of endogenous autosomal genes must be determined by an analysis of the history of methylation at their strategic regulatory sites.

What are the identities of other endogenous imprinted genes, and how many genes are imprinted? An extensive genetic analysis of parent-specific chromosome duplication/ deficiency phenomena has been carried out using reciprocal and Robertsonian translocations, revealing regions of the mouse genome that harbor developmentally important imprinted genes (Searle and Beechy 1978, 1985; Cattanach 1986; for review, see Surani 1986; Pedersen 1988; Solter 1988). In all, five autosomes were identified that showed developmental perturbations (ranging from midgestation to perinatal death) as a result of maternal duplication/paternal deficiency (chromosomes 2, 6, 7, 8, and 11), and four autosomes were similarly identified as paternal duplication/maternal deficiency syndromes (chromosomes 2, 7, 11, and 17); i.e., individuals inheriting both homologs of these autosomes from the specified parent and none from the other parent developed abnormally. Because the available translocation stocks encompassed large autosomal regions, there is presently no detailed mapping of the boundaries between imprinted and nonimprinted autosomal regions. At the minimum, at least ten imprinted genes could account for the observed mapping data; at the other extreme, perhaps 10% of the genome could be involved in imprinting, based on the extent of the translocations used to identify noncomplementing phenomena. Numerous genes involved in growth and morphogenesis have been mapped to the noncomplementing regions of the autosomes (Searle et al. 1989). The identification of two endogenous imprinted genes involved in the regulation of cell growth (DeChiara et al. 1990, 1991; Barlow et al. 1991) has been interpreted as the evolutionary consequence of conflicting parental investments in the growth of individual embryos at the expense of litter size (Haig and Graham 1991). An alternative view is that imprinting may have been involved in the evolution of placentation (Hall 1990). Whatever its origin, genomic imprinting appears to be involved in the etiology of several human diseases, including osteosarcoma (Toguchida et al. 1989), Angelman syndrome (Williams et al. 1990), and other degenerative or hyperplastic phenomena (for review, see Solter 1988; Hall 1990; Reik 1989). These observations strongly suggest that there are important developmental consequences of genomic imprinting among all mammals, including domestic species.

EARLY DEVELOPMENT IN DOMESTIC SPECIES

The study of embryonic and extraembryonic cell lineages in nonmurine mammals began in the 19th century. In an effort to understand embryonic development in agriculturally important mammals, many of these primarily descriptive studies utilized domestic ungulates: both artiodactyl (pig, cow [ox], sheep, and goat) and perissodactyl (horse). The similarity between early developmental events in the rabbit and those in domestic ungulates compels us to include the lagomorph embryo in this comparative analysis of pre- and peri-implantation development. We have also drawn from the limited information available for other mammals, particularly the insectivores *Elephantulus* and *Hemicentetes* and metatheria, in order to develop a sound perspective for understanding the early events in eutherian embryogenesis.

Cleavage

Among domestic species, cleavage divisions occur over a period of 2–8 days, generally while the embryos are in the oviduct. In some species, however, even the earlier cleavages occur in the uterus (Table 1). The pig embryo enters the uterus at the three- or four-cell stage (Assheton 1898a; Perry and Rowlands 1962; Oxenreider and Day 1965). In the horse, unfertilized eggs remain in the oviducts; only embryos reach the uterus (Betteridge et al. 1982). In embryos of domestic ungulates, divisions become increasingly asynchronous as cleavage proceeds (horse, Hamilton and Day 1945; pig, Assheton 1898a; Oxenreider and Day 1965; Hunter 1974; ox, Betteridge and Flechon 1988; goat, Amoroso et al. 1942; sheep, Assheton 1898b; Green and Winters 1945). Cleavage asynchrony would be expected to result in blastomeres of various sizes, as has been reported in particularly meticulous detail in the goat and sheep morula. However, in these and in some other domestic ungulates, blastomere size differences have also been reported at the two-cell stage (rabbit, Assheton 1895; pig, Assheton 1898a; Green and Winters 1946; ox, Hamilton and Laing 1946; Massip et al. 1983; goat, Amoroso et al. 1942; sheep, Assheton 1898b; Clark 1934). Clearly, such size discrepancies must arise from unequal cleavage and could be developmentally significant (Betteridge and Flechon 1988) if indeed size predisposes small blastomeres to occupy an interior location in the morula and thus assume an ICM fate, as it appears to in mice (Denker 1981a; Ziomek and Johnson 1981). On the other hand, blastomere size differences may be the inconsequential result of cleavage divisions rendered asymmetric by the large amount of yolk or yolk-like materials in the oocyte cytoplasm that has been reported, for instance, in the horse, pig, sheep, ox, cat, and guinea pig (Heuser and Streeter 1929; Green and Winters 1946; Boyd and Hamilton 1952). Individual sheep blastomeres from four- and eight-cell embryos have been shown to be totipotent (Willadsen 1979, 1980, 1981; Gatica et al. 1984) and ultrastructurally indistinguishable from one another (Calarco and McLaren 1976). Additionally, experimentally quartered 8-, 32-, and 64-cell ox morulae, as well as bisected ox blastocysts, have yielded viable twin or triplet calves (Willadsen and Polge 1981; Willadsen et al. 1981; Ozil et al. 1982; Ozil 1983; Lambeth et al. 1983; Williams et al. 1984). Hence, it is unlikely that size differences, whatever their cause, are responsible for determining the potency of blastomeres in ungulate embryos.

In the insectivore, *Elephantulus*, oocytes are shed in large numbers (60 per ovary); embryos undergo the first two cleavages in the oviduct and enter the uterus at approximately the four-cell stage. If they fail to reach the uterus, *Elephantulus* embryos degenerate at the four-cell stage in the oviduct (Tripp 1971). Of the dozen or so *Elephantulus* embryos that reach the uterine cornua, only two will implant, as the implantation site in each cornu is exceptionally small and typically accepts only one four-cell embryo. Before the embryo attaches to the uterine wall in this narrow cleft, it undergoes further cleavages and transforms into a blastocyst. All other embryos present in the cornua degenerate, although none appear to be abnormal (van der Horst 1942).

In contrast, the metatherian zygote traverses the oviduct quickly and reaches the uterus uncleaved. Metatherian cleavage patterns display minor species-specific variations (Selwood 1989a), but in at least three species, developmental arrest normally occurs at the four-cell stage (Selwood 1980, 1987).

Yolk elimination during cleavage, or deutoplasmolysis, has been described in a number of mammalian species. This process was first reported in 1909 by van der Stricht in the cleaving bat embryo (cited in Boyd and Hamilton 1952) but is perhaps best known in metatherian embryos, in which yolk is eliminated as many small particles (Hartman 1916, 1919; Hill 1918) or as a large mass into the intercellular spaces (Hill 1910; Selwood 1980; Selwood and Young 1983; Selwood and Sathananthan 1988). In the latter mode, the extruded yolk mass sometimes persists past the time when the primary endoderm cells form. In embryos of the metatherian *Antechinus stuartii*, the primary endoderm cells apparently originate in the embryonic hemisphere that contains the yolk mass (Selwood and Young 1983). However, the role of the extruded yolk mass in axis determination in metatherian embryos remains to be experimentally demonstrated.

TABLE 1 *DEVELOPMENTAL EVENTS IN SELECTED MAMMALS*

	Cleavage		*Morula compaction*		*Blastulation*		*Entry into uterus*		*Onset of implantation*	
Mammal	*2-cell*	*8-cell*	*days*	*cells*	*days*	*cells*	*days*	*stage*	*days*	*stage*
Lagomorph										
rabbit[a]	0.5	1.5	2	16–32	3	~128	3	blastocyst	6.5	blastocyst
Rodent										
guinea pig[b]	1	3.5	3.5	8	?	?	6	morula	6	blastocyst
mouse[c]	1.5	2.5	2	4–8	3.5	32	2	morula	4.5	blastocyst
Perissodactyl										
horse[d]	1	3	4.5	~16	6–8	64	5–6	blastocyst	34	past limb bud stage
Artiodactyl										
pig[e]	0.5	2–3	3	8–16	3.5–5	16	2.5–3.5	3–4-cell	14–15	early somite
ox[f]	1	2–4	4–5	16	5–6	32	3–4	morula	17–19	early somite
goat[g]	1	3–4	4	8	6	32	4–5	morula	?	?
sheep[h]	1	2	3	8	4.5–6	64	3–4	morula	15–16	early somite
Primate										
human[i]	2	3	3–5	16	5	64–107	3–4	morula	6–7	blastocyst

Insectivore										
Elephantulus[j]	?	?	none (?)	none (?)	?	4	2?	4-cell	?	blastocyst
Metatheria										
Didelphis[k]	0.5	2–3	none	none	4.5	32	<1	zygote	7–8	28 somites organogenesis
Antechinus[l]	1	3.5	none	none	6	~20	<1	zygote	23	8–12 somites, neurula
Sminthopsis[m]	0.5	2.5	none	none	4	~32	<1	zygote	?	organogenesis trilaminar
Trichosurus[n]	?	?	none	none	?	?	<1	zygote	?	blastocyst

The timing of events is indicated as days after fertilization.

References: [a]Assheton (1895), Anderson (1927), Boving (1959), Davies and Hesseldahl (1971); [b]Squier (1932), Blandau (1949); [c]Snell and Stevens (1966), Hogan et al. (1986); [d]Hamilton and Day (1945), Stevens and Morriss (1975), van Niekerk and Allen (1975), Allen (1982), Betteridge et al. (1982); [e]Assheton (1898a), Heuser and Streeter (1929), Green and Winters (1946), Boyd and Hamilton (1952), Perry and Rowlands (1962), Oxenreider and Day (1965), Hunter (1974), King et al. (1982); [f] Winters et al. (1942), Hamilton and Laing (1946), King et al. (1982), Lambeth et al. 1983), Betteridge and Flechon (1988), First and Barnes (1989); [g]Amoroso et al. (1942); [h]Assheton (1898b), Clark (1934), Green and Winters (1945), Chang and Rowson (1965), Calarco and McLaren (1976), Willadsen (1980), King et al. (1982), Moor et al. (1987), First and Barnes (1989); [i]Hertig and Rock (1945), Hertig et al. (1956), Hafez (1973), Brackett (1978), Edwards (1980); [j]van der Horst (1942), van der Horst and Gillman (1942); [k]Hartman (1919, 1923, 1928), McCrady (1938), New and Mizell (1972), Selwood (1989a); [l]Selwood (1980, 1981, 1986a,b, 1989a,b), Selwood and Young (1983); [m]Selwood (1989a,b); [n]Sharman (1961), Selwood (1989b).

Compaction and Blastocyst Formation

In the eutherian embryo, compaction occurs during the 8-, 16-, or 32-cell stage (see Table 1). Compaction appears to be initially reversible; indeed, eutherian embryos at these cleavage stages are portrayed in the literature as either loose aggregations of rounded blastomeres or subspherical masses of tightly packed, flattened cells. In mouse and hamster embryos, blastomeres have been observed to round up during mitosis (Bavister 1987; Garbutt et al. 1987; Skrecz and Karasiewicz 1987). The adhesion that occurs between blastomeres before compaction is mediated largely by uvomorulin, whose activity depends on the presence of a divalent cation such as Ca^{++} or Mg^{++} (Hyafil et al. 1980, 1981; Damsky et al. 1983; Gallin et al. 1983; Peyrieras et al. 1983; Shirayoshi et al. 1983; Vestweber and Kemier 1984; Yoshida-Noro et al. 1984; Johnson and Maro 1986; Johnson and Takeichi 1986; Johnson et al. 1986; Kemler et al. 1987). From the 32-cell stage, however, other cell interactions assume increasingly important roles in mediating cell-cell adhesions (for review, see Johnson and Maro 1986; Damsky 1989), and simply excluding Ca^{++} and Mg^{++} from culture media intended for 32-cell or older embryos no longer causes blastomere disaggregation. With mouse embryos, however, the omission of these cations increases the vulnerability of the inner blastomeres or ICMs to complement-mediated lysis (Y.P. Cruz et al., in prep.). Thus, compaction is a dynamic developmental event, its basis changing as the eutherian embryo transforms itself into a blastocyst.

Compaction is accompanied or quickly followed by events that presage cavitation. For instance, cell junctions form between the outer blastomeres of the compacted 16-cell embryo (morula) in many eutherian mammals (Enders 1971). The appearance of these junctions has been interpreted as the first sign of the differentiation of the trophoblast in sheep (Calarco and McLaren 1976) and rabbit embryos (van Blerkom et al. 1973). Another compaction-associated event well documented in mouse (Ziomek and Johnson 1981) and rat (Leis and Izquierdo 1984) embryos is polarization of the outer blastomeres. In bovine morulae, the apical surfaces of outer blastomeres become densely microvillous, whereas their lateral surfaces acquire apical tight junctions. Inner cells, meanwhile, become polygonal and closely apposed (Betteridge and Flechon 1988). In goat morulae, inner cells apparently result from the radial division of peripheral cells after the fifth cleavage division (Amoroso et al. 1942). Thus, the period from just before to just after the fifth cleavage division appears to be critical in determining the number of cells that come to occupy an inner position in goat morulae. This appears to be the case in pig embryos as well. Of 62 6–8-d.g. pig embryos examined, 2 lacked inner cells and 3 had at most two inner cells (Papaioannou and Ebert 1988). Although the absence of inner cells in this experiment could have been an artifact of the immunosurgical step required for the staining process used to differentiate between inner and outer cells, it could also indicate that some blastomeres had yet to attain an interior position even after the fifth cleavage division. Indeed, the same study reported that 2 of the 62 embryos studied had reached the early blastocyst stage (24 and 49 cells) but contained no inner cells (Papaioannou and Ebert 1988). Thus, although compaction precedes cavitation in the eutherian morula, it does not guarantee the establishment of inner and outer cell populations.

Compaction in eutherian embryos is quickly followed by cavitation. The compacted morula forms one or more cavities as a result of the active transport of Na^+, Cl^-, and HCO_3^- into its intercellular spaces (Lutwak-Mann et al. 1960; Daniel 1963, 1964; Borland 1977; Benos and Biggers 1981; Overstrom 1987). These cavities coalesce into the contiguous fluid-filled space, or blastocoel, which displaces the inner cells of the morula to an eccentric position. These displaced cells, which adhere to the inner surface of the blastocyst wall, the trophectoderm (trophoblast), constitute the ICM.

Unlike eutherians, metatherian embryos do not undergo compaction. The metatherian blastocyst forms without ionic transport into the intercellular spaces of the cleavage-stage embryo. The blastomeres in the cleavage-stage metatherian embryo establish superficial and extremely tenuous cell-cell contacts (Selwood 1986a). Indeed, enzymatic removal of embryonic investments of four-cell *A. stuartii* embryos results in blastomere dispersion (Selwood 1989b). In intact eight-cell metatherian embryos, blastomeres adhere to the inside surface of the zona pellucida (Hill 1910; Hartman 1916, 1919; Lyne and Hollis 1976, 1977; Selwood 1980; Selwood and Young 1983). As cells divide in this position, they also establish contact with each other (Lyne and Hollis 1976, 1977; Selwood 1987, 1989b). Eventually, an epithelium of flattened cells comes to line the zona pellucida. The cavity of this spherically configured epithelium is the blastocoel, formed passively from the intercellular spaces that become contiguous as a result of blastomere adhesion to the zona pellucida.

Two exceptional eutherians exhibit the metatherian manner of blastocyst formation. The insectivores *Hemicentetes* and *Elephantulus* form blastocysts without undergoing a compaction stage. The formation of the unilaminar blastocyst of *Hemicentetes* begins at the four- or eight-cell stage (Goetz 1937), with the blastomeres adhering to the interior surface of the zona pellucida in metatherian fashion (Bluntschli 1938). The unilaminar blastocyst is completed at the 16-cell stage (Goetz 1937). Blastocyst formation is even more unusual in *Elephantulus* because it occurs at the four-cell stage (van der Horst 1942; Tripp 1971). The two pairs of sister blastomeres at this stage become arranged in a crossed, typically eutherian configuration and proceed to form a cavity at the point where the blastomere pairs intersect. Two-cell blastocysts have also been reported in *Elephantulus*, apparently the result of blastocoel formation by the conjoined sister-cell pair arising from cleavage of a blastomere dislodged from a two-cell embryo. Blastomere disaggregation at the two- and four-cell stages occurs spontaneously and is not uncommon in *Elephantulus* embryos, since the zona pellucida disappears soon after the oocyte is fertilized (van der Horst 1942). Blastocoel formation in the *Elephantulus* embryo is thus atypical, differing significantly from that of the typical eutherian embryo, which requires compaction, and the metatherian, which requires a zona pellucida.

Blastocyst Expansion

Eutherian blastocysts undergo variable rates of expansion prior to implantation. Blastocyst expansion has been reported in the mammals listed in Table 1 with the exception of *Elephantulus*, in which blastocyst development has not been studied. The physiological dynamics of blastocyst expansion are well documented in the rabbit (Benos 1981; Benos et al. 1985; Biggers et al. 1988). The blastocysts of primates, dogs, and the ferret are also known to expand (Holst and Phemister 1971; Hafez 1972), but the physiological details of expansion have not been characterized in these embryos. Among eutherian mammals, the extent to which the blastocyst expands appears to be directly related to the superficiality of implantation. Thus, the pig blastocyst, which undergoes some expansion followed by tremendous elongation, forms a casual attachment to the uterine epithelium. In contrast, blastocysts of the mouse, rat, and hamster expand minimally (about fivefold) (Biggers 1972) but implant interstitially (Hafez 1972).

The mechanics of blastocyst expansion have been examined in several domestic ungulates. The pig blastocyst hatches on 6 d.g. (Perry and Rowlands 1962; Hunter 1977). Between days 9 and 14, a dramatic change in morphology transforms the blastocyst from a 6-mm sphere to a highly attenuated, filamentous chorionic vesicle measuring more than 100 mm in length (Heuser and Streeter 1929; Perry and Rowlands 1962; Anderson 1978). The rate of blastocyst elongation accelerates dramatically during this period from 0.25 mm/hr in blastocysts 4–9 mm in diameter to 30–45 mm/hr in blastocysts 10 mm or more in diameter (Geisert et al. 1982). Elongation from 10 to 150 mm requires only about 6 hours (Geisert et al. 1982), suggesting that this remarkable transformation requires cell deformation, not cell division (Perry 1981). A band of cells approximately 2 mm wide and spanning the distance from the edge of the ICM to the abembryonic pole along the wall of a 10-mm blastocyst has been identified as the elongation zone (Geisert et al. 1982). Within this band, trophectoderm cells undergo rapid remodeling of microfilaments and cell junctions (Mattson et al. 1990); densely packed endoderm cells extend numerous filopodia and migrate away from the elongation zone along the blastocoel-facing surface of the trophectoderm (Geisert et al. 1982). As a result of this activity, the blastocyst first becomes tubular and then filamentous.

Elongation of ox blastocysts probably occurs in the same

manner. Ox blastocysts hatch at 9 or 10 d.g (~200 cells, ~160 μm in diameter) and enlarge rapidly thereafter (~1000 cells, ~375 μm on days 11–12, ~10 mm on days 12–14). By 16 d.g., the blastocyst has transformed into an attenuated, bilaminar blastodermic vesicle approximately 100 mm long (Chang 1952; Betteridge et al. 1980; Betteridge and Flechon 1988). Polyploid transformation of trophectoderm cells begins after 12 d.g. and proceeds to affect approximately 25% of trophectoderm cells by 18 d.g. (Hare et al. 1980). No elongation zone has been reported in ox blastocysts in these embryos; transformation into attenuated filamentous vesicles apparently requires uterine factors (Betteridge and Flechon 1988).

Expansion in the sheep blastocyst proceeds rapidly as well. At 9 d.g., the sheep blastocyst is a unilaminar sphere (~300 cells, <1 mm diameter) (Green and Winters 1945; Rowson and Moor 1966; Bindon 1971). By 10 d.g., the blastocyst is bilaminar (~3000 cells, ~1 mm diameter), by 12 d.g., the blastocyst is 10–22 mm long (Wintenberger-Torres and Flechon 1974), and by 14 d.g., the blastocyst is attenuated and filamentous (~70–100 mm long) (Rowson and Moor 1966; Bindon 1971; Wintenberger-Torres and Flechon 1974; Carnegie et al. 1985). Ultrastructural documentation of the changes in cell shape that occur during this transformation reveals a flurry of activity in trophectoderm cells: Microvillar density and pinocytotic activity increase, the cytoplasmic tonofibrillar system becomes more prominent, and glycogen and lipids accumulate (Wintenberger-Torres and Flechon 1974; Guillomot et al. 1978). These activities notwithstanding, well-developed spot desmosomes persist in both the trophectoderm and endoderm, although widely scattered individual cells in the trophectoderm die and proceed to disintegrate (Carnegie et al. 1985). Uterine factors apparently also influence elongation in the sheep blastocyst, but these remain uncharacterized (Flechon et al. 1986).

Blastocyst expansion in metatherians occurs following completion of the unilaminar blastocyst. The blastocyst wall, or protoderm (McCrady 1938), is complete in several species by the 32-cell stage (4–6 d.g., depending on the species) (Table 1). In the dasyurid *A. stuartii*, the unilaminar blastocyst is formed at the 50-cell stage (9 d.g.) (Selwood 1980). In the bandicoots,

Perameles nasuta and *Isoodon macrourus*, the protoderm is not fully formed until the 75-cell stage (Lyne and Hollis 1976). Blastocyst formation in the native cat (*Dasyurus viverrinus*) is completed at an even later time, between the 108- and 130-cell stage (Hill 1910). The unilaminar blastocyst stage is achieved by embryos of approximately 0.5 mm diameter (Hill 1910; Hughes 1974). Blastocyst expansion in metatherians is associated with disappearance of the yolk mass in all species (Selwood and Young 1983) except *D. viverrinus*, in which the yolk mass persists until the blastocyst attains a diameter of between 4 and 6 mm (Hill 1910).

Trophoblast Differentiation

Differentiation in most eutherian blastocysts is evident from the earliest stages of blastocyst formation (Fig. 1). The trophectoderm (or trophoblast) cells that overlie the ICM constitute the polar trophoctoderm, whereas those that overlie the blastocoel make up the mural trophectoderm. In blastocysts that undergo considerable expansion, the polar trophectoderm disappears as the blastocysts enlarge, thus exposing the ICM, which by this time has differentiated into an epiblast and a hypoblast. The term Rauber's layer (Koelliker 1880) has been used to distinguish this dehiscent polar trophectoderm from that found in minimally expanding blastocysts such as that of the mouse and rat. Rauber's layer is found in blastocysts of the rabbit (Koelliker 1880; Assheton 1895; Flechon 1978; Mootz 1979; Williams and Biggers 1990), horse (Enders et al. 1988), and pig (Assheton 1898a; Heuser and Streeter 1929; Geisert et al. 1982; Stroband et al. 1984; Barends et al. 1989). Other embryos in which Rauber's layer occurs include those of the cat, dog, sheep, deer, mole, shrew, tree-shrew, and western spotted skunk (Mossman 1937; Boyd and Hamilton 1952; Enders et al. 1986, 1988).

The disappearance of Rauber's layer is well documented in ungulate embryos. In rabbit and pig blastocysts, the initially contiguous cells of Rauber's layer dissociate, disintegrate, and are apparently phagocytosed by cells of the underlying epiblast. Simultaneously, epiblast cells develop new junctions among themselves and with adjacent mural trophectoderm

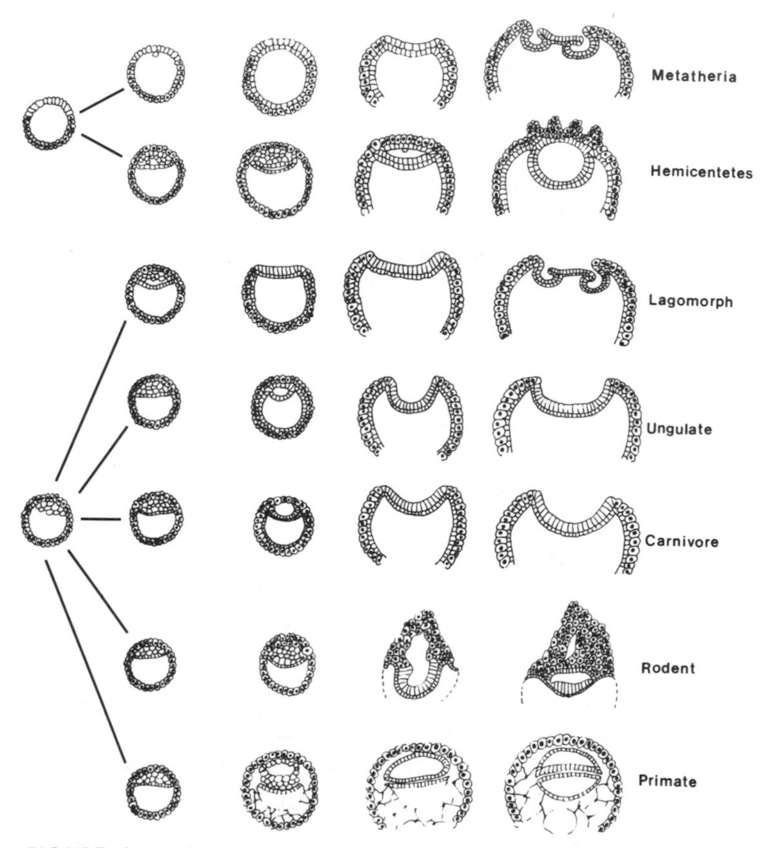

FIGURE 1 Schematic representation of differentiation of embryonic and extraembryonic lineages in different mammals. Trophoblast-derived cells are shown with solid nuclei. The epiblast is shown as tall, columnar cells; the hypoblast and its visceral and parietal endoderm derivatives are shown as cuboidal cells. (*Top*) In metatheria and *Hemicentetes*, a unilaminar protoderm forms first, and the epiblast and hypoblast originate from cells originally found in the protoderm. In eutherians, the ICM originates from blastomeres enclosed during cleavage, and compaction then gives rise to epiblast and hypoblast after blastocyst formation. (*Bottom*) In the rabbit (lagomorph), ungulates, and carnivores, the polar trophoblast (Rauber's layer) dehisces, exposing the germinal disc. The polar trophoblast in *Hemicentetes*, rodents, and primates does not dehisce but instead contributes to invasive derivatives of the chorionic ectoderm. (Based on Boyd and Hamilton 1952.)

cells, thus preserving the permeability seal originally provided by the intact trophectoderm (Barends et al. 1989; Williams and Biggers 1990). Disintegration of Rauber's layer occurs on day 6 in rabbit blastocysts (Williams and Biggers 1990) and 1 day later in pig blastocysts (Assheton 1895). In horse blastocysts, Rauber's layer becomes visibly disrupted on day 7. On day 10, cellular debris apparently resulting from this disintegration is evident in epiblast cells, which therefore must have phagocytosed cellular fragments from Rauber's layer (Enders et al. 1988).

The fate of Rauber's layer is an unmistakable instance of embryonic cell death. This stands in stark contrast with the fate of the polar trophectoderm, which is clearly its homolog in nonexpanding eutherian blastocysts. The polar trophectoderm has been shown in preimplantation mouse embryos to contribute to the mural trophectoderm (Copp 1979; Cruz and Pedersen 1985) and, following implantation, to the ectoplacental cone (Rossant 1986) and extraembryonic ectoderm (Papaioannou 1982). The significance of this clear divergence in fate by two patently homologous embryonic tissues awaits further investigation.

Gastrulation

Gastrulation occurs during the late preimplantation blastocyst (chorionic vesicle) stage in embryos of metatherian and domestic eutherians. This contrasts with gastrulation in rodents and primates, which occurs after implantation (see Table 1). Gastrulation consists of cell proliferation and rearrangement in the germinal (or germ) disc, which in eutherian embryos consists of the ICM (the future epiblast and hypoblast) and its overlying polar trophectoderm, when present. The germinal disc is typically flat or faintly convex in mammalian embryos that undergo blastocyst expansion. In rodent embryos, however, the germinal disc is steeply invaginated into the blastocoel, hence the name, egg cylinder (Snell and Stevens 1966). The conformation of the egg cylinder cannot be attributed to either early implantation or minimal expansion of the blastocyst. These characteristics are shared by rodent and primate embryos, but the germinal disc in the latter is flat or nearly so (Heuser and Streeter 1941; Hertig et al. 1956). Therefore, any analysis of mammalian gastrulation modeled on the rodent embryo should take into account these topological differences.

Differentiation of the eutherian ICM begins with hypoblast formation. The hypoblast arises from cells that migrate or delaminate from the blastocoel-facing surface of the ICM (Boyd and Hamilton 1952; Hafez 1972). These cells typically spread along the inner surface of the mural trophectoderm and so come to line the blastocyst cavity. The remainder of the ICM becomes the epiblast. At this time, Rauber's layer dehisces or is lost (Enders et al. 1988; Barends et al. 1989; Williams and Biggers 1990). In contrast, the nondehiscent polar trophectoderm of rodent and primate embryos persists and later actively participates in implantation by invading the uterine wall (Heuser and Streeter 1941; Hertig et al. 1956; Snell and Stevens 1966; Gardner et al. 1973; Papaioannou 1982). In mouse embryos, cell allocation between the ICM epiblast and polar trophectoderm is largely but not entirely completed at or immediately after the fifth cleavage (Nichols and Gardner 1984; Rossant and Croy 1985; Pedersen et al. 1986; Fleming 1987; for review, see Cruz 1990). Whether this occurs also in domestic animals remains to be demonstrated.

The most obvious manifestation of gastrulation is the formation of the primitive streak. In the pig embryo, this proliferative zone originates as a crescent-shaped region just beneath the surface of the future posterior region of the germinal disc (Boyd and Hamilton 1952). Elongating anteriorly into an oval area, the crescent gradually narrows as its margins dissociate into motile mesenchyme (mesoblast) cells, which later give rise to the mesoderm (Streeter 1927). In avian embryos, analysis of orthotopic grafts indicates that the mesenchyme associated with the primitive streak likewise produces embryonic endoderm (Nicolet 1971). Contrary to its name, the primitive endoderm (hypoblast) does not contribute to the embryonic body. The germ layers of the definitive embryo therefore have a gastrular origin. Hensen's node forms later, and thus separately from, the primitive streak. In the pig, Hensen's node first appears after mesenchyme formation as a new area of proliferation on the surface of the epiblast. At this time, the mesenchyme has just coalesced into a contiguous sheet subjacent to the epiblast but only in the future cranial region of the embryo. It is here in the most cephalic, mesenchyme-free region of the germinal disc closest to the anterior end of the primitive streak that Hensen's node appears (Streeter 1927). The cells arising from Hensen's node become exclusively chordamesoderm and later notochord (Slack 1983). Because both Hensen's node and the primitive streak are proliferative zones, new cells are intercalated into the steadily elongating germinal disc. The primitive streak eventually shortens, retreating posteriorly in tandem with Hensen's node. In the wake of this retreat, the neural tube, somites, and other principal parts of the embryonic body make their appearance.

Implantation and Placentation

The extraembryonic membranes are formed soon after gastrulation. Shortly after its formation through the primitive streak, the pig mesenchyme condenses into mesoderm and expands in the plane of the trophectoderm, beyond the confines of the germinal disc. The future coelom and exocoelom now appear, first as a cavity or split, and later as a widening gap in the embryonic (germinal disc) area and extraembryonic (beyond the germinal disc) mesoderm, respectively. The outer somatic layer of mesoderm lines the interior of the former mural trophectoderm (now the extraembryonic ectoderm); these closely apposed epithelia become the somatopleure. The inner splanchnic layer of mesoderm closely invests the underlying hypoblast (now extraembryonic endoderm); these constitute the splanchnopleure (Perry 1981). The extraembryonic portion of the somatopleure forms the amnion and chorion; those of the splanchnopleure form the yolk sac and allantois. The amnion is chronologically the first fetal membrane to form. In the pig, this closed pouch forms by fusion of the cephalic and caudal folds of somatopleure over the dorsal axis of the embryo (Perry 1981). In contrast, the rodent and primate amniotic cavity arises by transformation of a delaminated sinus in the epiblast-originated extraembryonic ectoderm. This sinus, or pro-amniotic cavity (Snell and Stevens 1966), acquires an exterior layer of extraembryonic mesoderm that originates from the cephalic and caudal folds of the embryo. At this time, the embryo constitutes the distal half of a steeply invaginated egg cylinder. Recent cell-marking studies reinforce the epiblast origin of the mouse amniotic ectoderm (K.A. Lawson and R.A. Pedersen, in prep.). This contrasts with the apparent mural trophectoderm (extraembryonic ectoderm) origin of the pig amniotic ectoderm (Heuser and Streeter 1941). To reconcile this apparent inconsistency, cell contribution between the mouse epiblast and mural trophectoderm (via a polar trophectoderm route) would have to be invoked. However, physical constraints imposed by the compression of the mouse germinal disc into a narrow egg cylinder should prevent the entry of epiblast-derived cells into the mural trophectoderm (Copp 1979). Moreover, the epiblast contributes to the chorion or ectoplacental cone in studies of mouse chimeras (Gardner and Rossant 1979; for review, see Rossant 1986). Therefore, arguments about homology lead us to question the conclusions of the pig and human studies, which are based solely on morphological observations.

The ungulate chorion arises from the subset of extraembryonic somatopleure that does not become the amnion. The arrangement of ectoderm and mesoderm (outer and inner layers, respectively, in the chorion) results from the simultaneous establishment of homotopic contiguity between the ectodermal and mesodermal components of the cephalic and caudal amniotic folds (Perry 1981). Chorion formation in primates (Heuser and Streeter 1941) and rodents (Snell and Stevens 1966; Perry 1981) occurs in a very different manner. In these early-implanting embryos, the chorion arises from the trophectoderm-derived cytotrophoblast, which acquires an internal lining of extraembryonic (somatic) mesoderm. The rodent and primate chorion forms not by folding, but by delamination.

The yolk sac is formed from the splanchnopleure. In ungulates, the cavity defined by the hypoblast persists as the cavity of the yolk sac. In other words, the cavity of the yolk sac is the former blastocoel. The extent of exocoelom formation, however, varies. Thus, the exterior of the parietal region of the yolk sac may remain contiguous with the interior of the chorion. This portion of contiguity is the bilaminar omphalopleure (Steven and Morriss 1975; Perry 1981), which fails to vascularize due to the poor development of its mesodermal component. The parietal portion of the yolk sac, however, becomes vascularized in carnivores and most ungulates, and its outer limits are defined by an annular blood vessel, the sinus terminalis. A vascularized yolk sac is frequently closely apposed to the chorion, with which it forms the trilaminar omphalopleure and serves a placental function. Metatherian embryos typically form a trilaminar omphalopleure and therefore a placenta (Sharman 1961; Renfree 1982). The metatherian placenta, however, is typically noninvasive. The notable exception is the bandicoot (Peramelidae) placenta, which forms not from the trilaminar omphalopleure but from the chorioallantoic membrane.

The yolk sac is a transient structure in certain embryonic mammals. In rodents, Reichert's membrane forms between parietal endoderm and the trophoblast giant cell layer. This largely acellular membrane is a modified basal lamina produced by the parietal endoderm (Hogan et al. 1980). As the mural trophectoderm invades the uterine epithelium during implantation, Reichert's membrane becomes the sole barrier between the blastocoel (yolk sac cavity) and the uterine milieu. The rodent yolk sac appears to have an endodermally derived exterior layer and therefore an "inside-out" configuration (Morriss 1975; Steven and Morriss 1975; Perry 1981). This inversion of the germ layers is yet another topographical peculiarity of the rodent conceptus. In primate embryos, the yolk sac makes an even more transitory appearance during development. Because the extraembryonic (primitive endoderm-derived) endoderm never completely lines the blastocoel prior to implantation (Steven and Morriss 1975), no Reichert's membrane forms and a bilaminar blastocyst stage is never attained (Wimsatt 1975). However, extraembryonic mesoderm appears precociously and completely lines the mural trophectoderm, which simultaneously transforms into the syncytial and cellular cytotrophoblast (extraembryonic ectoderm) component of the chorion. The yolk sac cavity (former blastocoel) persists, however, lined at first by extraembryonic mesoderm (Heuser's membrane). The hypoblast then belatedly enlarges and completely lines the blastocoel-facing surface of Heuser's membrane. The resulting yolk sac, however, soon degenerates but not before blood islands have formed in its mesoderm and hematopoietic cells have colonized the embryo (Heuser and Streeter 1941; Steven and Morriss 1975; Perry 1981). Although the yolk sac serves a primarily trophic function in amniotes with yolky eggs, it serves a purely hematopoietic function in eutherian mammals with yolkless eggs.

The allantois is chronologically the last and developmentally the most variable fetal membrane to form in mammals. Like the yolk sac, the allantois is derived from the splanchnopleure. It first arises, however, as an outpocketing of the extraembryonic endoderm that defines the ventral region of the embryo caudal to the base of the yolk sac (Perry 1981; Renfree 1982). The extent to which this endodermal pouch protrudes into the extraembryonic mesoderm varies among mammals. In eutherian mammals and peramelid metatherians, the endodermal component of the allantois forms precociously, enlarges, and encounters the extraembryonic splanchnic mesoderm. It is the mesodermal component that typically vascularizes the allantois and permits it, if sufficiently enlarged, to displace and supplant the yolk sac as the gas-, nutrient-, and waste-diffusion site in the placenta (Morriss 1975; Padykula and Taylor 1976; Perry 1981; Renfree 1982).

The allantois does not always develop as early or as extensively, however. In most metatherians, for instance, the allantois is poorly developed and remains enfolded by the wall of the yolk sac and is thus prevented from contacting the chorion (Sharman 1955; Hughes 1974; Fleming and Harder 1981; Renfree 1982). In certain ungulates such as the pig, the allantois does not vascularize but nevertheless enlarges to form the chorioallantoic sheet that comes to enclose the embryo (Corner 1921; Heuser 1927; Perry 1981). This sheet later establishes points of contact with the mouths of the uterine glands by means of button-like areolar projections (Amoroso 1952; Chang 1952; Steven 1975, 1983; King et al. 1982). Fetomaternal contact in the pig is thus extremely superficial; the placenta that forms is called diffuse.

The chorionic ectoderm is invasive in many mammals. In ungulates other than the pig, specialized regions of the chorioallantoic sheet (placenta) exhibit various degrees of invasiveness. For instance, the placentas in sheep, goats, and ox form chorionic projections, variously called cotyledons, cotyledonary burrs, caruncles, or placentomes (Assheton 1906; Steven 1975, 1983; Morgan and Wooding 1983; Wooding and Wathes 1980). These projections establish close connections with crypts or swellings in the uterine wall and later become extravasated as pregnancy progresses (Wimsatt 1951; Amoroso 1952; Steven 1975; Wooding and Staples 1981). In carnivores, the chorionic projections, or villi, are aggregated into a girdle-like band that establishes a vascular connection with the uterine wall. This band is later called the hemophagous organ, and the carnivore placenta is frequently described as zonary (Amoroso 1952; Steven 1975; Wimsatt 1975). In the horse omphalopleure, a belt of invasive chorionic cells (the chorionic girdle) coincides with the sinus terminalis, which otherwise is noninvasive and forms a diffuse placenta with the chorioallantoic sheet (Allen et al. 1973; Steven 1975, 1983). Trophoblast-derived cells from the chorionic girdle migrate from their native position in the chorion, invade the uterine wall, and phagocytose endometrial cells (Allen et al. 1973). These migratory cells establish themselves in the uterus and, as so-called equine endometrial cups, secrete pregnant mare serum gonadotropin (Allen et al. 1973; Van Niekerk and Allen 1975; Steven 1983; Whyte and Allen 1985; Whyte et al. 1986). Invasive, binucleate cells have also been reported in the placenta of the sheep (Wimsatt 1951; Davies and Wimsatt 1966; Lawn et al. 1969; Carnegie et al. 1985), goat (Lawn et al. 1969), and ox (Wimsatt 1951; Greenstein et al. 1958; Bjorkman 1969; King et al. 1979, Wooding 1982). Although typically noninvasive, the trophoblast component of the pig chorioallantoic membrane displays invasive properties in ectopic sites (Samuel and Perry 1972). This clearly underscores the homology among the various kinds of invasive cells and tissues derived from the chorion.

The invasiveness of the chorion is perhaps best demonstrated in rodents and primates. Implantation occurs early in these mammals, as the trophectoderm is instrumental in establishing attachment to the uterine wall (Schlafke and Enders 1975; Enders 1976; for review, see Parr and Parr 1989). The rodent embryo initiates attachment with its mural trophectoderm; the polar trophectoderm and extraembryonic component of the epiblast produces the intrusive ectoplacental cone, or trager (Perry 1981). In primate embryos, initial attachment is effected by the polar trophectoderm, which quickly transforms into the cellular and syncytial cytotrophoblast that later becomes the chorionic ectoderm (Heuser and Streeter 1941; Hertig and Rock 1945; Hertig et al. 1956). Trophoblast invasiveness in rodents and primates, however, is later confined to a small sector on the surface of the fully differentiated chorion. Here, the underlying extraembryonic (somatic) mesoderm induces proliferation of chorionic villi, and these, in turn, induce massive extravasation in the overlying maternal tissue (Steven and Morriss 1975; Perry 1981). As a result, a localized swelling forms, consisting of a highly vascularized endometrium and a large number of chorioinic villi. This is the discoid placenta, and it possesses neither allantois- nor yolk-sac-derived tissue. Although the allantois develops in rodents and primates, it does not establish contact with the chorion. This major topographical difference also needs to be considered when the rodent or primate placenta is used as a model to study the placenta of metatheria, lagomorphs, domestic ungulates, and carnivores.

Extracellular Coats

The mammalian embryo typically has at least one extracellular coat, the zona pellucida. First visible in ovarian follicles containing developing oocytes, the zona pellucida has traditionally been considered a product of ovarian follicular cells (see, e.g., Boyd and Hamilton 1952). Recent work shows, however, that the major glycoproteins of the mouse zona pellucida are produced by the oocyte itself (for review, see Wassarman 1990). One of the glycoproteins of the mouse zona pellucida, ZP3, has been found to bind specifically to a sperm surface protein and thus plays a significant role in fertilization (Bleil and Wassarmann 1980; Florman and Wassarmann 1985). The structural similarity of ZP3 to a glycoprotein component of the pig zona pellucida (Hedrick and Wardrip 1987) suggests that mammalian zonae pellucidae play similar, if not identical, roles in fertilization. An ultrastructural examination of the bovine zona pellucida reveals channels traversing its thickness. These channels were previously occupied by cytoplasmic projections emanating from the cells of the corona radiata (Flechon and Gwatkin 1980), the cellular accompaniment of recently ovulated oocytes. The zona pellucida is typically shed in the uterus or uterine cornua prior to implantation, although in the guinea pig, embryonic attachment to the uterine lining is established by trophoblast-originated ectoplasmic processes that extend through the zona pellucida at the abembryonic pole (Blandau 1949; Enders and Schlafke 1965, 1969; Parr 1973). In contrast, in *Elephantulus* oocytes, the zona pellucida is shed prior to the completion of the second meiotic division, coincident with the dispersion of the cumulus cells in the oviduct (van der Horst 1942). The lack of any extracellular coat in *Elephantulus* cleavage embryos apparently allows blastomere dissociation at the two- and four-cell stages, which in turn contributes to the high preimplantation mortality rate typical of this insectivore (van der Horst and Gillman 1942).

Other extracellular coats, or embryonic investments, are known in the rabbit and horse. Each rabbit oocyte and early-cleavage embryo receives an albuminous coat, or mucoprotein layer (Denker and Gerdes 1979; Denker 1981b), as it travels through the oviduct (Assheton 1894, 1895; Hammond 1934). In the uterus, the rabbit zona pellucida is dissolved completely but the mucoprotein coat remains (Denker 1981b). This coat is reinforced by the addition of a uterine secretion that congeals on its outer surface as the gloiolemma (Boving 1957, 1963; Davies and Hesseldahl 1971). The mucoprotein coat acquires an internally reinforcing layer as well, the neozona, which appears to be at least partly trophoblastic in origin (Denker and Gerdes 1979). The horse early-cleavage embryo has an outer gelatinous coat that imparts a roughened and bilaminar appearance to the underlying zona pellucida (Betteridge et al. 1982). This gelatinous coat is found in the newly ovulated oocyte (Hamilton and Day 1945; van Niekerk and Gerneke 1966) and becomes less prominent with time (Betteridge et al. 1982), suggesting that it is of oviductal origin (Flood et al. 1982). In addition, the horse embryo acquires a dense, acellular capsule in the uterus (Krolling 1937). This capsule is not an attenuated zona pellucida, but rather consists of material deposited between the trophoblast and the zona pellucida, which is eventually shed (Gygax et al. 1979; Betteridge et al. 1982).

The methaterian embryo possesses extracellular coats as well. The zona pellucida is the most internally located of these and forms while the oocyte is in the ovary (Hughes 1974). As the fertilized oocyte traverses the oviduct, a mucoid coat, or albumen layer, is deposited onto the outer surface of the zona pellucida. The mucoid coat varies in thickness among different species and like the zona pellucida, has a transitory existence, disappearing during gastrulation (Hughes 1974). The third extracellular coat of the marsupial embryo is a resilient shell (Hill 1910, 1918; Hartman 1916; McCrady 1938) formed from uterine secretions. This shell has the histochemical properties of the ovokeratin and ruptures in different species at different times during development, ranging from the early primitive streak to the fetal stage (McCrady 1938; Sharman 1961; Tyndale-Biscoe 1963; Ewer 1968; Renfree 1973; Hughes 1974).

CONCLUSIONS

The striking diversity of embryonic form among laboratory, domestic, and feral mammals raises the question of whether morphogenetic studies in any one species are relevant to such studies in other species. In particular, the differences between metatherian and eutherian embryos during cleavage and blastocyst formation suggest that different strategies exist for determining the first embryonic axis, dorso-ventral polarity, that is established by the animal-vegetal axis of the egg in amphibians, fish, reptiles, and birds (see Boyd and Hamilton 1952). The relationship between egg polarity and embryo polarity in some marsupials may suggest that similar mechanisms exist for determining their dorso-ventral axis as in nonmammalian vertebrate classes, although other explanations are possible (Selwood and Smith 1990). The available evidence for eutherian embryos provides no indication that embryonic polarity is determined in any way by egg polarity; rather, the pluripotent embryonic blastomeres appear to respond to epigenetic information in the form of their neighboring blastomeres, which determines the identity of trophectoderm and ICM progenitors by affecting cleavage planes. It is still unclear, however, why the ICM adheres to the trophectoderm at one pole of the blastocyst and not the other, or why the expanding eutherian blastocoel does not disperse the inner cells evenly around the inner surface of the trophoblast. The affinity of inner cells for each other is undoubtedly a significant factor in the morphogenesis of this cell population. Similarly, the peculiar egg cylinder morphology of rodent embryos after implantation may be explained by strong adherence between like cells of a lineage, in this case, the ectoplacental cone cells. The polar trophoblast cells of rabbit and several domestic species, in contrast, dehisce and move off the germinal disc, leaving the latter directly exposed as the outer embryonic cell layer, as it is in metatherian embryos, up to the time of amnion formation. Regardless of their basis, these anatomical differences between the embryos of rodents and other eutherian species show that early developmental events occur within distinct physical constraints (Snell and Stevens 1966; Kaufman 1990). This, in turn, indicates that the rodents have limited use as a morphogenetic model for domestic species during this period of development. It is in this regard that other eutherian mammalian embryos, such as those of the rabbit, as well as marsupial embryos, such as those of the opossum, may prove valuable. On the other hand, there is a remarkable similarity in the fates of extraembryonic and embryonic lineages in mouse and chick embryos. In both species, the hypoblast contributes descendants to extraembryonic, but not embryonic, structures, and the epiblast fate map reflects the temporal sequence of migration through the primitive streak. In both species, also, the anterior-posterior and left-right axes of the body plan are established during gastrulation and are organized by mesodermal descendants of migrating cells much as they are in amphibian embryos. In view of this strong evolutionary conservation of basic topological patterns of early embryogenesis within the vertebrates, it seems likely that cellular and molecular mechanisms for establishing the body plan would likewise be similar among these animals. These and other similarities compel us to conclude the following about embryos of eutherian mammals.

1. The pluripotency of cleavage-stage blastomeres is associated with their capacity to contribute descendants to both embryonic and extraembryonic cell lineages, up to the time that allocation of progenitors to trophectoderm and primitive endoderm occurs at the blastocyst stage.
2. The trophectoderm and primitive endoderm lineages contribute exclusively to extraembryonic structures, namely, the chorionic/trophoblastic ectoderm and the visceral/parietal yolk sac endoderm, respectively.
3. All other structures of the conceptus are derived from the epiblast lineage of the ICM, including the entire fetus, amniotic ectoderm and mesoderm, yolk sac and allantoic mesoderm, and chorionic mesoderm.
4. The fate of epiblast cells is determined during gastrulation and reflects the proximity of progenitors to the primitive streak, as well as other epigenetic factors whose identities are unknown. The gastrula fate map is probably evolutionarily conserved among mammals, because it is known to be conserved among amniotes and perhaps vertebrate species in general.
5. The functional consequences of genomic imprinting are unique to mammals and may reveal the identities of genes essential for differentiation and development of both extraembryonic and embryonic lineages. Analysis of genomic imprinting could contribute to an understanding of the etiology of genetic disease, chromosomal anomalies, and growth disturbances in mammals in general.

ACKNOWLEDGMENTS

We thank Ms. Liana Hartanto for her excellent assistance in preparing the manuscript. Work in the laboratory of Y.P.C. was supported by National Institutes of Health grant HD-24245; work in the laboratory of R.A.P. was supported by NIH grants HD-23651 and HD-25387, and by USDOE/OHER contract no. DE-AC03-76-SF01012.

REFERENCES

Allen, W.R. 1982. Embryo transfer in the horse. In *Mammalian egg transfer* (ed. C.E. Adams), p. 135. CRC Press, Boca Raton, Florida.

Allen, W.R., D.W. Hamilton, and R.M. Moor. 1973. The origin of equine endometrial cups. II. Invasion of the endometrium by trophoblast. *Anat. Rec.* **177:** 485.

Amoroso, E.C. 1952. Placentation. In *Marshall's physiology of reproduction*, 3rd edition (ed. A.S. Parkes), vol. II, p. 127. Longmans, Green, London.

Amoroso, E.C., W.F.B. Griffiths, and W.J. Hamilton. 1942. The early development of the goat (*Capra hircus*). *J. Anat.* **76:** 377.

Anderson, D. 1927. The rate of passage of the mammalian ovum through various portions of the fallopian tube. *Am. J. Physiol.* **82:** 557.

Anderson, L.L. 1978. Growth, protein content, and distribution of early pig embryos. *Anat. Rec.* **190:** 143.

Assheton, R. 1894. On the causes which lead to the attachment of the mammalian embryo to the walls of the uterus. *Q. J. Micros. Sci.* **37:** 173.

———. 1895. A re-investigation into the early stages of the development of the rabbit. *Q. J. Micros. Sci.* **37:** 113.

———. 1898a. The development of the pig during the first ten days. *Q. J. Micros. Sci.* **41:** 329.

———. 1898b. The segmentation of the ovum of the sheep, with observations on the hypothesis of a hypoblastic origin for the trophoblast. *Q. J. Micros. Sci.* **41:** 205.

———. 1906. VI. The morphology of the ungulate placenta, particularly the development of that organ in the sheep, and notes upon the placenta of the elephant and hyrax. *Philos. Trans. R. Soc. Lond.* **198:** 143.

Balakier, H. 1978. Induction of maturation in small oocytes from sexually immature mice by fusion with meiotic or mitotic cells. *Exp. Cell Res.* **112:** 137.

Balakier, H. and R.A. Pedersen. 1982. Allocation of cells to inner cell mass and trophectoderm lineages in preimplantation mouse embryos. *Dev. Biol.* **90:** 352.

Barends, P.M.G., H.W.J. Stroband, N. Taverne, G. teKronnie, M.P.J.M. Leen, and P.C.J. Blommers. 1989. Integrity of the preimplantation pig blastocyst during expansion and loss of polar trophectoderm (Rauber cells). *J. Reprod. Fertil.* **87:** 715.

Barlow, D.P., R. Stoger, B.G. Herrmann, K. Saito, and N. Schweifer. 1991. The mouse insulin-like growth factor type-2 receptor is imprinted and closely linked to the *Tme* locus. *Nature* **349:** 84.

Barton, S.C., M.A.H. Surani, and M.L. Norris. 1984. Role of paternal and maternal genomes in mouse development. *Nature* **311:** 374.

Bavister, B.D. 1987. Studies on the developmental blocks in cultured hamster embryos. In *The mammalian preimplantation embryo* (ed. B.D. Bavister), p. 219. Plenum Press, New York.

Beatty, R.A. 1967. Parthenogenesis in vertebrates. In *Fertilization* (ed. C.B. Metz and A. Monroy), vol. 1, p. 413. Academic Press, New York.

Beddington, R.S.P. 1981. An autoradiographic analysis of the potency of embryonic ectoderm in the 8th day postimplantation mouse embryo. *J. Embryol. Exp. Morphol.* **64:** 87.

———. 1986. Analysis of tissue fate and prospective potency in the egg cylinder. In *Experimental approaches to mammalian development* (ed. J. Rossant and R.A. Pedersen), p. 121. Cambridge University Press, United Kingdom.

Benos, D.J. 1981. Developmental changes in epithielial transport characteristics of preimplantation rabbit blastocysts. *J. Physiol.* **316:** 191.

Benos, D.J. and J.D. Biggers. 1981. Blastocyst fluid formation. In *Fertilization and embryonic development in vitro* (ed. L. Mastroianni and J.D. Biggers), p. 287. Plenum Press, New York.

Benos, D.J., J.D. Biggers, R.S. Balaban, J.W. Mills, and E.W. Overstrom. 1985. Developmental aspects of sodium-dependent transport processes of pre-implantation rabbit embryos. In *Regulation and development of membrane transport processes* (ed. J.S. Graves), p. 211. Wiley, New York.

Betteridge, K.J. and J.-E. Flechon. 1988. The anatomy and physiology of pre-attachment bovine embryos. *Theriogenology* **29:** 155.

Betteridge, K.J., M.D. Eaglesome, G.C.B. Randall, and D. Mitchell. 1980. Collection, description and transfer of embryos from cattle 10–16 days after estrus. *J. Reprod. Fertil.* **59:** 205.

Betteridge, K.J., M.D. Eaglesome, D. Mitchell, P.F. Flood, and R. Beriault. 1982. Development of horse embryos up to twenty two days after ovulation: Observations on fresh specimens. *J. Anat.* **135:** 191.

Biggers, J.D. 1972. Mammalian blastocyst and amnion formation. In *The water metabolism of the fetus* (ed. A.C. Barnes), p. 3. Charles C. Thomas, Springfield, Illinois.

———. 1987. Pioneering mammalian embryo culture. In *The mammalian preimplantation embryo: Regulation of growth and differentiation in vitro* (ed. B.D. Bavister), p. 1. Plenum Press, New York.

Biggers, J.D., J.E. Bell, and D.J. Benos. 1988. Mammalian blastocyst: Transport functions in a developing epithelium. *Am. J. Physiol.* **255:** C419.

Bindon, B.M. 1971. Systematic study of preimplantation stages of pregnancy in the sheep. *Aust. J. Biol. Sci.* **24:** 131.

Bjorkman, N.H. 1969. Light and electron microscopic studies on cellular alterations in the normal bovine placentome. *Anat. Rec.* **163:** 17.

Blandau, R.J. 1949. Observations on implantation of the guinea pig ovum. *Anat. Rec.* **103:** 19.

Bleil, J.P. and P.M. Wassarman. 1980. Mammalian sperm-egg interaction: Identification of a glycoprotein in mouse egg zonae pellucidae possessing receptor activity for sperm. *Cell* **20:** 873.

Bluntschli, H. 1938. Le developpement primaire et l'implantation chez un Centetine (Hemicentetes). *C.R. Assoc. Anat.* **44:** 39.

Borland, R.M. 1977. Transport processes in the mammalian blastocyst. In *Development in mammals* (ed. M.H. Johnson), vol. 1, p. 31. North-Holland Publishing, Amsterdam.

Boving, B. 1957. Rabbit egg coverings. *Anat. Rec.* **127:** 270.

———. 1959. Implantation. *Ann. N.Y. Acad. Sci.* **74:** 700.

———. 1963. Implantation mechanisms. In *Mechanisms concerned with conception* (ed. C.G. Hartman), p. 321. Macmillan, New York.

Boyd, J.D. and W.J. Hamilton. 1952. Cleavage, early development and implantation of the egg. In *Marshall's physiology of reproduction*, 3rd edition (ed. A.S. Parkes), vol. II, p. 1. Longmans, Green, London.

Brackett, B.G. 1978. Experimentation involving primate embryos. In *Methods in mammalian reproduction* (ed. J.C. Daniel, Jr.), p. 333. Academic Press, New York.

Calarco, P.G. and A. McLaren. 1976. Ultrastructural observations of preimplantation stages of the sheep. *J. Embryol. Exp. Morphol.* **36:** 609.

Carnegie, J.A., M.E. McCully, and H.A. Robertson. 1985. The early development of the sheep trophoblast and the involvement of cell death. *Am. J. Anat.* **174:** 471.

Cattanach, B.M. 1986. Parental origin effects in mice. *J. Embryol. Exp. Morphol.* **97:** 137.

Chang, M.C. 1952. Development of bovine blastocyst with a note on implantation. *Anat. Rec.* **113:** 143.

Chang, M.C. and L.E.A. Rowson. 1965. Fertilization and early development of dorset horn sheep in the spring and summer. *Anat. Rec.* **152:** 303.

Chisholm, J.C. 1988. Analysis of the fifth cell cycle of the mouse development. *J. Reprod. Fertil.* **84:** 29.

Clark, R.T. 1934. Studies on the physiology of reproduction in the sheep. *Anat. Rec.* **60:** 135.

Clarke, H.J., S. Varmuza, V.R. Prideaux, and J. Rossant. 1988. The developmental potential of parthenogenetically derived cells in chimeric mouse embryos: Implications for action of imprinted genes. *Development* **104:** 175.

Copp, A.J. 1979. Interaction between inner cell mass and trophectoderm of the mouse blastocyst. II. The fate of the polar trophectoderm. *J. Embryol. Exp. Morphol.* **51:** 109.

Copp, A.J. and D.C. Cockroft. 1990. *Postimplantation embryos: A practical approach*. IRL Press, Oxford, United Kingdom.

Corner, G.W. 1921. Cyclic changes in the ovaries and uterus of the sow, and their relation to the mechanism of implantation. *Carnegie Inst. Contrib. Embryol.* **13:** 117.

Cruz, Y.P. 1990. Mosaicism in the mouse trophectoderm. *Tissue & Cell* **22:** 103.

Cruz, Y.P. and R.A. Pedersen. 1985. Cell fate in the polar trophectoderm of mouse blastocysts as studied by microinjection of cell lineage tracers. *Dev. Biol.* **112:** 73.

Damsky, C. 1989. Cell-to-cell contact. *Curr. Opinion Cell Biol.* **1:** 881.

Damsky, C., C. Richa, D. Solter, K. Knudsen, and C.A. Buck. 1983. Identification and purification of a cell surface glycoprotein mediating intercellular adhesion in embryonic and adult tissue. *Cell* **34:** 455.

Daniel, J.C., Jr. 1963. Some kinetics of blastocyst formation as studied by the process of reconstitution. *J. Exp. Zool.* **154:** 231.

———. 1964. Early growth of rabbit trophoblast. *Am. Nat.* **98:** 85.

Davies, J. and H. Hesseldahl. 1971. Comparative embryology of mammalian blastocysts. In *The biology of the blastocyst* (ed. R.J. Blandau), p. 27. University of Chicago Press, Illinois.

Davies, J. and W. Wimsatt. 1966. Observations on the fine structure of the sheep placenta. *Acta Anat.* **65:** 182.

DeChiara, T.M., A. Efstratiadis, and E.J. Robertson. 1990. A growth-deficiency phenotype in heterozygous mice carrying an insulin-like growth factor II gene disrupted by targeting. *Nature* **345:** 78.

DeChiara, T.M., E.J. Robertson, and A. Efstratiadis. 1991. Parental imprinting of the mouse insulin-like growth factor II gene. *Cell* **64:** 849.

Denker, H.-W. 1981a. The determination of trophoblast and embryoblast cells during cleavage in the mammal: New trends in the interpretation of the mechanisms. *Verh. Anat. Ges.* **75:** 445.

———. 1981b. Proteinases and implantation. *J. Reprod. Fertil.* (suppl.) **29:** 183.

Denker, H.-W. and H.-J. Gerdes. 1979. The dynamic structure of rabbit blastocyst coverings. I. Transformation during regular preimplantation development. *Anat. Embryol.* **157:** 15.

Dickson, A.D. 1966. The form of the mouse blastocyst. *J. Anat.* **100:** 335.

Ducibella, T. and E. Anderson. 1975. Cell shape and membrane changes in the eight-cell mouse embryo; prerequisites for morphogenesis of the blastocyst. *Dev. Biol.* **47:** 45.

Dyce, J., M. George, and T.P. Fleming. 1987. Do trophectoderm and inner cell mass cells in the mouse blastocyst maintain discrete lineages? *Development* **100:** 685.

Edwards, R.G. 1980. *Conception in the human female.* Academic Press, New York.

Enders, A.C. 1971. The fine structure of the blastocyst. In *The biology of the blastocyst* (ed. R.J. Blandau), p. 71. University of Chicago Press, Illinois.

———. 1976. Anatomical aspects of implantation. *J. Reprod. Fertil.* (suppl.) **25:** 1.

Enders, A.C. and S.J. Schlafke. 1965. The fine structure of the blastocyst: Some comparative studies. In *Ciba symposium on preimplantation stages of pregnancy* (ed. G.E.W. Wolstenholine and M. O'Connor), p. 29. J. & A. Churchill, London.

———. 1969. Cytological aspects of trophoblast-uterine interaction in early implantation. *J. Anat.* **125:** 1.

Enders, A.C., S. Schlafke, N.E. Hubbard, and R.A. Mead. 1986. Morphological changes in the blastocyst of the western spotted skunk during activation from delayed implantation. *Biol. Reprod.* **34:** 423.

Enders, A.C., K.C. Lantz, I.K.M. Liu, and S. Schlafke. 1988. Loss of polar trophoblast during differentiation of the blastocyst of the horse. *J. Reprod. Fertil.* **83:** 447.

Ewer, R.F. 1968. A preliminary survey of the behavior in captivity of the dasyurid marsupial, *Sminthopsis crassicaudata* (Gould). *Z. Tierpsychol.* **25:** 319.

First, N.L. and F.L. Barnes. 1989. Development of preimplantation mammalian embryos. In *Development of preimplantation embryos and their environment* (ed. K. Yoshinaga and T. Mori), p. 151. Alan R. Liss, New York.

Flechon, J.-E. 1978. Morphological aspects of embryonic disc at the time of its appearance in the blastocyst of farm animals. In *Scanning electron microscopy* (ed. R.P. Becker and O. Johari), vol. II, p. 541. SEM, O'Hare, Illinois.

Flechon, J.-E. and R.B.L. Gwatkin. 1980. Immunocytochemical studies on the zona pellucida of cow blastocysts. *Gamete Res.* **3:** 141.

Flechon, J.-E., M. Guillomot, M. Charlier, B. Flechon, and J. Martal. 1986. Experimental studies on the elongation of the ewe blastocyst. *Reprod. Nutr. Dev.* **26:** 1017.

Fleming, M.W. and J.D. Harder. 1981. Uterine histology and reproductive cycles in pregnant and nonpregnant opossums, *Didelphis virginiana. J. Reprod. Fertil.* **63:** 21.

Fleming, T. 1987. A quantitative analysis of cell allocation to trophectoderm and inner cell mass in the mouse blastocyst. *Dev. Biol.* **119:** 520.

Fleming, T.P. and M.H. Johnson. 1988. From egg to epithelium. *Annu. Rev. Cell Biol.* **4:** 459.

Fleming, T.P. and S.J. Pickering. 1985. Maturation and polarization of the endocytotic system in outside blastomeres during mouse preimplantation development. *J. Embryol. Exp. Morphol.* **89:** 175.

Fleming, T.P., P. Cannon, S.J. Pickering. 1986. The cytoskeleton, endocytosis and cell polarity in the mouse preimplantation embryo. *Dev. Biol.* **113:** 406.

Fleming, T.P., J. McConnell, M.H. Johnson, and B.R. Stevenson.

1989. Development of tight junctions *de novo* in the mouse early embryo: Control of assembly of the tight-junction-specific protein, ZO-1. *J. Cell Biol.* **108:** 1407.

Flood, P.F., K.J. Betteridge, and M.S. Diocee. 1982. Transmission electron microscopy of horse embryos 3–16 days after ovulation. *J. Reprod. Fert.* (suppl.) **32:** 319.

Florman, H.M. and P.M. Wassarman. 1985. O-linked oligosaccharides of mouse ZP-3 account for its sperm receptor activity. *Cell* **41:** 313.

Fundele, R., M.L. Norris, S.C. Barton, W. Reik, and M.A. Surani. 1989. Systematic elimination of parthenogenetic cells in mouse chimeras. *Development* **106:** 29.

Gallin, W.J., G.M. Edelman, and B.A. Cunningham. 1983. Characterization of L-CAM, a major cell adhesion molecule from liver cells. *Proc. Natl. Acad. Sci.* **80:** 1038.

Garbutt, C.L., J.C. Chisholm, and M.H. Johnson. 1987. The establishment of the embryonic-abembryonic axis in the mouse embryo. *Development* **100:** 125.

Gardner, R.L. 1982. Investigation of cell lineage and differentiation in the extraembryonic endoderm of the mouse embryo. *J. Embryol. Exp. Morphol.* **68:** 175.

Gardner, R.L. and V.E. Papaioannou. 1975. Differentiation in the trophectoderm and inner cell mass. In *The early development of mammals* (ed. M. Balls and A.E. Wild), p. 107. Cambridge University Press, United Kingdom.

Gardner, R.L. and J. Rossant. 1979. Investigation of the fate of 4.5 day *post coitum* mouse inner cell mass cells by blastocyst injection. *J. Embryol. Exp. Morphol.* **52:** 141.

Gardner, R.L., V.E. Papaioannou, and S.C. Barton. 1973. Origin of the ectoplacental cone and secondary giant cells in mouse blastocysts reconstituted from isolated trophoblast and inner cell mass. *J. Embryol. Exp. Morphol.* **30:** 561.

Gatica, R., M.P. Borland, T.F. Crosby, and I. Gordon. 1984. Micromanipulation of sheep morulae to produce monozygotic twins. *Theriogenology* **21:** 555.

Geisert, R.D., J.W. Brookbank, R.M. Roberts, and F.W. Bazer. 1982. Establishment of pregnancy in the pig. II. Cellular remodeling of the porcine blastocyst during elongation on day 12 of pregnancy. *Biol. Reprod.* **27:** 941.

Goetz, R.H. 1937. Studien zur placentation der centetiden. II. Die implantation und fruhentwicklung von *Hemicentetes semispinosus* (Cuvier). *Z. Anat. Entrwickl.* **107:** 274.

Grant, S.G. and V.M. Chapman. 1988. Mechanisms of X-chromosome regulation. *Annu. Rev. Genet.* **22:** 199.

Green, W.W. and L.M. Winters. 1945. Prenatal development of the sheep. *Univ. Minn, Agric. Expt. Sta. Tech. Bull.* **169:** 1.

———. 1946. Cleavage and attachment stages of the pig. *J. Morphol.*

78: 305.

Greenstein, J., R. Murray, and R. Foley. 1958. Observations on the morphogenesis and biochemistry of the bovine pre-attachment placenta between 16 and 33 days of gestation. *Anat. Rec.* **132:** 321.

Guillomot, M., J.-E. Flechon, and S. Wintenberger-Torres. 1978. Cellular contact between the trophoblast and the endometrium at implantation in the ewe. In *Cellular and molecular aspects of implantation* (ed. S.R. Glaser and D.W. Bullock), p. 446. Plenum Press, New York.

Gygax, A.P., V.K Gangam, and R.M. Kenney. 1979. Clinical, microbiological and histological changes associated with uterine involution in the mare. *J. Reprod. Fertil.* (suppl.) **27:** 571.

Hafez, E.S.E. 1972. Differentiation of mammalian blastocysts. In *Biology of mammalian fertilization and implantation* (ed. K.S. Moghissi and E.S.E. Hafez), p. 296. Charles C. Thomas, Springfield, Illinois.

———. 1973. Gamete transport. In *Human reproduction* (ed. E.S.E. Hafez and T.N. Evans), p. 85. Harper and Row, Hagerstown, Maryland.

Haig, D. and C. Graham. 1991. Genomic imprinting and the strange case of the insulin-like growth factor II receptor. *Cell* **64:** 1045.

Hall, J.G. 1990. Genomic imprinting: Review and relevance to human diseases. *J. Hum. Genet.* **46:** 857.

Hamilton, W.J. and F.T. Day. 1945. Cleavage stages of the ova of the horse, with notes on ovulation. *J. Anat.* **79:** 127.

Hamilton, W.J. and J.A. Laing. 1946. Development of the egg of the cow up to the stage of blastocyst formation. *J. Anat.* **80:** 194.

Hammond, J. 1934. The fertilization of rabbit ova in relation to time: A method of controlling the litter size, the duration of pregnancy and the weight of young at birth. *J. Exp. Biol.* **11:** 40.

Handyside, A.H. 1980. Distribution of antibody- and lectin-binding sites on dissociated blastomeres from mouse morulae: Evidence of polarization at compaction. *J. Embryol. Exp. Morphol.* **60:** 99.

Hare, W.C.D., E.L. Singh, K.J. Betteridge, M.D. Eaglesome, D. Mitchell, R.J. Breton, and A.O. Trounson. 1980. Chromosomal analysis of 159 bovine embryos collected 12–18 days after oestrus. *Can. J. Genet. Cytol.* **22:** 615.

Hartman, C.G. 1916. Studies on the development of the opossum, *Didelphys virginiana*. I. History of early cleavage. II. Formation of the blastocyst. *J. Morphol.* **27:** 1.

———. 1919. Studies on the development of the opossum, *Didelphys virginiana*. III. Description of new material on maturation, cleavage and endoderm formation. IV. The bilaminar blastocyst. *J. Morphol.* **32:** 1.

———. 1923. The oestrus cycle in the opossum. *Am. J. Anat.* **32:** 353.

———. 1928. The breeding season of the opossum (*Didelphys vir-*

giniana) and the rate of intra-uterine and post-natal development. *J. Morphol.* **46:** 143.

Hedrick, J.L. and N.J. Wardrip. 1987. On the macromolecular composition of the zona pellucida from porcine oocytes. *Dev. Biol.* **121:** 478.

Hertig, A.T. and J. Rock. 1945. Two human ova of the previllous stage, having a developmental age of about seven and nine days, respectively. *Carnegie Inst. Contrib. Embryol.* **31:** 65.

Hertig, A.T., J. Rock, and E.C. Adams. 1956. A description of thirty-four human ova within the first seventeen days of development. *Am. J. Anat.* **98:** 435.

Heuser, C.H. 1927. A study of the implantation of the ovum of the pig from the stage of the bilaminar blastocyst to the completion of the fetal membranes. *Carnegie Inst. Contrib. Embryol.* **719:** 229.

Heuser, C.H. and G.L. Streeter. 1929. Early stages in the development of pig embryos, from the period of initial cleavage to the time of the appearance of limb-buds. *Carnegie Inst. Contrib. Embryol.* **20:** 1.

———.1941. Development of the macaque embryo. *Carnegie Inst. Contrib. Embryol.* **29:** 15.

Hill, J.P. 1910. The early development of the marsupialia, with special reference to the native cat (*Dasyurus viverrinus*). Contribution to the embryology of the marsupialia, IV. *Q. J. Micros. Sci.* **56:** 1.

———. 1918. Some observations on the early development of *Didelphys aurita*. (Contribution to the embryology of the marsupialia, V.). *Q. J. Micros. Sci.* **63:** 91.

Hogan, B.L.M. and R. Tilly. 1981. Cell interactions and endoderm differentiation in cultured mouse embryos. *J. Embryol. Exp. Morphol.* **62:** 379.

Hogan, B.L.M., A.R. Cooper, and M. Kurkinen. 1980. Incorporation into Reichert's membrane of laminin-like extracellular proteins synthesized by parietal endoderm cells of the mouse embryo. *Dev. Biol.* **80:** 289.

Hogan, B., F. Costantini, and E. Lacy. 1986. *Manipulating the mouse embryo: A laboratory manual*. Cold Spring Harbor Laboratory, New York.

Holst, P.J. and R.D. Phemister. 1971. The prenatal development of the dog: Preimplantation events. *Biol. Reprod.* **5:** 194.

Houliston, E., S.J. Pickering, and B. Maro. 1987. Redistribution of microtubules and pericentriolar material during the development of polarity in mouse blastomeres. *J. Cell Biol.* **104:** 1299.

Hughes, R.L. 1974. Morphological studies on implantation in marsupials. *J. Reprod. Fertil.* **39:** 173.

Hunter, R.H.F. 1974. Chronological and cytological details of fertilization and early embryonic development in the domestic pig, *Sus scrofa*. *Anat. Rec.* **178:** 169.

———. 1977. Physiological factors influencing ovulation, fertilization, early embryonic development and establishment of pregnancy in pigs. *Brit. Vet. J.* **133**:461.

Hyafil, F., C. Babinet, and F. Jacob. 1981. Cell-cell interaction in early embryogenesis: A molecular approach to the role of calcium. *Cell* **26:** 447.

Hyafil, F., D. Morello, C. Babinet, and F. Jacob. 1980. A cell surface glycoprotein involved in compaction of embryonal carcinoma cells and cleavage stage embryos. *Cell* **21:** 927.

Johnson, M.H. and B. Maro. 1984. The distribution of cytoplasmic actin in mouse 8-cell blastomeres. *J. Embryol. Exp. Morphol.* **82:** 97.

———. 1985. A dissection of the mechanisms generating and stabilising polarity in mouse 8- and 16-cell blastomeres; the role of cytoskeletal elements. *J. Embryol. Exp. Morphol.* **90:** 311.

———. 1986. Time and space in the early embryo: A cell biological approach to cell diversification. In *Experimental approaches to mammalian embryonic development* (ed. J. Rossant and R.A. Pedersen), p. 35. Cambridge University Press, United Kingdom.

Johnson, M.H. and M. Takeichi. 1986. The role of cell adhesion in the synchronization and orientation of polarization in 8-cell mouse blastomeres. *J. Embryol. Exp. Morphol.* **93:** 239.

Johnson, M.H. and C.A Ziomek. 1981. Induction of polarity in mouse 8-cell blastomeres: Specificity, geometry and stability. *J. Cell. Biol.* **91:** 431.

Johnson, M.H., B. Maro, and M. Takeichi. 1986. The role of cell adhesion in the synchronization and orientation of polarization in 8-cell mouse blastomeres. *J. Embryol. Exp. Morphol.* **93:** 239.

Kaslow, D.C. and B.R. Migeon. 1987. DNA methylation stabilizes X chromosome inactivation in eutherians but not in marsupials: Evidence for multistep maintenance of mammalian X dosage compensation. *Proc. Natl. Acad Sci.* **84:** 6210.

Kaufman, M.H. 1990. Morphological stages of postimplantation embryonic development. In *Postimplantataion mammalian embryos: A practical approach* (ed. A.J. Copp and D.L. Cockroft), p. 81. IRL Press, Oxford.

Kemler, R., B. Babinet, H. Eisen, and F. Jacob. 1987. Surface antigen in early differentiation. *Proc. Natl. Acad. Sci.* **74:** 4449.

King, G.J., B.A. Atkinson, and H.A. Robertson. 1979. Development of the bovine placentome during the second month of gestation. *J. Reprod. Fertil.* **53:** 173.

———. 1982. Implantation and early placentation in domestic ungulates. *J. Reprod. Fertil.* (suppl.) **31:** 17.

Koelliker, O. 1880. Die Entwicklung der Keimblatter des Kanischens. *Zool. Anz.* **3:** 370.

Krolling, O. 1937. Uber eine Keimblase in Stadium der Gastrula beim Pferd. *Z. Mikrosk-anat. Forsch.* **42:** 124.

Lambeth, V.A., C.R. Looney, S.A. Vrelkel, D.A. Jackson, K.G. Hill, and

R.A. Godke. 1983. Microsurgery on bovine embryos at the morula stage to produce monozygotic twin calves. *Theriogenology* **20:** 85.

Lawn, A.M., A.D. Chiquoine, and E.C. Amoroso. 1969. The development of the placenta in the sheep and goat: An electron microscope study. *J. Anat.* **105:** 557.

Lawson, K.A. and R.A. Pedersen. 1987. Cell fate, morphogenetic movement and population kinetics of embryonic endoderm at the time of germ layer formation in the mouse. *Development* **101:** 627.

Lawson, K.A., J.J. Meneses, and R.A. Pedersen. 1966. Cell fate and cell lineage in the endoderm of the presomite mouse embryo, studied with an intracellular tracer. *Dev. Biol.* **115:** 325.

Lewis, W.H. and E.S. Wright. 1935. On the early development of the mouse egg. *Carnegie Inst. Contrib. Embryol.* **25:** 113.

Lock, L.F., N. Takagi, and G.R. Martin. 1987. Methylation of the *Hprt* gene on the inactive X occurs after chromosome inactivation. *Cell* **48:** 39.

Lois, P. and L. Izquierdo. 1984. Cell membrane regionalization and cytoplasm polarization in the rat early embryo. *Roux's Arch. Dev. Biol.* **193:** 205.

Lutwak-Mann, C., J.B. Boursnell, and J.P. Bennett. 1960. Blastocyst-uterine relationship: Uptake of radioactive ions by the early rabbit embryo and its environment. *J. Reprod. Fertil.* **1:** 169.

Lyne, A.G. and D.E. Hollis. 1976. Early embryology of the marsupials *Isoodon macrourus* and *Perameles nasuta*. *Aust. J. Zool.* **24:** 361.

———. 1977. The early development of marsupials with special reference to bandicoots. In *Reproduction and evolution* (ed. J.N. Calaby and C.H. Tyndale-Biscoe), p. 293. Australian Academy of Science, Canberra.

Mann, J.R., I. Gadi, M.L. Harbison, S.J. Abbondanzo, and C.L. Stewart. 1990. Androgenetic mouse embryonic stem cells are pluripotent and cause skeletal defects in chimeras: Implications for genetic imprinting. *Cell* **62:** 251.

Maro, B., J. Kubiak, C. Gueth, H. De Pennart, E. Houliston, M. Weber, C. Antony, and J. Aghion. 1990. Cytoskeletal organization during oogenesis, fertilization and preimplantation development of the mouse. *Int. J. Dev. Biol.* **34:** 127.

Massip, A., W. Zwijsen, and J. Mulnard. 1983. Cinematographic analysis of the cleavage of the cow egg from 2-cell to 16-cell stage. *Arch. Biol.* **94:** 99.

Mattson, B.A., E.W. Overstrom, and D.F. Albertini. 1990. Transitions in trophectoderm cellular shape and cytoskeletal organization in the elongating pig blastocyst. *Biol. Reprod.* **42:** 195.

McConnell, J. and M. Lee. 1989. Presence of cdc 2(+)-like proteins in the preimplantation mouse embryo. *Development* **107:** 481.

McCrady, E. 1938. The embryology of the opossum. *Am. Anat. Mem.* **16:** 1.

McGrath and D. Solter. 1983. Nuclear transplantation in the mouse embryo by microsurgery and cell fusion. *Science* **220:** 1300.

McNeill, H., M. Ozawa, R. Kemler, and W.J. Nelson. 1990. Novel function of the cell adhesion molecule uvomorulin as an inducer of cell surface polarity. *Cell* **62:** 309.

Monk, M., M. Boubelik, and S. Lehnert. 1987. Temporal and regional changes in DNA methylation in the embryonic, extraembryonic and germ cell lineages during mouse embryo development. *Development* **99:** 371.

Moor, R.N., A.M. Glew, and F. Gandolfi. 1987. Cell cycle modification in the early ovine embryo. *Proc. Soc. Study Fertil.* (Abstr. 31).

Mootz, U. 1979. Rasterelektronenmikroskopische untersuchungen an der implantations bereiten Blastozyste des Kaninchens. *Verh. Anat. Ges.* **73:** 435.

Morgan, G. and F.B.P. Wooding. 1983. Cell migration in the ruminant placenta: A freeze-fracture study. *J. Ultrastruct. Res.* **83:** 148.

Morriss, G. 1975. Placental evolution and embryonic nutrition. In *Comparative placentation: Essays in structure and function* (ed. D.H. Steven), p. 87. Academic Press, New York.

Mossmann, H.W. 1937. Comparative morphogenesis of the fetal membranes and accessory uterine structures. *Carnegie Inst. Contrib. Embryol.* **26:** 129.

Nagy, A., M. Sass, and M. Markkula. 1989. Systematic non-uniform distribution of parthenogenetic cells in adult mouse chimaeras. *Development* **106:** 321.

New, D.A.T. and M. Mizell. 1972. Opossum fetuses grown in culture. *Science* **75:** 533.

Newport, J. and M. Kirschner. 1982. A major developmental transition in early *Xenopus* embryos. II. Control of the onset of transcription. *Cell* **30:** 687.

Nichols, J. and R.L. Gardner. 1984. Heterogeneous differentiation of external cells in individual isolated mouse inner cell masses in culture. *J. Embryol. Exp. Morphol.* **80:** 225.

Nicolet, G. 1971. Avian gastrulation. *Adv. Morphog.* **9:** 231.

O'Gorman, S., D.T. Fox, and G.M. Wahl. 1991. Recombinase-mediated gene activation and site-specific integration in mammalian cells. *Science* **251:** 1351.

Overstrom, E. 1987. In vitro assessment of blastocyst differentiation. In *The mammalian preimplantation embryo: Regulation of growth and differentiation in vitro* (ed. B.D. Bavister), p. 95. Plenum Press, New York.

Oxenreider, S.L. and B.N. Day. 1965. Transport and cleavage of ova in swine. *J. Anim. Sci.* **24:** 413.

Ozil, J.P. 1983. Production of identical twins by bisection of blastocysts in the cow. *J. Reprod. Fertil.* **69:** 463.

Ozil, J.P., Y. Heyman, and J.P. Reynaud. 1982. Production of monozygotic twins by micromanipulation and cervical transfer in the cow. *Vet. Rec.* **110:** 126.

Padykula, H.A and J.M. Taylor. 1976. Ultrastructural evidence for the loss of the trophoblastic layer in the chorio-allantoic placenta of Australian bandicoots (Marsupialia: Peramelidae). *Anat. Rec.* **186:** 357.

Paldi, A., A. Nagy, M. Markkula, I. Barna, and L. Dezso. 1989. Postnatal development of parthenogenetic<—>fertilized mouse aggregation chimeras. *Development* **105:** 115.

Papaioannou, V.E. 1982. Lineage analysis of inner cell mass and trophectoderm using microsurgically reconstituted mouse blastocysts. *J. Embryol. Exp. Morphol.* **68:** 119.

Papaioannou, V.E. and K.M. Ebert. 1988. The preimplantation pig embryo: Cell number and allocation to trophectoderm and inner cell mass of the blastocyst *in vivo* and *in vitro. Development* **102:** 793.

Parr, E.L. 1973. Shedding of the zona pellucida by guinea pig blastocysts: An ultrastructural study. *Biol. Reprod.* **8:** 531.

Parr, M.B. and E.L. Parr. 1989. The implantation reaction. In *Biology of the uterus*, 2nd edition (ed. R.M. and W.P. Jollie), p. 233. Plenum Press, New York.

Pedersen, R.A. 1986. Potency, lineage, and allocation in preimplantation mouse embryos. In *Experimental approaches to mammalian embyonic development* (ed. J. Rossant and R.A Pedersen), p. 3. Cambridge University Press, United Kingdom.

———. 1988. Early mammalian embryogenesis. In *The physiology of reproduction* (ed. E. Knobil et al.), p. 187. Raven Press, New York.

Pedersen, R.A., K. Wu, and H. Balakier. 1986. Origin of the inner cell mass in mouse embryos: Cell lineage analysis by microinjection. *Dev. Biol.* **117:** 581.

Perry, J.S. 1981. The mammalian fetal membranes. *J. Reprod. Fertil.* **62:** 321.

Perry, J.S. and I.W. Rowlands. 1962. Early pregnancy in the pig. *J. Reprod. Fertil.* **4:** 175.

Peyrieras, N., F. Hyafil, D. Louvard, N.D. Ploegh, and F. Jacob. 1983. Uvomorulin: A nonintegral membrane protein of early mouse embryos. *Proc. Natl. Acad. Sci.* **80:** 6247.

Pratt, H.P. 1989. Marking time and making shape: Chronology and topography in the early mouse embryo. *Int. Rev. Cytol.* **117:** 99.

Pratt, H.P. and A.C. Muggleton-Harris. 1988. Cycling cytoplasmic factors that promote mitosis in the cultural 2-cell mouse embryo. *Development* **104:** 115.

Rauber. 1875. Die erste entwicklung des kaninchens. *Sitzungber Naturfor. Gesell. Leipzig* No. 10, p. 103.

Reeve, W.J.D. 1981. Cytoplasmic polarity develops at compaction in rat and mouse embryos. *J. Embryol. Exp. Morphol.* **62:** 351.

Reik, W. 1989. Genomic imprinting and genetic disorders in man. *Hum. Genet. Dis.* **5:** 331.

Renfree, M.B. 1982. Implantation and placentation. In *Reproduction in mammals*, Book 2, *Embryonic and fetal development*, 2nd edition (ed. C.R. Austin and R.V. Short), p. 26. Cambridge University Press, United Kingdom.

Rossant, J. 1986. Development of extraembryonic cell lineages in the mouse embryo. In *Experimental approaches to mammalian embryonic development* (ed. J. Rossant and R.A. Pedersen), p. 97. Cambridge University Press, United Kingdom.

Rossant, J. and B.A. Croy. 1985. Genetic identification of tissue of origin of cellular populations within the mouse placenta. *J. Embryol. Exp. Morphol.* **86:** 177.

Rossant, J. and W. Tamura-Lis. 1981. Effect of culture conditions on diploid to giant-cell transformation in postimplantation mouse trophoblast. *J. Embyol. Exp. Morphol.* **62:** 217.

Rossant, J., R.L. Gardner, and H.L. Alexandre. 1978. Investigation of the potency of cells from the postimplantation mouse embryo by blastocyst injection: A preliminary report. *J. Embryol. Exp. Morphol.* **48:** 239.

Rossant, J., M. Vijh, L.D. Siracusa, and V.M. Chapman. 1983. Identification of embryonic cell lineages in histological sections of *M. musculus<—>M. caroli* chimaeras. *J. Embryol. Exp. Morphol.* **73:** 179.

Rowson, L.E.A. and R.M. Moor. 1966. Development of the sheep conceptus during the first fourteen days. *J. Anat.* **100:** 777.

Samuel, C.A. and J.S. Perry. 1972. The ultrastructure of pig trophoblast transplanted to an ectopic site in the uterine wall. *J. Anat.* **113:** 139.

Sanford, J.P., V.M. Chapman, and J. Rossant. 1985. DNA methylation in extraembryonic lineages of mammals. *Trends Genet.* **1:** 89.

Sanford, J.P., H.J. Clark, V.M. Chapman, and J. Rossant. 1967. Differences in DNA methylation during oogenesis and spermatogenesis and their persistence during early embryogenesis in the mouse. *Genes Dev.* **1:** 1039.

Sapienza, C., J. Paquette, T.H. Tran, and A. Peterson. 1969. Epigenetic and genetic factors affect transgene methylation imprinting. *Development* **107:** 165.

Schlafke, S. and A.C. Enders. 1975. Cellular basis of interaction between trophoblast and uterus at implantation. *Biol. Reprod.* **12:** 41.

Schultz, G.A 1986. Utilization of genetic information in the preimplantation mouse embryo. In *Experimental approaches to mammalian embryonic development* (ed. J. Rossant and R.A. Pedersen), p. 239. Cambridge University Press, United Kingdom.

Searle, A.G. and C.V. Beechey. 1978. Complementation studies with mouse translocations. *Cytogenet. Cell Genet.* **20:** 282.

———. 1985. Noncomplementation phenomena and their bearing on nondisjunctional effects. In *Aneuploidy, aetiology and mechanisms* (ed. V.L. Dellarco et al.), p. 363. Plenum Press, New York.

Searle, A.G., J. Peters, M.F. Lyon, J.G. Hall, E.P. Evans, J.N. Edwards, and V.J. Buckle. 1989. Chromosome maps of man and mouse. IV. *Ann. Hum. Genet.* **53:** 89.

Selwood, L. 1980. A timetable of embryonic development of the dasyurid marsupial *Antechinus stuartii* (Macleay). *Aust. J. Zool.* **28:** 649.

———. 1981. Delayed embryonic development in the dasyurid marsupial, *Antechinus stuartii. J. Reprod. Fertil.* (suppl.) **29:** 79.

———. 1986a. Cleavage in vitro following destruction of some blastomeres in the marsupial *Antechinus stuartii* (Macleay). *J. Embryol. Exp. Morphol.* **92:** 71.

———. 1986b. The marsupial blastocyst—A study of the blastocysts in the Hill Collection. *Aust. J. Zool.* **34:** 177.

———. 1987. Embryonic development in culture of two dasyurid mammals, *Sminthopsis crassicaudata* (Gould) and *Sminthopsis macroura* (Spencer) during cleavages and blastocyst formation. *Gamete Res.* **16:** 355.

———. 1989a. Marsupial pre-implantation embryos in vivo and in vitro. In *Development of preimplantation embryos and their environment* (ed. K. Yoshinaga and T. Mori), p. 225. Alan R. Liss, New York.

———. 1989b. Development *in vitro* of investment-free marsupial embryos during cleavage and early blastocyst formation. *Gamete Res.* **23:** 399.

Selwood, L. and A.H. Sathananthan. 1988. Ultrastructure of early cleavage and yolk extrusion in the marsupial *Antechinus stuartii. J. Morphol.* **195:** 327.

Selwood, L. and D. Smith. 1990. Time-lapse analysis and normal stages of development of cleavage and blastocyst formation in the marsupials the brown Antechinus and the stripe-faced Dunnart. *Mol. Reprod. Dev.* **26:** 53.

Selwood, L. and G.J. Young. 1983. Cleavage in vivo and in culture in the dasyurid marsupial *Antechinus stuartii* (Macleay). *J. Morphol.* **176:** 43.

Sharman, G.B. 1955. Studies on marsupial reproduction. III. Normal and delayed pregnancy in *Setonix brachyurus. Aust. J. Zool.* **3:** 56.

———. 1961. The embryonic membranes and placentation in five genera of diprotodont marsupials. *Proc. Zool. Soc. Lond.* **137:** 197.

Shirayoshi, Y., T.S. Okada, and M. Takeichi. 1983. The calcium-dependent cell-cell adhesion system regulates inner cell mass formation and cell surface polarization in early mouse development. *Cell* **35:** 631.

Skrecz, I. and J. Karasiewicz. 1987. Decompaction and recompaction of mouse preimplantation embryos. *Roux's Arch. Dev. Biol.* **196:** 397.

Slack, J.M.W. 1983. *From egg to embryo*. Cambridge University Press, United Kingdom.

Snell, G.D. and L.C. Stevens. 1966. Early embryology. In *Biology of the laboratory mouse*, 2nd edition (ed. E.L. Green), p. 205. McGraw Hill, New York.

Sobel, J.S. 1983. Cell-cell contact modulation of myosin organisation in the early mouse embryo. *Dev. Biol.* **100:** 207.

———. 1990. Membrane-cytoskeletal interactions in the early mouse embryo. *Cell Biol.* **1:** 341.

Sobel, J.S., E.G. Goldstein, J.M. Venuti, and M.J. Welsh. 1988. Spectrin and calmodulln in spreading mouse blastomeres. *Dev. Biol.* **126:** 47.

Solter, D. 1988. Differential imprinting and expression of maternal and paternal genomes. *Annu Rev. Genet.* **22:** 127.

Sorensen, R.A., M.S. Cyert, and R.A. Pedersen. 1985. Active maturation-promoting factor is present in mature mouse oocytes. *J. Cell Biol.* **100:** 1637.

Squier, R.R. 1932. The living egg and early stages of its development in the guinea pig. *Carnegie Inst. Contrib. Embryol.* **23:** 223.

Steven, D. 1975. Anatomy of the placental barrier. In *Comparative placentation: Essays in structure and function* (ed. D.N. Steven), p. 25. Academic Press, New York.

———. 1983. Interspecific differences in the structure and function of trophoblast. In *Biology of trophoblast* (ed. Y.W. Loke and A. Whyte), p. 111. Elsevier Science Publishers B.V., Amsterdam.

Steven, D. and G. Morriss. 1975. Development of fetal membranes. In *Placentation: Essays in structure and function* (ed. D.N. Steven), p. 58. Academic Press, New York.

Streeter, G.L. 1927. Development of the mesoblast and notochord in pig embryos. *Carnegie Inst. Contrib. Embryol.* **19:** 73.

Stroband, H.W.J., N. Taverne, and M.V.D. Bogaard. 1984. The pig blastocyst: Its ultrastructure and the uptake of protein macromolecules. *Cell Tissue Res.* **235:** 347.

Surani, M.A.H. 1986. Evidences and consequences of differences between maternal and paternal genomes during embryogenesis in the mouse. In *Experimental approaches to mammalian embryonic development* (ed. J. Rossant and R.A. Pedersen), p. 401. Cambridge University Press, United Kingdom.

Surani, M.A.H., S.C. Barton, and M.L. Norris. 1984. Development of reconstituted mouse eggs suggests imprinting of the genome during gametogenesis. *Nature* **308:** 548.

———. 1986. Nuclear transplantation in the mouse: Heritable differences between parental genomes after activation of the embryonic genome. *Cell* **45:** 127.

Surani, M.A., S.C. Barton, S.K. Howlett, and M.L. Norris. 1988. Influence of chromosomal determinants on development of androgenetic and parthenogenetic cells. *Development* **103:** 171.

Surani, M.A., N.D. Allen, S.C. Barton, R. Fundele, S.K. Howlett, M.L. Norris, and W. Reik. 1990. Developmental consequences of imprinting of parental chromosomes by DNA methylation. *Philos. Trans. R. Soc. Lond. B Biol. Sci.* **326:** 313.

Sutherland, A.E. and P.G. Calarco-Gillam. 1983. Analysis of compaction in the preimplantation mouse embryo. *Dev. Biol.* **100:** 328.

Sutherland, A.E., P.G. Calarco, and C.H. Damsky. 1989. Expression and function of cell surface extracellular matrix receptors in mouse blastocyst attachment and outgrowth. *J. Cell Biol.* **106:** 1331.

Sutherland, A.E., T.P. Speed, and P.G. Calarco. 1990. Inner cell allocation in the mouse morula: The role of oriented division during fourth cleavage. *Dev. Biol.* **137:** 13.

Swain, J.L., T.A. Stewart, and P. Leder. 1987. Parental legacy determines methylation and expression of an autosomal transgene: A molecular mechanism for parental imprinting. *Cell* **50:** 719.

Tarkowski, A.K and J. Wroblewska. 1967. Development of blastomeres of mouse eggs isolated at the 4- and 8-cell stage. *J. Embryol. Exp. Morphol.* **18:** 155.

Thomson, J.A. and D. Solter. 1988. The developmental fate of androgenetic, parthenogenetic, and gynogenetic cells in chimeric gastrulating mouse embryos. *Genes Dev.* **2:** 1344.

———. 1989. Chimeras between parthenogenetic or androgenetic blastomeres and normal embryos: Allocation to the inner cell mass and trophectoderm. *Dev. Biol.* **131:** 580.

Toguchida, J., K. Ishizaki, M.S. Sasaki, Y. Nakamura, M. Ikenaga, M. Kato, M. Sugimoto, Y. Kotoura, and T. Yamamuro. 1989. Preferential mutation of paternally derived RB gene as the initial event in sporadic osteosarcoma. *Nature* **338:** 156.

Tripp, H.R.H. 1971. Reproduction in elephant-shrews (Macroscelididae) with special reference to ovulation and implantation. *J. Reprod. Fertil.* **26:** 149.

Tyndale-Biscoe, C.H. 1963. *Life of marsupials*. Edward Arnold, London.

Vakaet, L. 1984. Early development of birds. In *Chimeras in developmental biology* (ed. N. LeDouarin and A. McLaren), p. 71. Academic Press, London.

Van Blerkom, J., C. Manes, and J.C. Daniel, Jr. 1973. Development of preimplantation rabbit embryos *in vivo* and *in vitro*. *Dev. Biol.* **35:** 262.

van der Horst, C.J. 1942. Early stages in the embryonic development of *Elephantulus*. *S. Afr. J. Med. Sci.* (biol. suppl.) **7:** 55.

van der Horst, C.J. and J. Gillman. 1942. Pre-implantation phenomena in the uterus of *Elephantulus*. *S. Afr. J. Med. Sci.* **7:** 47.

van Niekerk, C.H. and W.R. Allen. 1975. Early embryonic development in the horse. *J. Reprod. Fertil.* (suppl.) **23:** 495.

van Niekerk, C.H. and W.H. Gerneke. 1966. Persistence and parthenogenetic cleavage of tubal ova in the mare. *Onderstepoort. J. Vet. Res.* **33:** 195.

Vestweber, D. and R. Kemler. 1984. Rabbit antiserum against purified surface glycoprotein decompacts mouse preimplantation embryos and reacts with specific adult tissues. *Exp. Cell Res.* **152:** 169.

Vestweber, D., A. Gossler, K. Boller, and R. Kemler. 1987. Expression and distribution of cell adhesion molecule uvomorulin in mouse preimplantation embryos. *Dev. Biol.* **124:** 451.

Wassarman, P.M. 1990. Profile of a mammalian sperm receptor. *Development* **108:** 1.

Whyte, A. and W.R. Allen. 1985. Equine endometrium at pre-implantation stages of pregnancy has specific glycosylated regions. *Placenta* **6:** 537.

Whyte, A., C.D. Ockleford, F.B.P. Wooding, M. Hamon, W.R. Allen, and S. Kellie. 1986. Characteristics of cells derived from the girdle region of the preimplantation blastocyst of the donkey. *Cell Tissue Res.* **246:** 343.

Willadsen, S.M. 1979. A method for culture of micromanipulated sheep embryos and its use to produce monozygotic twins. *Nature* **277:** 298.

———. 1980. The viability of early cleavage stages containing half the normal of blastomeres in the sheep. *J. Reprod. Fertil.* **59:** 357.

———. 1981. The developmental capacity of blastomeres from 4- and 8-cell sheep embryos. *J. Embryol. Exp. Morphol.* **65:** 165.

Willadsen, S.M. and C. Polge. 1981. Attempts to produce monozygotic quadruplets in cattle by blastomere separation. *Vet. Rec.* **108:** 211.

Willadsen, S.M., H. Lehn-Jensen, C.B. Fehilly, and R. Newcomb. 1981. The production of monozygotic twins of preselected parentage by micromanipulation of non-surgically collected cow embryos. *Theriogenology* **15:** 23.

Williams, B.S. and J.D. Biggers. 1990. Polar trophoblast (Rauber's layer) of the rabbit blastocyst. *Anat. Rec.* **227:** 211.

Williams, C.A., R.T. Zori, J.W. Stone, B.A. Gray, E.S. Cantu, and H. Ostrer. 1990. Maternal origin of 15q11-13 deletions in Angelman syndrome suggests a role for genomic imprinting. *Am. J. Med. Genet.* **35:** 350.

Williams, T.J., R.P. Elsden, and G.E. Seidel, Jr. 1984. Pregnancy rates with bisected bovine embryos. *Theriogenology* **22:** 521.

Wimsatt, W.A. 1951. Observations on the morphogenesis, cytochemistry, and significance of the binucleate giant cells of the placenta of ruminants. *Am. J. Anat.* **89:** 233.

———. 1975. Some comparative aspects of implantation. *Biol. Reprod.* **12:** 1.

Winkel, G.K. and R.A. Pedersen. 1988. Fate of the inner cell mass in mouse embryos as studied by microinjection of lineage tracers. *Dev. Biol.* **127:** 143.

Wintenberger-Torres, S. and J.-E. Flechon. 1974. Ultrastructural evolution of the trophoblast cells of the pre-implantation sheep blastocyst from day 8 to day 18. *J. Anat.* **118:** 143.

Winters, L.M., W.W. Green, and R.E. Comstock. 1942. Prenatal development of the bovine. *Univ. Minn. Agric. Expt. Sta. Tech. Bull.* **151:** 1.

Wooding, F.B.P. 1982. The role of the binucleate cell in ruminant placental structure. *J. Reprod. Fertil.* (suppl.) **31:** 31.

Wooding, F.B.P. and L.D. Staples. 1981. Functions of the trophoblast papillae and binucleate cells in implantation in the sheep. *J. Anat.* **133:** 110.

Wooding, F.B.P. and D.C. Wathes. 1980. Binucleate cell migration in the bovine placenta. *J. Reprod. Fertil.* **59:** 425.

Yoshida-Noro, C., N. Suzuki, and M. Takeichi. 1984. Molecular nature of the calcium-dependent cell-cell adhesion system in mouse teratocarcinoma and embryonic cells studied with a monoclonal antibody. *Dev. Biol.* **101:** 19.

Ziomek, C.A. and M.H. Johnson. 1980. Cell surface interactions induce polarization of mouse 8-cell blastomeres at compaction. *Cell* **21:** 935.

———. 1981. Properties of polar and apolar cells from the 16-cell mouse morula. *Roux's Arch. Dev. Biol.* **190:** 287.

Cloning by Nuclear Transfer and Embryo Splitting in Laboratory and Domestic Animals

R.S. Prather[1] and J.M. Robl[2]
[1]Department of Animal Science, University of Missouri
Columbia, Missouri 65211
[2]Department of Veterinary and Animal Science
University of Massachusetts
Amherst, Massachusetts 01003

HISTORY OF CLONING

Techniques for the cloning of embryos, either by splitting into equal parts or by transplanting nuclei, arose out of attempts to solve fundamental questions in developmental biology. These questions related to how blastomeres of an early embryo give rise to differentiated cell types.

One hypothesis for the differentiation of cell types involves the segregation of specific cytoplasmic determinants. Classic studies (Roux 1888; Driesch 1892) have shown that in a variety of lower vertebrates and invertebrates, maternal cytoplasmic determinants that are prelocalized in the egg become partitioned differentially among the cleaving blastomeres. These determinants then direct the development of the various cell lineages in the embryo. As the cells cleave and the cytoplasmic determinants are partitioned among cells, they lose their potency or ability to give rise to all the cell types in the adult. Therefore, assessing the developmental potential of individual blastomeres or split embryos is a definitive way of testing this hypothesis. Early work by Nicholas and Hall (1942) in the rat, Tarkowski (1959) in the mouse, and Seidel (as referenced by Kelly 1977) in the rabbit showed that individual two-cell-stage blastomeres are capable of giving rise to live offspring. Moore et al. (1968) extended this work to show that

Animal Applications of Research in Mammalian Development
 0-87969-333-9/91 $3.00 + 00

even isolated eight-cell blastomeres are capable of giving rise to normal offspring. This work, along with subsequent studies, demonstrated that prelocalized cytoplasmic determinants are probably not involved in mammalian embryo development, because a new individual can arise from a single isolated blastomere.

Monitoring the development of split embryos also introduced the possibility of producing identical offspring by splitting the embryos into several parts. Willadsen (1979; Willadsen et al. 1981) first used this approach to obtain identical offspring from farm animal (sheep, cattle, goats, and pigs) embryos. This procedure involved the isolation of blastomeres from early-stage embryos and the transfer of individual blastomeres or pairs of blastomeres to evacuated zonae. These were then embedded in agar and transferred to a sheep oviduct as a temporary incubator. After several days of development in the oviduct, the embryos were recovered, dissected free of agar, and transferred to recipients. Although the procedure was successful, it was cumbersome and not practical because it required the use of early-stage oviductal embryos. In addition, cow oviductal embryos are difficult to recover and handle after manipulation, thereby limiting the use of this technique in this species. However, the development of techniques for easily and efficiently splitting embryos at the morula and blastocyst stage (Ozil et al. 1982; Williams et al. 1982; Lambeth et al. 1983) became a commercially viable method of cloning for the bovine embryo transfer industry.

A second mechanism by which the various cell types of the body may arise is through nuclear differentiation. This hypothesis suggests that specific genes may be physically lost in the differentiation of specific cell types (Weismann 1893). Spemann (1938) suggested that one way to test this hypothesis would be to transfer nuclei from differentiated adult tissues to unfertilized eggs. If these nuclei could support development and differentiation of an embryo into a normal adult, it could be assumed that the nuclei retained full potency and that all nuclei in an individual are equivalent. Spemann's experiment was first carried out by Briggs and King (1952), who showed that nuclei from blastula stage *Rana pipiens* embryos could be transferred to enucleated and activated meiotic metaphase II oocytes and that the resulting embryo could develop to the blastula stage. They also alluded to the fact that nuclei from more advanced stages might not be able to direct development at similar rates. Their conclusions were subsequently borne out by additional investigators (for review, see Gurdon 1986; DiBerardino 1980, 1987, 1989).

Methods to transfer nuclei in mammalian cells were subsequently acquired from a variety of different sources (Briggs and King 1952; Graham 1969) and put into practice by McGrath and Solter (1983) in the mouse. This method, in a modified form, was used to duplicate the amphibian experiments in the cow (Prather et al. 1986) and sheep (Willadsen 1986). Studies in these species and in the rabbit (Stice and Robl 1988) and pig (Prather et al. 1989a) demonstrated not only that nuclei from early mammalian embryos are totipotent, but also that they can be used to produce genetically identical individuals. In this way, the current methods for cloning by nuclear transfer in mammals were developed.

BIOLOGICAL BASIS AND THEORY OF CLONING

To grasp the biological basis of the two methods of cloning, basic developmental events must be understood. The importance of this is discussed in context with the corresponding procedures below.

Splitting

Methods

Two general procedures, depending on the stage of the embryos, have been used for splitting embryos. Methods for splitting cleavage-stage sheep, swine, and cow embryos have been described in detail by Willadsen et al. (1981; Willadsen 1982). The procedure involves microsurgically removing the zona pellucida, mechanically separating the blastomeres, and then injecting them singly or in groups into previously evacuated zonae pellucidae. Because reliable culture systems were not available for the domestic species, it was necessary to transfer the split embryos to either definitive recipients or temporary incubator animals. Unfortunately, cleavage-stage embryos with damaged zonae pellucidae did not survive well following transfer. Embedding the embryos in agar cylinders and transferring them to an incubator species (Willadsen 1979) helped to overcome survival problems.

Methods for splitting late-morula- and early-blastocyst-stage embryos arose out of interest in applying this technique to bovine embryo transfers. Because bovine embryos are recovered and transferred nonsurgically from the uterus at the late morulae and blastocyst stage, a method for commercial embryo splitting would have to be applicable to these stages. Such a procedure would have the advantage of not requiring transfer to an incubator species.

Several methods for splitting nonsurgically recovered uterine cow embryos have been developed. Williams et al. (1982) used a fine surgical blade mounted on a glass micropipette to cut through the zona pellucida and the embryo. After bisection, one of the halves was removed from the zona pellucida using a 40–50-μm micropipette and placed in an evacuated surrogate zona pellucida. The procedure used by Lambeth et al. (1982) was similar, except that a finely drawn glass needle was used to bisect the embryo after it had been removed from the zona pellucida. The method of Ozil et al. (1982) involved the use of five glass microinstruments: a holding pipet, two glass needles for opening the zona pellucida and bisecting the embryo, and two other instruments for extracting and replacing the embryos in the zona pellucida. The most recent method reported involves a single blade slice of an embryo in a culture dish. This procedure is simple, quick, and results in high rates of development (Williams and Moore 1988). All of these methods have been used widely to produce identical twins and to increase the number of offspring obtained from a set of embryos.

Results

Split embryos have a smaller than normal number of cells present during early cleavage, but the offspring that are born are normal in size because regulation of cell numbers occurs after the time of blastocyst formation (Buehr and McLaren

1974; O'Brien et al. 1984; Rands 1985). Studies of this process show that blastocyst formation occurs on a temporal schedule and is independent of cell number (Surani et al. 1980; Dean and Rossant 1984). Further studies evaluating the time at which cell number regulation occurs suggest that in the mouse, regulation occurs between days 5 and 6 (Lewis and Rossant 1982). Furthermore, this regulation of cell number arises from a decreased rate of cell division that occurs in normal, but not in manipulated, embryos during this time period. Thus, split embryos result in animals that are similar in size, cell number, and resulting performance as compared with control embryos.

In principle, embryo splitting should yield as many descendant embryos as there are blastomeres. However, the maximum number of offspring that result from the splitting of a single embryo is limited in practice to a much smaller number. The greatest survival rate comes from bisection of the blastocyst. For example, although it is theoretically possible to produce four identical embryos from one by splitting, the efficiency of the technique limits the use of splitting to the production of identical twins. Because single members of twin pairs, or both half embryos, may be lost following transfer, between 50% (Leibo and Rall 1987) and 78% (Willadsen et al. 1981; Williams et al. 1982) of the split embryos surviving bisection at the blastocyst stage will be twin pairs.

The reduced yield is further illustrated by the rate of survival from time of splitting to birth for separated blastomeres of two-, four-, and eight-cell-stage rabbit embryos. Whereas 30% of the two-cell-stage blastomeres survive to term, only 19% and 11% of the four- and eight-cell-stage blastomeres survive (Moore et al. 1968). In the cow, 75% (21/28) of half embryos and 41% (9/22) of quarter embryos survive to day 50 of gestation (Willadsen et al. 1981). The decreased rate of development appears to be a result of an inadequate number of cells present at the time of compaction and blastulation (Tarkowski and Wroblewska 1967; Willadsen 1982). When the blastocyst begins to form, cells that are in the interior of the compacted embryo give rise to the inner cell mass. If too few cells are present during blastocyst formation and if no inner cells are present, a blastocyst composed of trophectoderm

TABLE 1 *PREGNANCY RATES FROM WHOLE AND SPLIT EMBRYOS*

	Type of embryo[a]	
Study	*whole*	*split*
Baker and Shea (1985)	74/98 (76)	159/248 (64)
Heyman (1985)	26/45 (58)	4/11 (36)
Leibo and Rall (1987)	291/515 (56)	441/842 (52)

[a]Values given reflect number of pregnancies/total number of attempted implants. Percentage of successful pregnancies indicated in parentheses.

forms without an inner cell mass (Tarkowski and Wroblewska 1967; Menino and Wright 1983). This type of conceptus will not result in development to term, but yields only trophoblastic vesicles that implant without undergoing further cell division (Gardner et al. 1973).

Although the efficiency of the technique is low and split embryos are less viable than whole embryos, it is possible to increase the total number of offspring produced from a group of embryos by splitting. Comparisons of results obtained by several investigators indicate that pregnancy rates with split embryos are 4–22% lower than whole embryos (Table 1). In all cases, splitting increased the total number of calves obtained from the original group of embryos, and in one case, it was nearly doubled.

A variety of factors may affect the survival rate of split embryos. Blastocyst-stage embryos from day 7.0 to 7.5 have a higher survival rate than earlier-stage embryos (Williams et al. 1984). Pregnancy rates decline dramatically with lower-quality embryos (Brem et al. 1984). Further manipulations such as freezing or culturing (Heyman 1985) also are more harmful to split embryos than to whole embryos.

Nuclear Transfer

Methods

Current methods of nuclear transfer in domestic species are derived from the method developed by McGrath and Solter (1983). The donor embryos and unfertilized recipient oocytes are treated with cytoskeletal inhibitors, a micropipette is inserted into the oocyte, and the metaphase chromosomes are

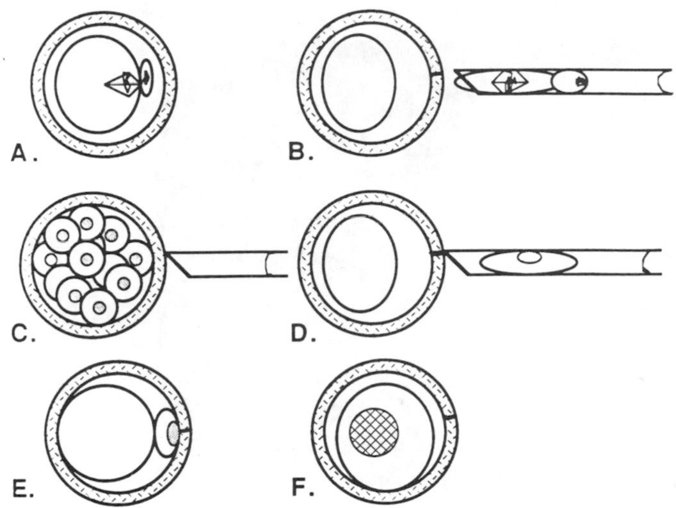

FIGURE 1 Scheme of nuclear transfer procedure. (*A*) Meiotic metaphase II oocyte. (*B*) Meiotic metaphase II oocyte after removal of polar body and metaphase chromosomes. (*C*) Sixteen-cell-stage donor embryo. (*D*) Before transfer of 16-cell-stage blastomere to enucleated oocyte from *B*. (*E*) After transfer of donor blastomere to enucleated oocyte. (*F*) After exposure to electric pulse and subsequent fusion. Note that swelling of the transferred nucleus as well as chromatin reorganization is depicted. (Reprinted, with permission, from Prather and First 1990b.)

removed in a portion of membrane-bounded cytoplasm (Fig. 1). Successful enucleation is monitored by observing the removal of the chromosomes directly (Stice and Robl 1988), by indirect staining using the DNA-specific fluorescent dye bisbenzimide (Tsunoda et al. 1988; Prather and First 1990a; Westhusin et al. 1990), or by mounting a portion of the enucleated oocytes and assuming an equal efficiency of enucleation in the remaining eggs (Willadsen 1986; Prather et al. 1987; Smith and Wilmut 1989). A single blastomere from the donor embryo (or portions thereof) is then aspirated into the micropipette and expelled into the perivitelline space, adjacent to the enucleated oocyte (Fig. 2). The next step is the fusion of the two cells within the perivitelline space (see Fig. 3). This can be accomplished in some species with Sendai virus (Graham 1969) or with electrofusion (Berg 1982).

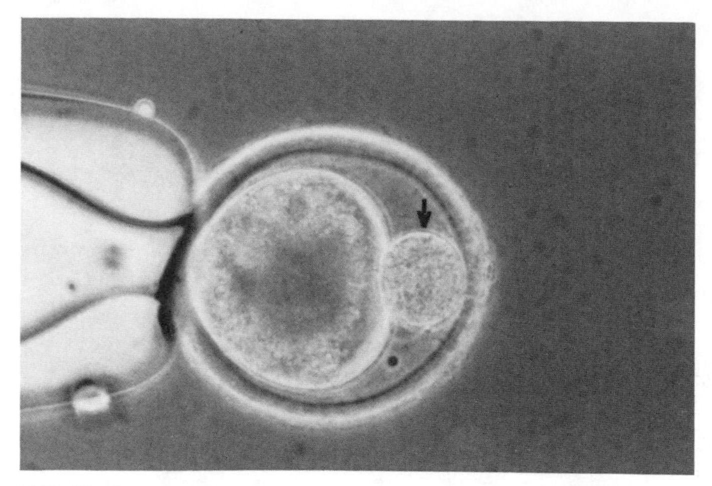

FIGURE 2 An eight-cell-stage bovine blastomere (arrow) transferred to an enucleated oocyte.

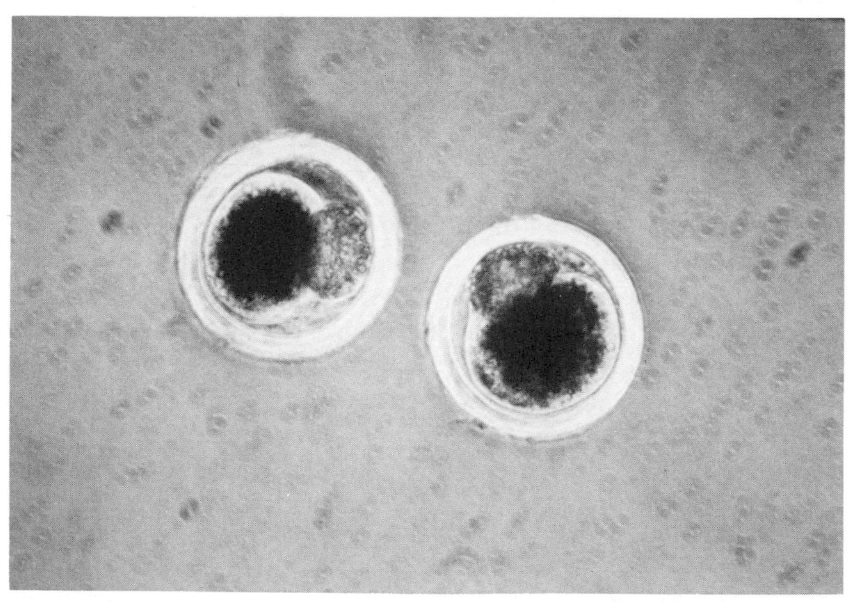

FIGURE 3 Four-cell-stage porcine blastomeres transferred to enucleated oocytes and pulsed with electricity. Note that the nuclear transfer embryo on the right has begun the fusion process but that cell fusion is not evident in the nuclear transfer embryo on the left. (Reprinted, with permission, from Prather and First 1990a.)

The efficiencies of the enucleation procedures can reach 100% when the chromosomes are directly or indirectly observed (Tsunoda et al. 1988; Stice and Robl 1988; Prather and First 1990a), whereas the percentage of enucleated oocytes is lower when chromosomal removal is based solely on the location of the first polar body (Willadsen 1986; Prather et al. 1987, 1989a).

Activation is thought to occur coincident with electrofusion. It has been known for many years that electrical pulses are an effective parthenogenetic agent in the mouse (Whittingham 1980). The specific mechanism of electrical activation is not known, but it may be related to membrane depolarization and calcium leakage after electrically induced pore formation (Whittingham 1980). As with fusion, electrically induced activation varies greatly from study to study. Factors that affect activation rates are many and include age and species of oocyte, type of chamber and medium in which the pulse is given, and type of pulse (Collas et al. 1989; Ozil 1990).

Genetic Uniformity of Clones

Defining the genetic composition of embryos resulting from nuclear transfer is not as straightforward as it may appear. With current technology, it appears that all of the nuclei in an early embryo are identical; however, there may be both nuclear and cytoplasmic exceptions. Some cells exhibit chromosomal changes, including translocations (King and Linares 1983), diminution (Beerman 1977), gene rearrangements (Malissen et al. 1984; Alt et al. 1987), gene amplification (Tobler 1975), and mutation. If any of these events occur in only a portion of the cells in an early embryo (i.e., if the donor embryo is mosaic), the resulting clones may be composed of two or more populations of nuclear genomes. If chromosomal rearrangements are replicated with serial nuclear transfers, the effect could be magnified.

Another possible source of genetic variation is the cytoplasm. There are numerous mitochondrial genomes within each cell, and the distribution during cleavage is thought to follow population genetics. The mitochondrial genotypes of a single maternal line have been observed to change (Hauswirth and Laipis 1982; Laipis et al. 1982; Michaels et al. 1982; Olivo et al. 1983), suggesting that it will be important to evaluate nuclear cytoplasmic interactions.

Another organelle that may have its own genome is the centriole. It has recently been determined that basal bodies have their own genome (Hall et al. 1989). In mammals, centrioles are absent during the first few cleavages (Szollosi et al. 1972) and it would be interesting to determine the fate of centrioles that are transferred to oocytes. Do they disappear or are they degraded? Do they participate in cleavage, changing the shape of the mitotic spindle from a barrel shape with broad poles to a fusiform shape with narrow mitotic poles during the early cleavages (Schatten et al. 1985)? Thus, the genetic uniformity of nuclear transfer clones remains to be determined.

Phenotypic Uniformity of Clones

Because of their uniformity, genetically identical monozygotic twins have been of interest as experimental models for many years. However, identical twins are not necessarily phenotypically identical. Phenotype is determined by epigenetic phenomena, uterine, neonatal, and postnatal environments, in addition to nuclear and cytoplasmic genetics (Seidel 1983). The uniformity of members of a twin set can be estimated by the intraclass correlation coefficient (p_I) (Biggers 1986), with a value of 1 indicating a perfect correlation. For natural monozygotic twins in cattle, estimates of p_I have been calculated for a variety of physiological and biochemical parameters, including milk production, behavior, and growth. Values ranged from 0.17 to 0.99, but nearly half were above 0.90 (Biggers 1986), indicating a high degree of similarity between monozygotic twins. In human monozygotic twins, intraclass correlation coefficients for various traits were slightly lower than those for cattle, ranging between 0.53 and 0.84, with 0.65 as the median. The coefficients are always considerably higher for monozygotic twins than for dizygotic twins (Biggers 1986).

Artificially cloned animals may not necessarily be as uniform as natural monozygotic twins. Natural twins develop within the same uterus and in some cases with common circulation, and the neonatal and postnatal environments are usually similar. Artificially produced twins are usually transferred to different recipients and consequently the environment in which the animal develops may be considerably different. In addition, there may be differences resulting from the methods of producing the individual clones. To date, however, no information is available to conclude definitively that artificially produced monozygotic twins are either less uniform or more uniform than natural twins.

Nuclear Reorganization as Compared with Pronuclear Development

The process of fertilization dramatically alters sperm chromatin. This chromatin reorganization allows the sperm to participate in the early cleavages and to direct subsequent development. In this section, the development of the male pronucleus is compared with the development of the transferred nucleus during the cloning process.

Morphological changes. After fertilization, the sperm chromatin decondenses and acquires a nuclear envelope derived mainly from the egg cytoplasm. Decondensation begins at the nuclear periphery and spreads to the interior, resulting in a large increase in the volume of the sperm chromatin. The process of sperm decondensation probably results in part from the disruption of disulfide bonds, because decondensation can be mimicked in vitro with reducing agents (Calvin and Bedford 1971). Following sperm nuclear decondensation, the nuclear envelope forms around both male and female pronuclei, and the nuclei increase dramatically in size. During this time, multiple nucleoli appear in the developing pronuclei and gradually aggregate to form a single predominate nucleolus (Prather et al. 1990b).

Events similar to pronuclear formation also occur in nuclei transplanted into the cytoplasm of an oocyte. After transfer to an activated oocyte, an eight-cell-stage nucleus swells in diameter from 9.6 μm to 22 μm in the rabbit (Stice and Robl 1988) and from 14.3 μm to 27.3 μm in the pig (Prather et al. 1990a). When four-cell-stage blastomeres or four-cell-stage karyoplasts (equal in volume to an eight-cell-stage blastomere) are transferred to one-half oocytes or to relatively intact oocytes, the transferred nucleus swells to a similar volume (Prather et al. 1990a). This suggests that cytoplasmic components from the oocyte and four-cell-stage blastomere responsible for nuclear swelling are not competing with each other to cause swelling. If that were the case, the four-cell karyoplast transferred to the relatively intact oocyte would be expected to swell more than the four-cell-stage blastomere transferred to the one-half oocyte because it would have the greatest supply of components from the oocyte and the least from the transferred nucleus. Interestingly, 4-, 8- and 16-cell-stage nuclei swell to a similar volume, although this represents different degrees of swelling for the nuclei of different cell stages.

The time of fusion in relation to activation is important for full development as a pronucleus. Nuclear transfer experiments in the mouse suggest that for maximum swelling and transformation into a pronucleus-like structure, the nucleus must be transferred in a 90-minute window around the time of activation. When transferred too early, the nuclei undergo premature condensation with individualization of the chromosomes. However, when transferred too late, the nuclei fail to develop as a pronucleus would (Czolowska et al. 1984). In addition, there appears to be a limit to the decondensation factors present. Multiple nuclei or polyspermy results in only a portion of the nuclei swelling and remodeling into a pronuclei-like structure. Nucleolar changes after nuclear transfer are also time-dependent. When they occur synchronous with activation, these changes suggest that there is inactivation of rRNA genes as the nucleoli become compact and agranular (Szollosi et al. 1988).

Biochemical changes. One type of biochemical change seen during nuclear reorganization is that of *protein exchange*. During spermatogenesis, histones are replaced by protamines (Bloch 1969). Coincident with chromosome decondensation and disulfide bond reduction is the release of these basic protamines from the fertilizing sperm. The protamines are replaced with basic proteins similar to those found within the female pronucleus (Poccia et al. 1981), and this exchange may be a result of phosphorylation of the protamines (Wiesel and Schultz 1981).

Overall protein movement into the developing sperm pronucleus is illustrated with the movement of [^{3}H]lysine-labeled oocyte proteins accumulating in the developing pronuclei of rabbits (Motlik et al. 1980). After nuclear transfer in amphibians, there is an accumulation of cytoplasmic egg nonhistone proteins by the transplanted nuclei and a major loss of nonhistone proteins from the transplanted nucleus into the cytoplasm (DiBerardino and Hoffner 1975; Hoffner and DiBerardino 1977). This protein exchange precedes the swelling observed and may even facilitate the process of swelling (Merriam 1969).

The nuclear lamins are an example of specific protein exchange in mammalian embryos. The nuclear lamin composition changes during early development in a variety of species (for review, see Stricker et al. 1989). In the pig, the nuclear lamin A/C epitope (as defined by the monoclonal antibody J9) is lost from the nuclear envelope after the four-cell stage, but it can be reacquired if nuclei are transferred to an enucleated, activated meiotic metaphase II oocyte (Fig. 4) (Prather et al. 1989b). This is interpreted to result from acquisition of the lamin proteins from a cytoplasmic pool in the recipient oocyte. A similar change in epitope is seen in the mouse, but the lamin antigens are not acquired if the recipient is at either the pronuclear stage or enucleated pronuclear stage (Prather et al. 1991).

The second type of biochemical change is that involved with *DNA synthesis*. DNA synthesis in pronucleus-stage mouse eggs has been shown by autoradiography to begin about 5–6 hours after pronuclear formation (Luthardt and Donahue 1973; Kirshna and Generoso 1977). The duration of S phase is about 4 hours, and the male pronucleus begins and ends DNA synthesis earlier than the female pronucleus (Luthardt and Donahue 1973). Replication of DNA at the pronuclear stage permits cleavage to occur while maintaining the diploid state of the embryonic cells.

DNA synthesis occurs after nuclear transfer in both amphibians (Graham et al. 1966) and hamsters (Naish et al.

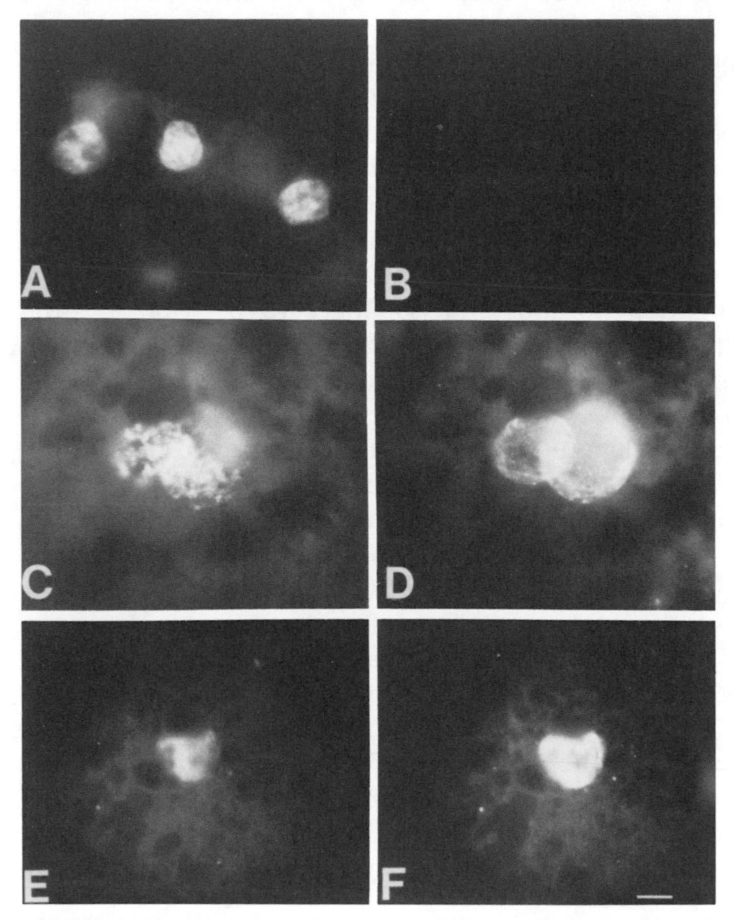

FIGURE 4 Nuclear lamin epitope after transfer of a 16-cell stage porcine nucleus to an activated enucleated meiotic metaphase II oocyte. DNA (Hoechst 33258 stain) in cells from 16-cell-stage embryo (*A*), pronucleus stage egg (*C*), an activated oocyte after nuclear transfer (*E*) (nucleus from 16-cell-stage embryo in *A*), and corresponding lamin A/C (J9) reactivity (*B*, *D*, *F*; respectively). Note the absence of lamin A/C (J9) reactivity in the 16-cell-stage blastomeres, whereas after transfer to an activated enucleated metaphase II oocyte, the nucleus acquires the antigen. (*B*, *D*, *E*) Photographed and developed under identical conditions. Cells in *A*, *B*, *E*, and *F* were mounted and stained on the same coverslip. Bar, 10 μm; size may not be indicative of in vivo size, because nuclear size is dependent on the degree of flattening under the coverslip. (Reprinted, with permission, from Prather et al. 1989b.)

1987). In *Xenopus*, DNA synthesis begins sooner and does not last as long in nuclear transfer embryos derived from developmentally immature nuclei as in nuclei from more advanced tissues. For example, nuclei from brain tissue require 40 minutes for a significant number of transferred nuclei to initiate DNA synthesis and more than 90 minutes to complete DNA synthesis, whereas nuclei from embryonic stages begin DNA synthesis within 20 minutes and are completed by 60 minutes after nuclear transfer (Graham et al. 1966). More importantly, DNA synthesis occurs only after the transfer of nuclei in G_1 or S phase and not in nuclei in G_2 stage (DeRoeper et al. 1977). Thus, in amphibians, which lack a G_2 phase during early cleavage, nuclei from rapidly dividing embryonic cells can be transferred at any stage of their cell cycle except during M and can direct later development at equal levels (Ellinger 1977). This is in contrast to more slowly dividing tissue culture cells, where nuclei in G_2 promote development at a higher level than nuclei in G_1 (von Bergoldington 1981). This may result from incomplete DNA synthesis after nuclear transfer and before cell division by the cells early in the cell cycle, whereas those nuclei in G_2 would not need to enter or complete DNA synthesis before the first cleavage division. Similar regulation is shown in mammalian cells (for review, see Rao 1980). However, the regulation of DNA synthesis in nuclei that are transferred using electrofusion has not been documented (see below).

Incomplete DNA replication is a major problem in the development of amphibian nuclear transfer embryos and causes chromosomal abnormalities that arise during the first cell cycle after nuclear transfer (DiBerardino and Hoffner 1970). As more advanced nuclei are transferred, incomplete DNA replication is more likely to occur, leading to chromosomal abnormalities after transfer (for review, sec DiBerardino 1979; Gurdon 1986). Interestingly, as nuclei become more and more developmentally advanced, the length of their cell cycle increases. After nuclear transfer in amphibians, nuclei are required to replicate and prepare for cleavage within 1 hour. Any incomplete replication observed may result from differences in the lengths of the cell cycles. Thus, a comparison between the differences in lengths of the cell cycle and degree of differentiation is confounded. These variables may be evaluated separately in mammals where the length of the first cell cycle is much longer (>15 hr) than in *Xenopus* (35 min).

Embryos derived from nuclear transfer show a progressive attrition during development. Nuclei transferred from late blastula stage *Xenopus* embryos into an oocyte develop to the blastula stage with a frequency of 62%, to a tail-bud tadpole with a frequency of 48%, to a swimming tadpole with a frequency of 38%, and to a young frog with a frequency of 35% (Gurdon 1964). The developmental restriction point of arrested embryos is stably inherited as shown with serial nuclear transfer and is likely due to karyotypic abnormalities (for review, see DiBerardino 1987). In mammalian nuclear transfer, a similar loss of pregnancies is observed. Nuclear transfer in cattle using cleavage-stage donor nuclei (i.e., 2–16-cell-stage embryos) results in 20% (Prather et al. 1987) to 35% (Bondioli et al. 1990) morula/blastocyst formation. Of the morula/blastocysts transferred to recipients, only 37% (7/19) established pregnancies as confirmed by ultrasound between days 21 and 30 or 22% (104/463) as confirmed on day 42 (Bondioli et al. 1990). The pregnancy rate for morphologically graded number-1 embryos was reported by Bondioli et al. (1990) to be 33%, a rate about half that reported for fresh embryo transfers. A parallel observation is seen in the pig using four-cell-stage embryos as donors (Prather et al. 1989a). Of three gilts each receiving ten or more nuclear transfer embryos on day 6, all had extended estrus cycles (28 days, 52 days, 72 days). This implies pregnancy establishment with later abortion (it should be noted that the pig is a litter-bearing species and four conceptuses are needed to maintain a pregnancy to term).

In both cattle and pigs, a reduced pregnancy rate is thus observed after the transfer of cloned embryos to the reproductive tracts of recipient females. It is interesting to speculate that some of these concepti may have begun development, but encountered developmental restrictions like those found in amphibian nuclear transfer embryos, and that developmental restriction points are a result of chromosomal abnormalities.

The concept of incomplete DNA replication resulting in chromosomal abnormalities and ensuing abortions is especially intriguing considering the hypothesis of Blow and Laskey (1988) that the nuclear envelope must remain intact once DNA synthesis begins. These authors suggest that if the nuclear envelope is perturbed, a licensing factor enters the nucleus and promotes DNA synthesis (i.e., DNA synthesis occurs in previously replicated DNA). The nuclear transfer procedures developed for cloning incorporate a method of cell fusion that causes transient plasma membrane breakdown. The breakdown of the nuclear envelop may also occur. If this were the case, then the licensing factor entering the transferred nucleus could result in replication of previously replicated DNA, leading to polyploidy. Since the electrofusion procedure may allow nuclei in G_2 phase to undergo DNA synthesis, it may be important to transfer nuclei that have not yet undergone *any* DNA replication.

Nuclear Reprogramming

Reprogramming of morphological developmental events. In both mammals and amphibians, a reprogramming of the timing of morphological events has been reported following nuclear transfer. Although not precisely timed, mammalian nuclei recapitulate the early cleavage stages of development prior to compaction, blastocoel formation, implantation, and, in some cases, complete development to term. If the nuclei are not reprogrammed, some serious developmental consequences would result. If, for example, a 32-cell-stage cow nucleus were transferred to an enucleated activated meiotic metaphase II oocyte without undergoing reprogramming, it might lead to compaction and blastocyst formation at the two- to four-cell stage. Such an embryo would fail after implantation in the mouse, because a minimum of 8–16 cells are required to form a competent blastocyst, containing both an inner cell mass and a trophoblast (Tarkowski and Wroblewska 1967; Papaioannou and Ebert 1986). The best example of reprogramming after nuclear transfer is in the sheep embryo, as studied by Smith and Wilmut (1989). Sheep inner-cell-mass nuclei transferred to enucleated oocytes recapitulated the early cleavage stages before compaction and blastocoel formation. Evidence that reprogramming occurred is the observation that some of the resulting embryos developed to term. If the nuclei had not been reprogrammed to behave as a zygote nucleus, then their developmental program would not be expected to match that of the surrogate uterus, and the conceptus would have been aborted.

Interestingly, most nuclear transfers in the mouse result in poorer development than in the cow and sheep. Apparently, reprogramming does not occur when eight-cell-stage or later nuclei are transferred to enucleated two-cell blastomeres or to enucleated zygotes (Modlinski 1981; McGrath and Solter 1984; Robl et al. 1986; Barnes et al. 1987), in contrast to an earlier, unconfirmed report (Illmensee and Hoppe 1981). Although inner-cell-mass nuclei, when transferred to enucleated meiotic metaphase II oocytes, can direct development of some embryos to morphologically normal blastocysts, their subsequent developmental potential has not been tested (Tsunoda et al. 1988 1989). Primordial germ cells from 12.5- to 16.5-day-old fetuses can direct development of a trophoblastic vesicle, but without an apparent inner cell mass (Tsunoda et al. 1989). The successful development to term of two-cell embryos receiving nuclei from later blastomeres by Tsunoda et al. (1987) contrasts with the results of Robl et al. (1986), who could only obtain development to midgestation stages. The cooling of the donor embryos prior to transfer is the only observable difference in procedures between the two groups and is not without precedent, as cooling has a beneficial affect on development when used in *Rana pipiens* (Hennen 1970).

Reprogramming of specific genes. Although the reprogramming of specific genes has yet not been identified in mammals, precise genomic reprogramming in amphibians has been described. Muscle-specific actin is produced in the myotome cells of the developing *Xenopus* embryo during the gastrula stage. Myotome cell nuclei that are transferred stop transcribing the muscle-specific actin gene and do not again produce muscle-specific actin until the gastrula stage, and then only in the myotome cells (Gurdon et al. 1984). A final example of nuclear reprogramming is that of the $5S^{ooc}$ gene. This gene is transcribed for only a short period of time during the late blastula stage. Nuclei from the gastrula stage transplanted to oocytes do not transcribe this gene until the resulting embryo progresses to the late blastula stage (Wakefield and Gurdon 1983). Thus, where nuclear reprogramming has been described at the molecular level, it has been temporally specific.

Splitting versus Cloning

With a full understanding of the two procedures of cloning obtained by splitting and by nuclear transfer, a direct comparison is warranted. For successful development to term, cloning by nuclear transfer must result in reprogramming the transferred nucleus so that it behaves as a zygotic nucleus capable of recapitulating early developmental events. Serial nuclear transfer would result in the daughter nuclei repeatedly traversing these early developmental events, and an unlimited number of identical individuals could potentially be created. Recent information in cattle shows that serial nuclear transfer can be conducted at least six times with resulting development to the morula or blastocyst stage (S. Stice, pers. comm.) and with subsequent development to term in third-generation nuclear transfers (Stice et al. 1991; K. Bondioli, pers. comm.). In addition, the largest clone set born alive is a group of 11 calves at Granada Genetics (K. Bondioli, pers. comm.). This is in contrast to cloning by splitting, in which the developmental cascade of events is not altered in split embryos. Because there is a minimum number of cells required to be present at the morula/blastocyst stage, the embryo can be split into a maximum of four portions (see above). Other regulatory mechanisms operate in split embryos that compensate for the reduced cell number, but these occur after the blastocyst stage. Cloning by splitting thus results in a limited number of identical individuals, whereas cloning by nuclear transfer has the potential to produce an unlimited number of genomically identical individuals.

APPLICATIONS TO SCIENCE AND AGRICULTURE

The ability to produce large numbers of identical individuals would have many benefits to both science and agriculture. In scientific experimentation, researchers could be sure that all observed variation in cloned animals would be due to the environment, rather than the nuclear genotype of the individuals. This would greatly reduce the number of animals required for experimentation in species where inbred strains are not available. Basic research on the mechanisms of early differentiation could be conducted using nuclear transfer techniques as a tool.

Many segments of agriculture would also benefit from these approaches. Maintenance of a clonal line in frozen storage (see Van Blerkom, this volume) while clones are tested would facilitate rapid advances in genetic improvement of livestock species. Large numbers of identical individuals would be ideal for high-level management systems, as all animals would be under the same management system and would respond similarly to management changes. Clonal lines would produce a known product. This would benefit producers and processors, who would be able to predict their product. Finally, consumers would benefit from receiving a known product. With the increase in rate of genetic change possible, breeders could change their product as consumer demand fluctuates. Thus, the potential for benefits from the application of this technology extends across not only agriculture, but all of society.

FUTURE DIRECTIONS

The procedures for cloning embryos by nuclear transfer in some species are currently limited by the inability to activate the oocyte adequately. This has been overcome in the rabbit with multiple pulses of electricity and precise timing of the stage of the recipient oocyte (Collas and Robl 1989; Ozil 1990). Not only are activation rates increased, but the rate of parthenogenic development to blastocyst and implantation stages is also increased. Current methods of activation in the pig (Hagen et al. 1991) and cow (Ware et al. 1989) result in adequate levels of pronuclear formation, but the rates of development to the morula or blastocyst stage are low (R.S. Prather, unpubl.). Although the oocytes can form a single pronucleus, this activation does not necessarily mimic that of fertilization. These results in domestic animals are in contrast to those of the mouse, where oocytes can be parthenogenetically activated at high rates and can subsequently develop to mid-gestation stages (Surani et al. 1987).

Another area of needed attention is that of transferring more developmentally advanced nuclei. To date, one of the most advanced nuclei that has been transferred is an inner-cell-mass-stage nucleus in the sheep (Smith and Wilmut 1989). Experiments using more advanced nuclei need to be conducted. Embryonic stem cells need to be developed in agriculturally important animals and evaluated as nuclei donors for cloning procedures. If developed, stem cells would preclude serial nuclear transfers, or could themselves be used to incorporate novel genes into the germ line of chimeras, as described in Stewart (this volume).

A third area of emphasis is that of DNA synthesis after nuclear transfer. The transfer of nuclei in G_1 of the cell cycle should result in progression through S phase in the recipient oocyte, whereas G_2 nuclei would not progress into S phase. However, the results of Blow and Laskey (1988) suggest that the integrity of the nuclear membrane is important in regulating a single round of DNA synthesis. Using the current methods of nuclear transfer that incorporate electrofusion may result in a transient breakdown of the transferred nuclear envelope, thus (according to the model by Blow and Laskey 1988) permitting the transferred nucleus in G_2 to undergo another round of replication that would interfere with subsequent development. These areas of research are in the forefront of investigations needed to understand the basic mechanisms of the currently available cloning procedures.

ACKNOWLEDGMENTS

This manuscript was prepared while supported by the Cooperative State Research Service, U.S Department of Agriculture under agreement no. 88-37240-4197 (R.S.P.), Food for the 21st Century (R.S.P.), USDA grant no. 88-37240-410 (J.M.R.), and is a contribution from the Missouri Agricultural Experiment Station Journal Series Number 11,158.

REFERENCES

Alt, F.W., K. Blackwell, and G.D. Yancopoulos. 1987. Development of the primary antibody repertoire. *Science* **238:** 1079.

Baker, R.D. and B.F. Shea. 1985. Commercial splitting of bovine embryos. *Theriogenology* **23:** 3.

Barnes, F.L., J.M. Robl, and N.L. First. 1987. Nuclear transplantation in mouse embryos: Assessment of nuclear function. *Biol. Reprod.* **36:** 1267.

Beerman, S. 1977. The diminution of heterochromatic chromosomal segments in Cyclops. *Chromosoma* **60:** 297.

Berg, H. 1982. Biological implications of electric field effects. V: Fusion of blastomeres and blastocysts of mouse embryos. *Bioelectrochem. Bioenerg.* **9:** 223.

Biggers, J.D. 1986. The potential use of artificially produced monozygotic twins for comparative experiments. *Theriogenology* **26:** 1.

Bloch, D.P. 1969. A catalog of sperm histones. *Genetics* (suppl.) **61:** 93.

Blow, J.J. and R.A. Laskey. 1988. A role for the nuclear envelope in controlling DNA replication within the cell cycle. *Nature* **332:** 546.

Bondioli, K.R., M.E. Westhusin, and C.R. Looney. 1990. Production of identical bovine offspring by nuclear transfer. *Theriogenology* **33:** 165.

Brem, G., B. Kruff, B. Szilvassy, and H. Tenhumberg. 1984. Identical simmental twins through microsurgery of embryos. *Theriogenology* **21:** 225

Briggs, R. and T.J. King. 1952. Transplantation of living nuclei from blastula cells into enucleated frogs' eggs. *Proc. Natl. Acad. Sci.* **38:** 455.

Buehr, M. and A. McLaren. 1984. Size regulation in chimaeric mouse embryos. *J. Embryol. Exp. Morphol.* **31:** 229.

Calvin, H.I. and J.M. Bedford. 1971. Formation of disulfide bonds in the nucleus and accessory structures of mammalian spermatozoa during maturation in the epididymis. *J. Reprod. Fertil.* (suppl.) **13:** 67.

Collas, P. and J.M. Robl. 1989. Age of recipient oocyte for nuclear transplantation in the rabbit. *Biol. Reprod.* (suppl. 1) **40:** 110.

Collas, P., J.J. Balise, G.A. Hofmann, and J.M. Robl. 1989. Electrical activation of mouse oocytes. *Theriogenology* **32:** 835.

Czolowska, R., J.A. Modlinski, and A.K. Tarkowski. 1984. Behavior of thymocyte nuclei in nonactivated and activated mouse oocytes. *J. Cell Sci.* **69:** 19.

Dean, W.L. and J. Rossant. 1984. Effect of delaying DNA replication on blastocyst formation in the mouse. *Differentiation* **26:** 134.

DeRoeper, A., J.A. Smith, R.A. Watt, and J.M. Barry. 1977. Chromatin dispersal and DNA synthesis in G1 and G2 HeLa cell nuclei injected into *Xenopus* eggs. *Nature* **265:** 469.

DiBerardino, M.A. 1979. Nuclear and chromosomal behavior in amphibian nuclear transplants. *Int. Rev. Cytol.* (suppl.) **9:** 129.

———. 1980. Genetic stability and modulation of metazoan nuclei transplanted into eggs and oocytes. *Differentiation* **17:** 17.

———. 1987. Genomic potential of differentiated cells analyzed by nuclear transplantation. *Am. Zool.* **27:** 623.

———. 1989. Genomic activation in differentiated somatic cells. In *Developmental biology* (ed. M.A. DiBerardino and L.D. Etkin), vol. 6, p. 175. Plenum Press, New York.

DiBerardino, M.A. and N. Hoffner. 1970. Origin of chromosomal abnormalities in nuclear transplants—A reevaluation of nuclear differentiation and nuclear equivalence in amphibians. *Dev. Biol.* **23:** 185.

———. 1975. Nucleocytoplasmic exchange of nonhistone proteins in amphibian embryos. *Exp. Cell Res.* **94:** 235.

Driesch, H. 1892. The potency of the first two cleavage cells in echinoderm development. Experimental production of partial and double formations. In *Foundations of experimental embryology* (ed. B.H. Willier and J.M. Oppenheimer), p. 38. Hafner, New York.

Ellinger, M.S. 1977. The cell cycle and transplantation of blastula nuclei in *Bombina orientalis*. *Dev. Biol.* **65:** 81.

Gardner, R.L., V.E. Papaioannou, and S.C. Barton. 1973. Origin of the ectoplacental cone and secondary giant cells in mouse blastocysts reconstituted from isolated trophoblast and inner cell mass. *J. Embryol. Exp. Morphol.* **30:** 561.

Graham, C.F. 1969. The fusion of cells with one- and two-cell mouse embryos. *Wistar Inst. Symp. Monogr.* **9:** 19.

Graham, C.F., K. Arms, and J.B. Gurdon. 1966. The induction of DNA synthesis by frog egg cytoplasm. *Dev. Biol.* **14:** 349.

Gurdon, J.B. 1964. The transplantation of living cell nuclei. *Adv. Morphol.* **4:** 1.

———. 1986. Nuclear transplantation in eggs and oocytes. *J. Cell Sci.* (suppl.) **4:** 287.

Gurdon, J.B., S. Brennan, S. Fairman, and T.J. Mohun. 1984. Transcription of muscle-specific actin genes in early *Xenopus* development: Nuclear transplantation and cell dissociation. *Cell* **38:** 691.

Hagen, D.R., R.S. Prather, and N.L. First. 1991. Response of porcine oocytes to electrical and chemical activation during maturation in vitro. *Mol. Reprod. Dev.* **28:** 70.

Hall, J.L., Z. Ramanis, and D.J.L. Luck. 1989. Basal body/centriolar DNA: Molecular genetic studies in chlamydomonas. *Cell* **59:** 121.

Hauswirth, W.W. and P.J. Laipis. 1982. Mitochondrial DNA polymorphism in a maternal lineage of Holstein cows. *Proc. Natl. Acad. Sci.* **79:** 4686.

Hennen, S. 1970. Influence of spermine and reduced temperature on the ability of transplanted nuclei to promote normal development in eggs of *Rana pipiens*. *Proc. Natl. Acad. Sci.* **66:** 630.

Heyman, Y. 1985. Factors affecting the survival of transferred whole and half-embryos in cattle. *Theriogenology* **23:** 63.

Hoffner, N.J. and M.A. DiBerardino. 1977. The acquisition of egg cytoplasmic nonhistone proteins by nuclei during nuclear reprogramming. *Exp. Cell Res.* **108:** 421.

Illmensee, K. and P.C. Hoppe. 1981. Nuclear transplantation in mus musculus: Developmental potential of nuclei from preimplantation embryos. *Cell* **23:** 9.

Kelly, S.J. 1977. Studies of the developmental potential of 4- and 8-cell stage mouse blastomeres. *J. Exp. Zool.* **200:** 365.

King, W.A. and T. Linares. 1983. A cytogenetic study of repeat-breeder heifers and their embryos. *Can. Vet. J.* **24:** 112.

Kirshna, M. and W.M. Generoso. 1977. Timing of sperm penetration, pronuclear formation, pronuclear DNA synthesis and first cleavage in naturally ovulated mouse eggs. *J. Exp. Zool.* **202:** 245.

Laipis, P.J., C.J. Wilcox, and W.W. Hauswirth. 1982. Nucleotide sequence variation in mitochondrial deoxyribonucleic acid from bovine liver. *J. Dairy Sci.* **65:** 1655.

Lambeth, V.A., C.A. Looney, S.A. Voelkel, K.G. Hill, D.A. Jackson, and R.A. Godke. 1982. Micromanipulation of bovine morulae to produce identical twin offspring. In *Proceedings of the 2nd World Congress on Embryo Transfer in Mammals*. Sept. 20–22, p 55. Annecy, France.

———.1983. Microsurgery on bovine embryos at the morula stage to produce monozygotic twin calves. *Theriogenology* **20:** 85.

Leibo, S.P. and W.R. Rall. 1987. Increase in production of pregnancies by bisection of bovine embryos. *Theriogenology* **27:** 245.

Lewis, N.E. and J. Rossant. 1982. Mechanisms of size regulation in mouse embryo aggregates. *J. Embryol. Exp. Morphol.* **72:** 169.

Luthardt, F.W. and R.P. Donahue. 1973. Pronuclear DNA synthesis in mouse eggs. *Exp. Cell Res.* **82:** 143.

Malissen, M., K. Minard, S. Mjolsness, M. Kronenberg, J. Goverman, T. Hunkapiller, M.B. Prystowsky, Y. Yoshikai, F. Fitch, T.W. Mak, and L. Hood. 1984. Mouse T cell antigen receptor: Structure and organization of constant and joining gene segments encoding the β polypeptide. *Cell* **37:** 1101.

McGrath, J. and D. Solter. 1983. Nuclear transplantation in the mouse embryo by microsurgery and cell fusion. *Science* **220:** 1300.

———. 1984. Inability of mouse blastomere nuclei transferred to enucleated zygotes to support development in vitro. *Science* **226:** 1317.

Menino, A.R. and R.W. Wright, Jr. 1983. Effects of pronase treatment, microdissection and zona pellucida removal on the development of porcine embryos and blastomeres in vitro. *Biol. Reprod.* **28:** 433.

Merriam, R.W. 1969. Movement of cytoplasmic proteins into nuclei induced to enlarge and initiate DNA or RNA synthesis. *J. Cell Sci.* **5:** 333.

Michaels, G.S., W.W. Hauswirth, and P.J. Laipis. 1982. Mitochondrial DNA copy number in bovine oocytes and somatic cells. *Dev. Biol.* **94:** 246.

Modlinski, J.A. 1981. The fate of inner cell mass and trophectoderm nuclei transplanted to fertilized mouse eggs. *Nature* **292:** 342.

Moore, N.W., C.E. Adams, and L.E.A. Rowson. 1968. Developmental potential of single blastomeres of the rabbit egg. *J. Reprod. Fertil.* **17:** 527.

Motlik, J., V. Kopecny, J. Pivko, and J. Fulka. 1980. Distribution of proteins labeled during meiotic maturation in rabbit and pig eggs at fertilization. *J. Reprod. Fertil.* **58:** 415.

Naish, S., S.J. Perreault, and B. Zirkin. 1987. DNA synthesis following microinjection of heterologous sperm and somatic cell nuclei into hamster oocytes. *Gamete Res.* **18:** 109.

Nicholas, J.S. and B.V. Hall. 1942. Experiments on developing rats. II. The development of isolated blastomeres and fused eggs. *J. Exp. Zool.* **90:** 441.

O'Brien, M.J., E.S. Critser, and N.L. First. 1984. Developmental potential of isolated blastomeres from early mouse embryos. *Theriogenology* **22:** 601.

Olivo, P.D, M.J. Van de Walle, P.J. Laipis, and W.W. Hauswirth. 1983. Nucleotide sequence evidence for rapid genotype shifts in the bovine mitochondrial DNA D-loop. *Nature* **306:** 400.

Ozil, J.P. 1990. The parthenogenetic development of rabbit oocytes after repetitive pulsatile electrical stimulation. *Development* **109:** 117.

Ozil, J.P., Y. Heyman, and J.P. Renard. 1982. Production of monozygotic twins by micromanipulation and cervical transfer in the cow. *Vet. Rec.* **110:** 126.

Papaioannou, V.E. and K.M. Ebert. 1986. Comparative aspects of embryo manipulation in mammals. In *Experimental approaches to mammalian embryonic development* (ed. J. Rossant and R.A. Pederson), p. 67. Cambridge University Press, England.

Poccia, D., J. Salik, and G. Krystal. 1981. Transitions in histone variants of the male pronucleus following fertilization and evidence for a maternal store of cleavage-stage histones in sea urchin eggs. *Dev. Biol.* **82:** 287.

Prather, R.S. and N.L. First. 1990a. Cloning of embryos. *J. Reprod. Fertil.* (suppl.) **40:** 227.

———. 1990b. Cloning embryos by nuclear transfer. *J. Reprod. Fertil.* (suppl.) **41:** 125.

Prather, R.S., M.M. Sims, and N.L. First. 1989a. Nuclear transplantation in pig embryos. *Biol. Reprod.* **41:** 414.

———. 1990b. Nuclear transfer in the pig: Nuclear swelling. *J. Exp. Zool.* **255:** 355.

Prather, R.S., F.L. Barnes, J.M. Robl, and N.L. First. 1986. Multiplication of bovine embryos. *Biol. Reprod.* (suppl.) **34:** 192.

Prather, R.S., J. Kubiak, G.G. Maul, N.L. First, and G. Schatten. 1991. The expression of nuclear lamin A and C epitopes is regulated by the developmental stage of the cytoplasm in mouse oocytes or embryos. *J. Exp. Zool.* **257:** 110.

Prather, R.S., M.M. Sims, G.G. Maul, N.L. First, and G. Schatten. 1989b. Nuclear lamin antigens are developmentally regulated during porcine and bovine embryogenesis. *Biol. Reprod.* **41:** 123.

Prather, R.S., F.L. Barnes, M.M. Sims, J.M. Robl, W.H. Eyestone, and N.L. First. 1987. Nuclear transplantation in the bovine embryo: Assessment of donor nuclei and recipient oocyte stage. *Biol. Reprod.* **37:** 859.

Prather, R.S., C. Simerly, G. Schatten, D.R. Pilch, S.M. Lobo, W.F. Marzluff, W.L. Dean, and G.A. Schultz. 1990a. U3 snRNPs and nucleolar development during oocyte maturation, fertilization and early embryogenesis in the mouse: U3 snRNA and snRNPs are not regulated coordinate with other snRNAs and snRNPs. *Dev. Biol.* **138:** 247.

Rands, G.F. 1985. Cell allocation in half-and-quadruple-sized preimplantation mouse embryos. *J. Exp. Zool.* **236:** 67.

Rao, M.V.N. 1980. Nuclear proteins in programming cell cycles. *Int. Rev. Cytol.* **67:** 291.

Robl, J.M., B. Gilligan, E.S. Critser, and N.L. First. 1986. Nuclear transplantation in mouse embryos: Assessment of recipient cell stage. *Biol. Reprod.* **43:** 733.

Roux, W. 1888. Contributions to the developmental mechanics of the embryo. On the artificial production of half-embryos by destruction of one of the first two blastomeres and the later development (postgeneration) of the missing half of the body. In *Foundations of experimental embryology* (ed. B.H. Willer and J.M. Oppenheimer), p. 2. Hafner, New York.

Schatten, G., C. Simerly, and H. Schatten. 1985. Microtubule configurations during fertilization, mitosis, and early development in the mouse and the requirement for egg microtubule-mediated motility during mammalian fertilization. *Proc. Natl. Acad. Sci.* **82:** 4152.

Seidel, G.E., Jr. 1983. Production of genetically identical sets of mammals: Cloning? *J. Exp. Zool.* **228:** 347.

Smith, L.C. and I. Wilmut. 1989. Influence of nuclear and cytoplasmic activity on the development in vivo of sheep embryos after nuclear transfer. *Biol. Reprod.* **40:** 1027.

Spemann, H. 1938. *Embryonic development and induction*, p. 210. Hafner, New York.

Stice, S.L. and J.M. Robl. 1988. Nuclear reprogramming in nuclear transplant rabbit embryos. *Biol. Reprod.* **39:** 657.

Stice, S.L., C.L. Keefer, M. Maki-Laurila and P.E. Phillips. 1991.

Producing multiple generations of bovine nuclear transplant embryos. *Theriogenology* **35:** 273.

Stricker, S., R. Prather, C. Simerly, H. Schatten, and G. Schatten. 1989. Nuclear architectural changes during fertilization and development. In *The cellular biology of fertilization* (ed. H. Schatten and G. Schatten), p. 225. Academic Press, New York.

Surani, M.A.H., S.C. Barton, and A. Burling. 1980. Differentiation of 2-cell and 8-cell mouse embryos arrested by cytoskeletal inhibitors. *Exp. Cell Res.* **125:** 257.

Surani, M.A.H., S.C. Barton, and M.L. Norris. 1987. Experimental reconstruction of mouse eggs and embryos: An analysis of mammalian development. *Biol. Reprod.* **36:** 1.

Szollosi, D., P. Calarco, and R.P. Donahue. 1972. Absence of centrioles in the first and second meiotic spindles of mouse oocytes. *J. Cell Sci.* **11:** 521.

Szollosi, D., R. Czolowska, M.S. Szollosi, and A.K. Tarkowski. 1988. Remodeling of mouse thymocyte nuclei depends on the time of their transfer into activated, homologous oocytes. *J. Cell Sci.* **91:** 603.

Tarkowski, A.K. 1959. Experimental studies on regulation in the development of isolated blastomeres of mouse eggs. *Acta Theriol.* **3:** 191.

Tarkowski, A.K. and J. Wroblewska. 1967. Development of blastomeres of mouse eggs isolated at the 4- and 8-cell stage. *J. Embryol. Exp. Morphol.* **36:** 155.

Tobler, H. 1975. The occurrence and developmental significance of gene amplification. In *Biochemistry of animal development* (ed. R. Weber), vol. 3, p. 91. Academic Press, New York.

Tsunoda, Y., T. Tokunaga, H. Imai, and T. Uchida. 1989. Nuclear transplantation of male primordial germ cells in the mouse. *Development* **107:** 407.

Tsunoda, Y., Y. Shioda, M. Onodera, K. Nakamura, and T. Uchida. 1988. Differential sensitivity of mouse pronuclei and zygote cytoplasm to Hoechst staining and ultraviolet irradiation. *J. Reprod. Fertil.* **82:** 173.

Tsunoda, Y., T. Yasui, Y. Shioda, K. Nakamura, T. Uchida, and T. Sugie. 1987. Full-term development of mouse blastomere nuclei transplanted into enucleated two-cell embryos. *J. Exp. Zool.* **242:** 147.

Von Bergoldington, C.H. 1981. The developmental potential of synchronized amphibian cell nuclei. *Dev. Biol.* **81:** 115.

Wakefield, L. and J.B. Gurdon. 1983. Cytoplasmic regulation of 5S RNA genes in nuclear-transplant embryos. *EMBO J.* **2:** 1613.

Ware, C.B., F.L. Barnes, M. Maiki-Laurila, and N.L. First. 1989. Age dependence of bovine oocyte activation. *Gamete Res.* **22:** 265.

Weismann, A. 1893. *The germ-plasm: A theory of heredity*. (Translated by W. Newton Parker and H. Ronnfeld.) Walter Scott, Ltd., London.

Westhusin, M.E., M.J. Levanduski, R. Scarborough, C.R. Looney, and K.R. Bondioli. 1990. Utilization of fluorescent staining to identify enucleated demi-oocytes for utilization in bovine nuclear transfer. *Biol. Reprod.* (suppl.) **42:** 176.

Whittingham, D.G. 1980. Parthenogenesis in mammals. In *Oxford reviews in reproductive biology* (ed. C.A. Finn), vol. 2, p. 205. Oxford University Press, England.

Wiesel, S. and G.A. Schultz. 1981. Factors which may affect removal of protamine from sperm DNA during fertilization in the rabbit. *Gamete Res.* **4:** 25.

Willadsen, S.M. 1979. A method for culture of micromanipulated sheep embryos and its use to produce monozygotic twins. *Nature* **227:** 298.

———. 1982. Micromanipulation of embryos of the large domestic species. In *Mammalian egg transfer* (ed. C.E. Adams). p. 185. CRC Press, Boca Raton, Florida.

———. 1986. Nuclear transplantation in sheep embryos. *Nature* **320:** 63.

Willadsen, S.M., H. Lehn-Jensen, C.B. Fehilly, and R. Newcomb. 1981. The production of monozygotic twins of preselected parentage by micromanipulation of non-surgically collected cow embryos. *Theriogenology* **15:** 23.

Williams, T.J. and L. Moore. 1988. Quick-splitting of bovine embryos. *Theriogenology* **29:** 477.

Williams, T.J., R.P. Elsden, and G.E. Seidel, Jr. 1982. Identical twin bovine pregnancies derived from bisected embryos. *Theriogenology* **17:** 114.

———. 1984. Pregnancy rates with bisected bovine embryos. *Theriogenology* **22:** 521.

Changes in Domestic Livestock through Genetic Engineering

K.M. Ebert[1] and J.P. Selgrath[2]
Departments of [1]Anatomy and Cellular Biology and [2]Comparative Medicine, Tufts University Schools of [1,2]Veterinary Medicine, [1]Medicine and Dental Medicine North Grafton, Massachusetts 01536

INTRODUCTION

The classic mechanism used to increase the efficiency of food production in domestic species is a breeding program designed to select animals with superior qualities and to transmit these traits through propagation of these animals. Recently, the propagation of superior lines has been accelerated by developments in biotechnology including artificial insemination (AI), embryo transfer (ET) and in vitro fertilization (IVF). With the advent of the "new biology," i.e., the direct manipulation of genetic information by means of molecular biology techniques, selected genes can now be transferred directly to the genome of an organism. The foreign genes in these animals can be expressed in such a way as to produce biologically active molecules that alter the phenotype of the transformed animal. More importantly, these phenotypic changes can be transferred from one generation to another.

This innovative technology, however, is far in advance of the present identification of economically beneficial phenotypic changes that are regulated by a single gene product. The optimization of this technology is thus governed by coordinate research that identifies genetic elements as prime targets for manipulation to alter appropriately the domestic species in a more beneficial fashion. With this in mind, we outline some of the areas that have been changed by classic breeding programs and postulate ways of genetically manipulating domes-

Animal Applications of Research in Mammalian Development
 0-87969-333-9/91 $3.00 + 00

tic species by the process of genetic engineering. We illustrate the genetic engineering approach with appropriate references to the initial stages of research using animal models.

GENE INJECTION TECHNOLOGY

As a result of the first successful transfer of a foreign gene into the embryo of a mouse (Gordon et al. 1980), microinjection of DNA has become a widely used technique for investigations of physiological, genetic, and immunological systems (for reviews, see Palmiter and Brinster 1986; Cuthbertson and Klintworth 1988; Hanahan 1989; Westphal 1989). The animals that incorporate the foreign gene are termed transgenic. The incorporation of genes into transgenic mice is generally stable and is passed on to succeeding generations in a Mendelian fashion (Palmiter and Brinster 1986). The mouse has been the preferred animal for genetic manipulation because of the vast amount of knowledge of the animal's genetics as it relates to physiology, immunology, and developmental biology. More recently, investigations with the mouse have indicated the usefulness of the transgenic technique to alter or augment useful traits in farm animals (see Table 1) (Ebert 1989; Pursel et al. 1989b).

The production of transgenic farm animals involves unique solutions to difficulties that are not encountered when utilizing mice as models (Steele and Pursel 1990). Figure 1 illustrates the production of transgenic goats in our laboratory. Although the following description is limited to the production of transgenic goats, the same steps are required to produce transgenic swine (Ebert et al. 1988), sheep (Clark et al. 1989; Rexroad et al. 1989), and cattle (McEvoy and Sreenen 1990).

Fertilized embryos at the pronuclear stage are collected from superovulated and synchronized females 72–78 hours following progesterone removal (Ebert et al. 1988; Selgrath et al. 1990). A midventral incision is made, and the reproductive tract of the animal is exteriorized. A fluted cannula is inserted into the oviductal ostium through the fimbria, and 15–20 ml of phosphate-buffered saline solution is retrograde-flushed using a 27-gauge needle inserted into the oviductal lumen at the uterotubal junction.

Oviduct flushings are collected and placed in sterile petri dishes and immediately examined under a stereomicroscope for the presence of ova. Embryos are placed on a depression slide for examination using Nomarski interference contrast microscopy (160x). Upon visualization of pronuclei, the male pronucleus is injected with several hundred copies of the foreign gene resuspended in 2–3 pl of 10 μM Tris-EDTA buffer using standard microinjection technology (Fig. 2). Some farm animal embryos are opaque and no nuclear structures are visible with interference contrast microscopy; however, embryos can be centrifuged at 13,000*g* in an Eppendorf centrifuge in order to partition the lipid inclusions to reveal the pronuclei (Ebert et al. 1988). Following examination and injection, embryos are placed in Ham's F12 nutrient media (GIBCO) plus 10% fetal calf serum and cultured under paraffin oil (Fisher, 0-122) in an atmosphere of 5% CO_2 in air at 37°C until transfer. Ideally, the embryos should be transferred to recipient females as soon as possible (1–3 hr) to minimize the deleterious effects of long-term culture on embryo viability (Papaioannou and Ebert 1986). However, it is also possible to generate transgenic livestock from one-cell-injected embryos that have been cultured up to 72 hours prior to transfer to the uterus of synchronized recipient females (K.M. Ebert, unpubl.).

Recipients are anesthetized, and the reproductive tract is exteriorized as described for donors. Ovaries are examined for the presence of corpus luteum (CL). Injected embryos are transferred with as little fluid as possible to the oviduct ipsilateral to the CL using a sterilized 50-μl capillary tube inserted into the oviduct lumen via the fimbria.

Analysis of gene incorporation is performed using a probe specific for the gene injected. Once a potential transgenic animal is born, DNA is extracted from blood monocytes and a small segment of ear tissue and then analyzed by dot-blot (Feinberg and Vogelstein 1983) and Southern analyses (Goosens and Kan 1981) to determine the copy number of the transgene. Northern analysis of the mRNA is required in order to determine the tissue-specific expression of the foreign gene and the relative concentration of the transcript in different cell types. Those animals determined to be transgenic are used to develop a unique line of transgenic animals. It should be noted

TABLE 1 *PRODUCTION OF TRANSGENIC LIVESTOCK*

Species	Gene[a]	Fragment length (kb)	Integration[b]	Copy number	Expression[c]	References
Swine	mMT-hGH	2.6	10.4	1–490	61.1	Hammer et al. (1985)
	mMT-bGH	2.6	6.0	1–28	88.9	
Sheep	mMT-hGH	2.6	1.3	1	0.0	Miller et al. (1989)
Swine	mMT-hGRF	2.5	4.0	30–100	28.6	Purcel et al. (1989b)
	mMT-hIGF-I	3.8	11.8	10	25.0	
Sheep	mMT-HSV-TK	8.4	3.4	1	0.0	Simons et al. (1988)
	sBLG-αIAT	11.0	9.1	4	0.0	
Sheep	sBLG-hFIX	11.0	7.7	1–40	50.0	Clark et al. (1989)
Swine	mWAP-mWAP	7.0	2.6	several	20.0	Pursel et al. (1990)
Swine	bPRL-bGH	2.7	20.0	1–10	40.0	Polge et al. (1989)
Sheep	mMT-bGH	2.6	4.3	3	100.0	Rexroad et al. (1989)
	mMT-hGRF	2.5	14.3	1–6	11.1	
Swine	hMT-pGH	2.7	35.3	1–15	16.7	Vize et al. (1988)

Sheep	sMT-sGH5	5.26	4.9	–	0.0	Murray et al. (1989)
	sMT-sGH9	4.86	13.0	–	100.0	
Swine	mMLV-rGH	2.6	6.7	8	100.0	Ebert et al. (1988)
Swine	mMLV-pGH	2.3	10.2	1–20	16.7	Ebert et al. (1990)
	hCMV-pGH	4.7	46.9	–	13.3	
	mMLV-pGHs	2.8	30.3	–	0.0	
Bovine	mMT-rGH	–	0.0	–	0.0	McEvoy and Sreenan (1990)
	mMT-rGH-BE	–	0.0	–	0.0	
	mMT-β-gal	–	0.0	–	0.0	
Bovine	mMT-hGH	4.0	6.7	–	0.0	Brem et al. (1985)
Swine	rPEPCK-bGH	–	–	–	–	Weighart et al. (1990)

[a]Abbreviations for promoters and structural genes: (b) bovine; (h) human; (m) mouse; (p) porcine; (r) rat; (s) sheep; (αIAT) α1-antitrypsin; (BE) viral sequences; (β-gal) β-galactosidase; (BLG) β-lactoglobulin; (FIX) clotting factor IX; (GH) growth hormone; (GRF) growth hormone releasing factor; (HSV-TK) herpes simplex virus thymidine kinase; (IGF-I) insulin-like growth factor I; (MT) metallothionein; (PEPCK) phosphoenolpyruvate carboxykinase; (PRL) prolactin; (WAP) whey acid protein.

[b]Calculated as a percentage of transgenic animals born.

[c]Calculated as a percentage of transgenic animals exhibiting foreign gene expression.

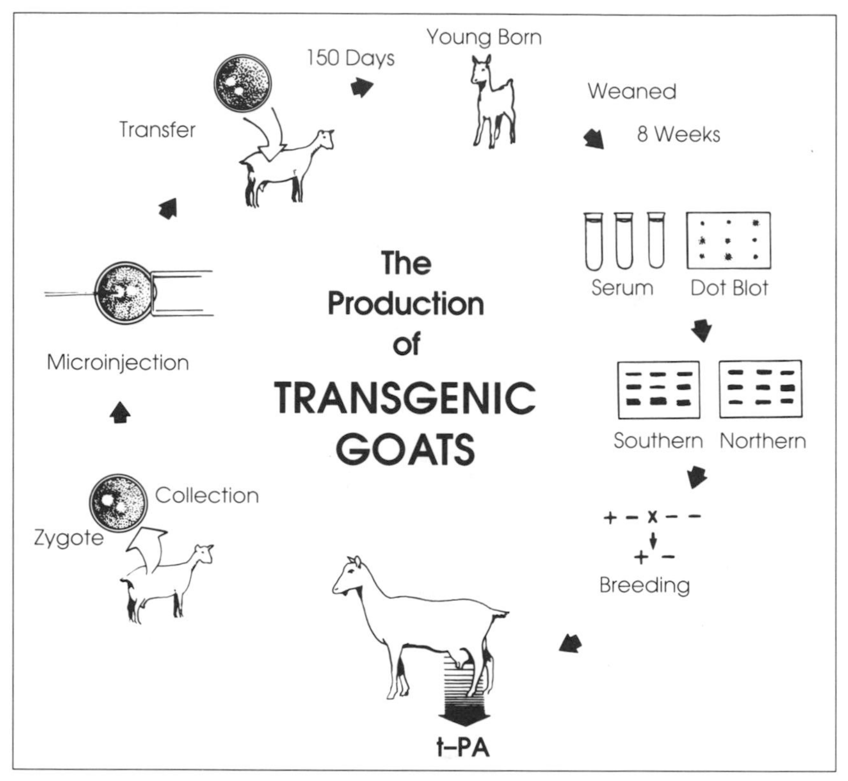

FIGURE 1 Schematic diagram outlining the methodologies used in the production of transgenic goats by microinjection of fusion genes.

that each transgenic animal produced by the conventional microinjection technique is unique in that integration is random and can occur in more than one location of the genome. Breeding schemes that are designed to produce homozygous transgenics often produce mutated animals, indicating that the integrated gene has disrupted developmentally important genes.

GROWTH PERFORMANCE

Almost all farm animal species have some aspect of their development or management based on growth performance. For the purpose of simplicity, we concentrate on animals whose economic production is directly related to growth (e.g., beef cattle, swine, and sheep).

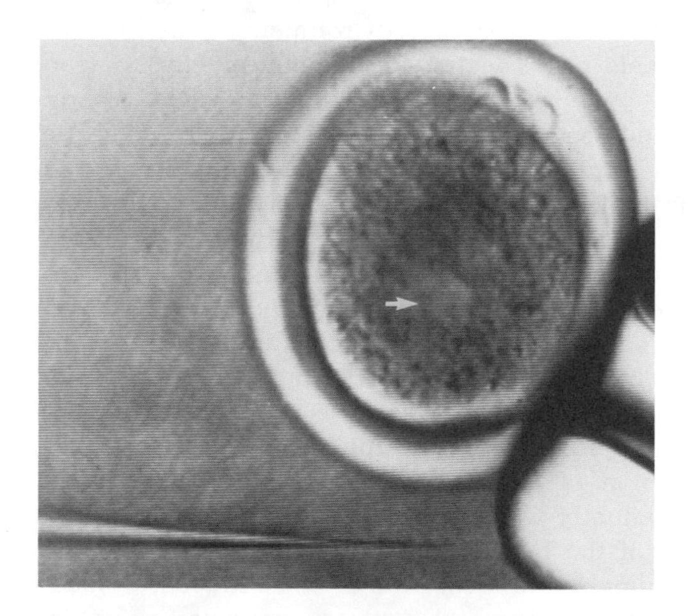

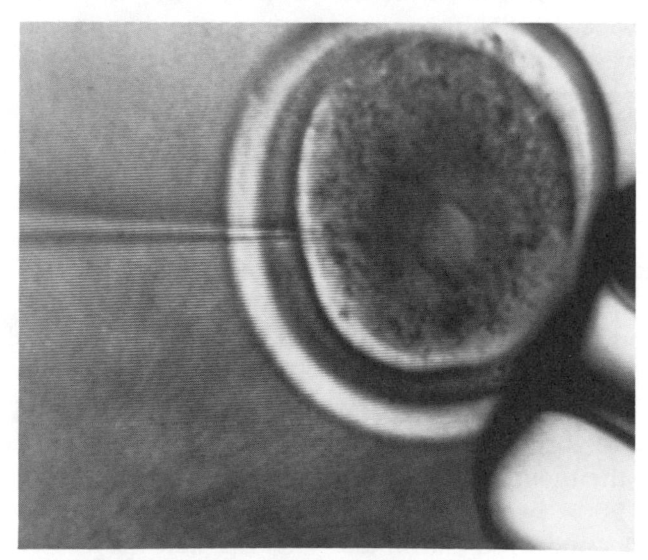

FIGURE 2 (*Top panel*) Goat zygote. The male pronucleus is the larger of the two pronuclei (arrow). (*Bottom panel*) Same zygote after injecting 2–3 pl of Tris-EDTA solution containing the foreign gene. Zygote is approximately 140 μm in diameter.

Cattle

Beef cattle are one of the few animals to produce a marketable product in dramatically different environments. The current production animals are selected for maximum efficiency in both permanent pastures, where stocking rates are several head per acre, and desert-like areas with stocking rates in the tens of acres per head (Koch and Algeo 1983). This diversity in environment is indicated by the number of different breeds and breed crosses examined for efficient beef production (Long 1980; Thrift and Aaron 1987). Even with the diversity in environment, animal type, or management methodology, the beef producer must meet market demands (consumer meat preference) within economic parameters (Thonney 1987). Consequently, we can draw conclusions as to the relative effect of historical breeding programs in the United States.

The introduction of the Research and Marketing Act in 1946 led to an industry-wide recognition of the need for genetic improvement of the beef animal in response to consumer demand. For example, the ideal animal of 1945 was a small-framed, early-maturing animal with a propensity for fat deposition (Warwick 1958), but current consumer demand is for a much different animal, with less fat and a more efficient feed-to-muscle conversion ratio. Selection for the optimum type of animal is often determined through performance evaluation programs (Willham 1982).

Performance traits are selected by identifying animals exhibiting superior qualities; however, the measurement of beef cattle performance is subject to individual interpretation (Gibson and Smith 1989). One unit of measure that is reasonably consistent is food conversion efficiency (lbs. fed to lbs. gained) (Davis 1987). The animal that produces the most meat for the lowest cost will be regarded as the superior line, independent of environment or breed. Consequently, beef-breeding programs select for the most efficient animal.

In the early 1900s, beef producers within the United States typically raised purebred animals of English ancestry (Hereford, Shorthorn, and Angus). With the development of technologies such as AI and ET, European breeds (exotics) were introduced to improve growth traits (Koch and Algeo

1983). The introduction of exotics illustrates a very important aspect of maximizing growth performance in the commercial cattle industry: crossbreeding versus breed selection. English breeds were selected to optimize available range resources, and in one comprehensive study, they had exhibited little change in basic growth parameters in a 20-year period previous to 1971 (Nadarajah et al. 1987). The introduction of European breeds, such as the Charolais and the Simmental, heavily influenced breeding programs in the years from 1970 to 1980.

Crossbreeding maximizes hybrid vigor, marked by the increase of complementary characteristics to produce offspring with the best qualities of both the exotic and English breeds. Instead of maintaining purebred lines, the modern cattle rancher maximizes his profits by selecting herd sires for growth (Nwakalor et al. 1986) and crossbreeding to cows with favorable maternal characteristics. Therefore, at the present time, selection within breeds has had little influence, and crossbreeding has had the major influence on the making of the modern beef animal. Future gains in beef growth performance must therefore come by maximizing growth characteristics within breeds, since the positive heterosis effects of crossbreeding are limited by the genetic potential of the initial purebred lines (Cunningham et al. 1987).

Swine

The swine industry in the United States also has a recent history of research-based improvement. Although the Regional Swine Research Breeding Laboratory was set up in 1937 to identify criteria to be used for selection, the commercial swine producer did not incorporate these criteria until the 1950s. The pork industry initially produced meat animals with a large amount of body fat. These "lard-type" animals were extremely inefficient in energy conversion because the production of fat requires substantial energy intake, thereby increasing the cost of raising swine (Fredeen and Harmon 1983). The need for a more energy-efficient animal to reduce production costs, combined with a consumer demand for leaner pork, forced the pork industry to select for leaner, more efficient market animals (Bereskin and Steele 1986). The carcass yield, which is the weight of an animal's carcass after slaughter divided by the live weight, has increased from 69.5% in 1960 to 71% in 1980. Lard, on the other hand, has decreased from 14.6% of body weight in 1960 to 5.3% in 1980 (Fredeen and Harmon 1983).

Selection for growth in market swine will continue to follow two related lines. An increase in feed efficiency will greatly decrease the cost per pound of pork produced, since feed is the greatest cost of pork production (Bereskin 1986). In addition, further growth rate increases will allow an animal to reach market weight earlier, thus decreasing both feed and housing care costs. As in beef cattle, improvements in growth and carcass composition are largely due to crossbreeding programs using selected breeds, rather than within-breed selection (Bereskin and Steele 1986; Buchanan 1987; McLaren et al. 1987).

Sheep and Goats

Since market lamb production in the United States is complementary to wool production, and market kid production is largely a by-product of the dairy goat business, very little selection has been done to improve the growth characteristics of purebred lines. Nevertheless, terminal crosses are used to produce marketable lambs selected from wool-breed ewes. The terminal cross-sire is usually selected from a meat breed. The use of terminal crosses is so important to the industry that breeds with superior wool characteristics are known as ewe breeds, and breeds with superior growth and muscle characteristics are known as ram breeds. The use of larger ram breeds has resulted in a doubling in market lamb weight since 1940 (Parker and Pope 1983). Again, this is mainly due to crossbreeding advantages, rather than selection for individual traits.

New Approaches

Although growth performance in domestic livestock is largely regulated by multigene product interactions, the direct administration of growth hormone (GH) alone can reduce carcass fat and increase feed efficiency in swine (Chung et al. 1985; Etherton et al. 1986, 1987), sheep (I.D. Johnson et al. 1985), and cattle (Brumby 1959). The expression of GH in transgenic mice from a fusion gene consisting of the metallothionein promoter linked to the rat or human GH gene increased the rate of growth and resulted in a doubling of the overall size of the animal (Palmiter et al. 1982). Gene transfer was immediately postulated as a viable approach toward changing the growth characteristics of domestic species to increase the efficiency of farm animals in an extremely short period of time (i.e., years) compared to the decades required by classic breeding regimes (Ward 1982; Ebert 1989).

The first promoter used to drive the expression of heterologous GH in pigs and sheep was the metallothionein promoter. Several laboratories have fused this promoter to either the rat or human GH structural genes and produced transgenic pigs and sheep (Hammer et al. 1985; Brem et al. 1985). Various themes of this basic construct have also been reported. For instance, Vize et al. (1988) used a gene construct consisting of the human metallothionein promoter linked to the porcine GH gene, and Miller et al. (1989) used the mouse metallothionein promoter and the bovine GH gene to create transgenic pigs. The metallothionein promoter may not be optimal for expression of foreign genes in transgenic livestock, and additional promoters were sought to direct expression in different tissues under different physiological conditions. Ebert et al. (1988) reported the successful expression of the rat GH gene with the use of the Moloney murine leukemia virus (Mo-MLV) promoter-enhancer. Subsequently, we have shown that both the Mo-MLV and the cytomegalovirus (CMV) promoters are effective in expressing porcine GH cDNA fusion genes in transgenic pigs (Ebert et al. 1990). Other attempts to regulate GH genes in transgenic livestock differentially used either the bovine prolactin promoter (Polge et al. 1989) or the rat phosphoenolypyruvate carboxykinase (PEPCK) promoter (Pinkert et al. 1990; Wieghart et al. 1990). Attempts were also made to obtain homologous regulatory elements combined with the homologous GH by producing transgenic sheep that contained the ovine metallothionein–ovine GH fusion gene (Murray et al.

1989). Although all of these experiments had varying degrees of expression, as indicated by the elevated levels of circulating GH, there was no control of the regulation of the gene, and thus GH was produced continuously.

Transgenic livestock that express heterologous (Hammer et al. 1985; Pursel et al. 1987, 1989a; Ebert et al. 1988; Miller et al. 1989; Rexroad et al. 1989) or homologous (Vize et al. 1988; Ebert et al. 1990) GH fusion genes have experienced adverse consequences of joint disease, lethargy, infertility, dermatitis, and renal disease. The concomitant failure to breed the original transgenic livestock successfully seriously challenges the application of this approach to alter efficiently and economically carcass fat or enhanced growth of domestic species. Alternate approaches will have to be developed to affect specific cell types, such as muscle and fat. Some of the approaches being considered are to increase circulating insulin-like growth factors specifically, to increase circulating levels of GH-binding proteins, and to increase growth hormone receptors in a tissue-specific manner. Alternate approaches are to formulate promoter elements that can be regulated by exogenous substances during the period of growth and development of the animal. Animals produced using any of these approaches could accrue economic benefits.

LACTATION

Dairy

The direct relationship between animal selection and economics is best illustrated by the dairy industry. In 1955, total United States milk production from slightly over 21 million dairy cows was almost 56 million kilograms. In 1980, 60 million kilograms of milk were produced by slightly under 11 million cows. In essence, milk production on a per cow basis doubled between 1955 and 1980 (Voelker 1981; Niedermeier et al. 1983). Improvements within the following three areas of the dairy industry are responsible for this increased milk production: farm management, nutrition, and artificial insemination. Although both nutrition and dairy farm management have had a major effect on the amount of milk produced by an individual dairy cow (Nott et al. 1981), they are not discussed here as these parameters are not likely to be manipulated through genetic engineering.

The direct effect of artificial insemination on the increase of milk production is related to the recognition of specific selection goals (Crowley and Niedermeier 1981; Miller 1981) by the industry. The recognition of specific desirable traits in milking animals led to the formation in 1927 of the Dairy Herd Improvement Association (DHIA) (Voelker 1981), which collects data on specific animals that are of economic importance to the dairy herder. The stringent records kept by the DHIA on individual cow performance are often used to determine cull and breeding choices. In addition, the DHIA record system has allowed criteria to be chosen and analyzed so that sires selected for performance can produce females with high lactation with a great degree of reliability (Miller 1981).

The first AI center was founded in New Jersey in 1938. Because early AI centers were without benefit of cryopreserved semen, their service was limited to farmers within relatively small geographical regions. The development of extenders increased the efficient use of bull semen, but it was not until sperm was frozen and thawed with reliable results did AI have a tremendous impact on the dairy industry (Miller 1981). Sires used for semen donors are chosen from young bulls following herdmate, daughter-dam, and progeny testing. A bull is used for artificial insemination only after the complete analysis of DHIA criteria for these comparisons is made. The animal is rated by expected progeny difference, or the difference between that bull's offspring and the herd average (Miller 1981; Voelker 1981; Niedermeier et al. 1983). Semen from these sires are collected, frozen, and available for purchase by producers. The improvements made by dairy cattle breeding are the result of within-breed selection rather than crossbreeding programs (Niedermeier et al. 1983), in contrast to improvements made in areas of animal management, such as beef and pork production.

The use of AI and sire selection based on DHIA records and daughter-dam comparisons has made a tremendous impact on the production of milk (Niedermeier et al. 1983). Very few farm animal production systems can match the tremendous increases in milk production on a per-animal basis. This reflects the change in consumer preferences for dairy products. Milk consumption has remained stable in the United States since 1960 (~236 kg per person per year), but dramatic decreases have occurred in whole milk consumption with a concurrent rise in the purchase of skim and lowfat milk (Niedermeier et al. 1983). The production of milk has increased approximately 40% since 1960; concurrently, the production of fat has increased only about 30% (Crowley and Niedermeier 1981). This reflects the increased prevalence of the Holstein breed that has been shown to have a larger milk production and a lower percentage of fat.

Consequences of Lactation for Beef Production

Weaning weight of calves is of great economic concern to the cow-calf producer, who is paid for the pounds of animals sold at weaning. Milk production by the dam is one of the main factors influencing weaning weight of calves (Boggs et al. 1980; Beal et al. 1990). At the present time, semen catalogs list beef bull expected progeny differences for daughter's milk production. But until AI becomes more prevalent in beef production, the rancher will probably continue to increase cow milk production through crossbreeding programs based on breed characteristics instead of by individual sire records as is done in the dairy industry.

New Approaches

The ability to increase lactational efficiency is also an area where genetic engineering methodologies are of use. Initial experiments by Machlin (1973) showed that exogenous bovine growth hormone, when given to lactating dairy cows, increased milk production by 18%. However, the expense of purifying bovine growth hormone from natural sources far outweighs the economical advantage of its use for increasing milk production. Seeburg et al. (1983) successfully used recombinant DNA technologies to produce large amounts of bovine growth hormone by fermentation. Several corporations are now scaling up this process in order to make the growth hormone, also known as bovine somatotropin (BST), commercially available.

The potential of effectively elevating endogenous levels of BST by the microinjection of fusion genes may be a competitive approach to the exogenous hormone injection regime. However, the major obstacle to this approach is the current lack of identified genetic regulatory elements required to increase the transcriptional activity of the fusion gene during lactation. Any endogenous control of lactation (i.e., using prolactin or steroids) would probably result in chronic elevated circulating growth hormone that would cause adverse side effects for these animals. One possible approach would be to develop a fusion gene containing the BST gene driven by an exogenous promoter that could be regulated by an artificial food additive. At present, this is a formidable task, and the exogenous delivery of growth hormone is the most attractive means of increasing milk production by biotechnology methods.

Apart from trying to increase milk production by hormone regulation, the mammary gland may also be a target for altering the protein constituency of the milk. Along these lines, transgenic dairy animals may provide us with an inexhaustible supply of valuable pharmaceutical polypeptides if we could construct transcriptionally efficient fusion genes that are expressed only in the mammary gland and produce large amounts of the polypeptide coincident with lactation. Gene transfer can also be directed toward changing the milk constituents in domestic livestock, either by the introduction of more copies of existing genes or by designing genes that utilize the most active promoter elements in milk protein genes and fusing these elements to other structural genes. Milk constituents that alter the processing qualities of milk and its byproducts offer an additional target for transgenic livestock (Wilmut et al. 1990). Introduction of a gene into the mouse for the sheep β-lactoglobulin (BLG), which is not present in rodents, resulted in dramatic changes in milk constituents; the expression levels actually exceeded the relative concentration of BLG in sheep milk (Simons et al. 1987). Such experiments offer the promise of producing modified animals that will have enhanced nutritional and economic values. The cloning of mammary-specific regulatory genetic elements and the fusion of these elements to structural genes are presently the major approach in transgenic technology for the production of specific gene products in milk.

The transcription of the mouse mammary tumor virus (MMTV) in mammary carcinomas has been reported (Parks et al. 1974; Lasfargues et al. 1976). The presumptive regulatory long terminal repeat (LTR) sequences of the virus has been fused to the c-*myc* structural gene and microinjected into mouse zygotes. The resulting transgenic mice were shown to produce mammary adenocarcinomas following pregnancy, indicating that the LTR sequences were adequate to direct the expression of a fusion gene in mammary tissue (Stewart et al. 1984). Further dissection of the LTR region and analysis of the regulatory behavior of other fusion genes in transgenic animals showed that expression of a thymidine kinase reporter gene occurred in lactating mammary gland (Ross and Solter 1985).

The MMTV LTR sequences, however, were not tissue-specific for the mammary gland and therefore would not be appropriate for specific production of a pharmaceutical in lactating animals. A more tissue-specific gene product would be required to assure that the product under the control of the promoter would exclusively be released into milk and not other body fluids, where it might affect the physiological state of the transgenic animal. Genes coding for milk products may be better candidates for a mammary-tissue-specific promoter that would direct synthesis of the product by mammary epithelial cells selectively. Gertler et al. (1982) found that the casein promoter was an effective promoter for specific transcription in the mammary gland. This information became the impetus for creating mammary-specific fusion gene products.

Through transgenic technology, various milk protein gene promoters have been used to direct the production of biologically active peptides by mammary epithelial cells (Table 2). More importantly, the peptides were shown to be secreted into the milk effectively and to have biological activity following purification. Two laboratories concurrently reported the production of transgenic mice that were capable of producing foreign peptides in their milk. Simons et al. (1987) used the sheep BLG gene and showed that transgenic mice could secrete more BLG in their milk than is found in sheep milk.

TABLE 2 *PRODUCTION OF FOREIGN PROTEINS BY TRANSGENIC ANIMALS*

Species	Promoter	Gene	Fragment length (kb)	Copy number	Specificity	Expression levels [a]	References
Mouse	bovine α-lactalbumin	bovine α-lactalbumin	3.1	1–40	mammary	450 μg/ml	Vilotte et al. (1989)
Sheep	sheep β-lactoglobulin	human anti-hemophilic factor IX	12.1	10	mammary	25 ng/ml	Clark et al. (1989)
Mouse	mouse whey acid protein	human tissue plasminogen activator	2.6	3–50	mammary tongue sublingual gland kidney	400 ng/ml	Gordon et al. (1987); Pittius et al. (1988b)
Mouse	mouse whey acid protein	human CD4	3.4	1–several hundred	mammary	200 ng/ml	Yu et al. (1989)
Mouse	sheep β-lactoglobulin	sheep β-lactoglobulin	16.2	1–20	mammary	23 mg/ml	Simons et al. (1987)
Mouse	rat β-casein	bacterial CAT	2.8	1–40	mammary thymus	–	Lee et al. (1989)
Rabbit	rabbit β-casein	human interleukin-2	6.9	2–10	mammary	430 ng/ml	Buhler et al. (1990)
Swine	mouse whey acid protein	mouse whey acid protein	7.0	several	mammary	–	Pursel et al. (1990)
Mouse	bovine αS1-casein	human urokinase	45.0	2–100	mammary	2 mg/ml	Meade et al. (1990)

[a]Represents amount of protein secreted into milk.

Similarly, Gordon et al. (1987) used a fusion gene consisting of the murine whey acid protein gene promoter linked to the human tissue plasminogen activator (tPA) cDNA. Their transgenic mice were shown to produce biologically active tPA in their milk at concentrations as high as 50 $\mu g\ ml^{-1}$. It is important to note that neither gene construct was exclusively expressed in the mammary tissue, and this may be cause for concern if the biologically active peptides are produced outside the mammary tissue (Simons et al. 1987; Pittius et al. 1988a,b). Recently, experiments by Gordon and co-workers (pers. comm.) have not indicated any adverse effects of biologically active tPA in mice, as these mice have produced several generations of animals with no overt pathological problems.

Extrapolating from the mice experiments to domestic livestock, similar applications would obviously result in production of large volumes of milk, making the transgenic pharmaceutical animal approach economically viable. Transgenic sheep carrying fusion genes containing the BLG promoter sequences linked to either human clotting factor IX or human α1-antitrypsin genes have been reported (Simons et al. 1988; Clark et al. 1989). More recently, our laboratory, in collaboration with Genzyme Corporation, has successfully microinjected dairy goat embryos and produced transgenic goats containing a fusion gene consisting of the murine whey acid protein gene promoter linked to the human tPA cDNA. A female expressing the fusion gene at a concentration of 3.2 mg/liter has produced biologically active tPA throughout her lactation, with no effect on the lactational output (Fig. 3).

REPRODUCTION

The efficient production of farm animals must include some selection for reproduction. This selection criterion can include increased litter size/twinning rate, decreased interval to rebreeding, decreased reproductive disorders such as dystocia (complicated delivery) or premature abortion, and earlier puberty. The major control of reproductive efficiency is primarily regulated by complex physiological systems. The endocrine, nutritional, behavioral, and physical parameters all

FIGURE 3 (*Top panel*) Transgenic female Alpine goat that secreted tissue plasminogen activator (tPA) in her milk throughout lactation. (*Bottom panel*) Two of her offspring.

play major roles in optimizing reproductive potential. Again, our classic breeding programs have been selected for the optimization of some of these interactions. More importantly, these physiological systems may be directly altered through genetic engineering.

Beef Cattle

The primary selection criteria in the beef breeds have been for an earlier puberty and an earlier return to estrus (Bourdon and Brinks 1983). Animals that have a short return to estrus postparturition have the economic benefit of decreasing the time a producer is paying for that animal without continued production. Earlier puberty is also important to the producer for a variety of reasons. First, a heifer that breeds early in its reproductive life produces more calves over its lifetime than a heifer that is bred later (Donaldson 1968). Second, an animal that conceives early in the breeding season maintains that capability throughout its lifetime (Lesmeister et al. 1973). Third, a heifer that breeds early in its first breeding season will produce heavier calves and will calve earlier than its late-breeding counterparts (Short and Bellows 1971). All of these traits have economic impacts.

Selection for these traits has not been within-breed but has been heavily influenced by crossbreed selection. The use of breed crosses is suitable to provide and maintain maximum heterosis (Urick et al. 1986). Breeding younger heifers, coincident with the use of sires selected for maximum gain, has resulted in an increase in dystocia (Norris et al. 1986). Artificial insemination companies have addressed this problem by assigning a "calving ease" score to potential sires (Elzo et al. 1987). However, fewer than 5% of all commercial beef cows are bred by AI (Koch and Algeo 1983) and the calving-ease rating generally does not affect the beef producer.

Dairy Cattle

Due to the high-intensity reproductive management of most dairies (of which 60% use AI exclusively), the records on return to estrus postbreeding are complete. The selection for reproductive performance is the same as that for beef cattle. However, since the measure of a dairy cow's economic worth is the amount of milk produced per year, and not pounds of beef, selection for reproduction traits is less important than selection for milk production.

Swine

The economics of swine production are based on the pounds of pork produced per sow per year and are directly related to the average number of offspring per litter and the growth performance of the offspring (Fredeen and Harmon 1983). For an increase in fecundity to be economically important, it must be measured as the result in an increased number of weaned piglets per sow. Weaning rates in litter-bearing animals are affected by three variables in reproductive/natal management. The three areas are (1) increased ovulation rates, leading to an increased number of viable fetuses; (2) increased postfertilization survival of embryos and fetuses, leading to higher litter size; and (3) increased mothering ability, as evidenced by a higher neonatal survival rate. In designing the optimum production operation, the most successful formula would be to maintain both a mothering breed to optimize numbers of offspring born per litter and a meat breed that is genetically superior for growth performance. Crossing of these two breeds would generate an F_1 hybrid that incorporates the most economically sound component of both breeds.

Sows can have 2.3 litters per year (Martinat-Botte' et al. 1985). Therefore, an increase of two piglets per litter is an increase of almost five pigs per year per sow going to market. Initial introduction of breeds, such as the Yorkshire and Landrace (Gaugler et al. 1984), into programs specifically designed to increase litter size has had a beneficial effect on the industry. However, the need for large litter size and the selection for this trait has not substantially changed the number of piglets per litter; selection for increased litter size has only accounted for 0.23 additional piglets per farrowing during the last 20 years (Fredeen and Harmon 1983). In a study to determine the effectiveness of selection to increase ovulation rates over ten generations, Johnson et al. (1984) found no change in ovulation rate in the selected lines versus the controls. To confound the selection for litter size, females raised in a large litter are at an environmental disadvantage during neonatal development. This disadvantage ultimately affects their reproductive potential, and a decrease in the size of the litters they produced (as compared to their mothers' litters) is often observed (Bichard and David 1985). Consequently, although selection for litter size is extremely important, it will not be until specific selection criteria are accepted and used by the pork industry, and confounding effects of environment are understood, that real change in litter size can be expected (R.K. Johnson et al. 1985). This may reflect the genetic limitation of the chosen meat lines for which we have historically selected. Yet the fact that the "Chinese" breeds of swine (e.g., the Meishan and Jiaxing) have an average of 13–17 pigs per litter demonstrates the actual genetic potential that can be realized (Legault 1985).

Sheep and Goats

Reproductive performance in sheep is one of the major factors controlling profitability (Bradford et al. 1986). The need for multiple offspring has led to the successful breeding for prolific sheep. Breeds such as the Romanov (Draincourt et al. 1988) and the Finnish Landrace (Vesely and Swierstra 1987) are known for high rates of twinning. The ability to select for twinning rate within breed lines indicates that a selectable physiological control mechanism exists. This is further indicated by the use of exogenous gonadotropins to increase ovulation rates. The use of specific proteins, such as exogenous follicle stimulating hormone (FSH), given immediately prior to estrus increases ovulation rates (Selgrath et al. 1983). Although high endogenous levels of FSH have been proposed as the reason for the high fecundity rates of specific prolific breeds (Lahlou-Kassi et al. 1984), there is evidence that high fecundity is not directly related to increasing endogenous levels of FSH in all highly prolific breeds (Draincourt et al. 1988).

The Booroola strain of Merino sheep are unique in that average ovulation (5.08) and lambing rates (2.29) are extremely high. There is also evidence that the increased fecundity is related to increased circulating FSH levels but not increased levels of luteinizing hormone (LH; Bindon and Piper 1986). In addition, the fecundity of the Booroola strain can be traced to a single allele, denoted as the *F* gene, which can be incorporated into crossbreeding programs designed to increase fecundity. The increased fecundity of animals with the *F* gene indicates that increasing fecundity is an area that is well suited to genetic manipulation.

New Approaches

Although major histocompatibility complex (MHC) antigens play essential roles in the immune tolerance of tissue from one animal to another, these surface antigens may also play a role in reproductive events. The preimplantation embryo development gene (PED), initially found in mice, appears to be associated with the H-2 complex and may affect the rate of early embryonic development (Warner 1986). Similar MHC genes associated with domestic species may be yet another target for genetically manipulating the reproductive efficiency of our livestock. Evidence exists which suggests that certain swine leukocyte antigen (SLA) gene clones can transfect mouse L cells, resulting in cell-surface expression of SLA antigens (Singer et al. 1983). Rothchild et al. (1984) have implicated the SLA complex as a correlate to high and low ovulation rates. A detailed investigation of the influence of the SLA haplotype on both ovulation rates and litter size was also reported using miniature pigs specifically bred for MHC haplotypes (Conley et al. 1988). Transgenic mouse models containing the porcine MHC antigens on the surface of cells have recently been produced (Frels et al. 1985). These models may provide the means to test the theory that MHC antigens influence reproductive performance.

IMMUNE RESPONSE

Until recently, the advantage of breeding for disease resistance was not recognized. The classic methods of preventing disease in herds was either to vaccinate or to prevent contact between infected and clean animals. It is now recognized that there are elements of disease resistance/susceptibility that are genetically based. A Nobel Prize was awarded in 1980 to Benacerraf, Snell, and Dausset for their recognition of the role of the MHC in the immune response. A single gene of the MHC was found to influence immune response levels (McDevitt and Benacerraf 1969; Benacerraf and McDevitt 1972). The susceptibility of chickens to leukosis virus is receptor-mediated, and breeding chickens without the receptor confers resistance (Crittendon 1974). The resistance of certain sheep to the nematode

Haemonchus contortus is also genetically linked and is inherited as a dominant characteristic (Wakelin 1978).

It should be possible to select for a resistance to a disease that is genetically determined. Selection for resistance has already occurred in some farm animal species. Swine have been selected for a natural resistance to *Brucella suis* with positive results (Warner et al. 1987). As the level of understanding of immune responses grows, so will the ability to use present breeding selection techniques to promote disease resistance.

New Approaches

Although disease resistance as conferred by the immunological system is very complex and involves both cell-mediated and humoral immunity, certain aspects of the immune process can be targeted for potential alterations by genetic engineering. As described in an excellent review by Hanahan (1989), various alterations within the complex immune system are being produced through the production of transgenic mice; this work underscores the possible new approaches to generating genetically modified domestic species that will have improved disease resistance. However, as there is very little information on the molecular etiology of economically important diseases, it is premature to identify the likely targets for genetic manipulation via transgenic livestock.

Some of the possible approaches toward modifying the immune system in transgenic livestock have been nevertheless exemplified through several mouse models. Functional immunoglobulin proteins are major defense mechanisms against diseases. It has been postulated that disease resistance to pathogenic organisms could be inherited if the foreign immunoglobulin genes could be formulated to produce biologically active antibodies.

For some time, transgenic technology has produced mice that make functional Igκ gene products in the form of authentic heavy and light chain proteins (Brinster et al. 1983). Subsequently, several laboratories have produced transgenic mice expressing immunoglobulin molecules (Storb 1987) and MHC genes (LeMeur et al. 1985; Pinkert et al. 1985; Yamamura et al. 1985). These types of experiments have been extended to develop an animal model that would produce a specific antibody that offers protection against a certain pathogenic organism. Pinkert et al. (1989) have generated transgenic mice that express functional rearranged heavy and light chain immunoglobulin against the antigen phosphorylcholine, an antigen found on the surface of bacteria and fungi. These experiments suggest that resistance to pathogens can be conferred in transgenic animals.

Viral oncogenesis is also an important area for disease control in domestic species. Lacey et al. (1986) produced transgenic mice that expressed the viral genome of the economically important bovine papillomavirus type 1 (BPV-1) DNA. These animals were susceptible to tumors of dermal origin, resulting in a tissue-specific phenotype similar to that observed in BPV-1 infection of cattle. This model system could be used in studies of BPV-1 etiology and therapy.

SUMMARY

Conventional breeding programs have continued to improve the domestic farm animal. However, breeding strategies require prolonged periods of time to develop and assess changes occurring over several generations. Changes in consumer preferences may require an ability to quickly change market animals by deviating from breeding schemes that are producing the present animals. For example, the pork and beef industry made concerted efforts to produce meat with substantial fat and marbling qualities to conform to consumer taste. However, the current emphasis on decreasing cholesterol intake may necessitate product change to satisfy consumer preference for extremely lean meat. Utilizing only conventional breeding strategies would require an extensive period of time (~20 years). Through genetic engineering, it may be possible to accomplish this change in animal carcass quality in a much shorter period of time. Significant changes in carcass composition can be realized either by hormone injections or by increased production of hormones within the animal using the transgenic approach.

The transfer of genetic material by recombinant DNA technologies is only a few years old, but the concept of producing transgenic agricultural species is generating momentum. Areas that have been emphasized in classic breeding programs might benefit from direct manipulation at the gene level. However, if the reality of modifying livestock species through genetic engineering is to be achieved, major efforts are required to learn more about the genes and the regulatory elements responsible for these potential changes. This new transgenic approach may eventually result in new forms of domestic livestock that can increase food production, improve the efficiency of feed conversion, increase disease resistance, and become bioreactors for the production of valuable pharmaceuticals for human and animal medicine.

ACKNOWLEDGMENTS

The authors gratefully acknowledge the contributions of the following people to the Transgenic Program at Tufts University: R. Godin, D. Paquin-Platts, K. Baldwin, and T. Smith for technical help; T. Smith for graphic design; B. Bannon and J. Ebert for helping in the preparation of this manuscript; and D. Willman for the goat photographs.

REFERENCES

Beal, W.E., D.R. Notter, and R.H. Akers. 1990. Techniques for estimation of milk yield in beef cows and relationships of milk yield to calf weight postpartum reproduction. *J. Anim. Sci.* **68:** 937.

Benacerraf, B. and H. McDevitt. 1972. Histocompatibility linked immune response genes. A new class of genes that controls the formation of species immune response has been identified. *Science* **175:** 273.

Bereskin, B. 1986. A genetic analysis of feed conversion efficiency and associated traits in swine. *J. Anim. Sci.* **62:** 910.

Bereskin, B. and N.C. Steele. 1986. Performance of Duroc and Yorkshire boars and gilts and reciprocal breed crosses. *J. Anim. Sci.* **62:** 918.

Bichard, M. and P.J. David. 1985. Effectiveness of genetic selection for prolificacy in pigs. *J. Reprod. Fertil.* (suppl.) **33:** 127.

Bindon, B.M. and L.R. Piper. 1986. Booroola (*F*) gene: Major gene affecting ovine ovarian function. In *Genetic engineering of animals, an agricultural perspective* (ed. J.W. Evans and A. Hollander), p. 67. Plenum Press, New York.

Boggs, D.L., E.F. Smith, R.R. Schalles, B.E. Brent, L.R. Corah, and R.J. Pruitt. 1980. Effects of milk and forage intake on calf performance. *J. Anim. Sci.* **51:** 550.

Bourdon R.M. and J.S. Brinks. 1983. Calving date versus calving interval as a reproductive measure in beef cows. *J. Anim. Sci.* **57:** 1412.

Bradford, G.E., J.F. Quirke, and T.R. Famula. 1986. Fertility, embryo survival and litter size in lines of Targhee sheep selected for weaning weight or litter size. *J. Anim. Sci.* **62:** 895.

Brem, G., B. Brenig, H.M. Goodman, R.C. Selden, F. Graf, B. Kruff, K. Springman, J. Hondele, J. Meyer, E.-L. Winnaker. and H. Krausslich. 1985. Production of transgenic mice, rabbits and pigs by microinjection into pronuclei. *Zuchthygiene* **20:** 251.

Brinster, R.L., K.A. Ritchie, R.E. Hammer, R.L. O'Brien, B. Arp, and U. Storb. 1983. Expression of a microinjected immunoglobulin gene in the spleen of transgenic mice. *Nature* **306:** 332.

Brumby, P.J. 1959. The influence of growth hormone on growth in young cattle. *N. Z. J. Agric. Res.* **21:** 683.

Buchanan, D.S. 1987. The crossbred sire: Experimental results for swine. *J. Anim. Sci.* **65:** 117.

Buhler, T.A., T. Bruyere, D.F. Went, G. Stranzinger, and K. Burki. 1990. Rabbit β-casein promoter directs secretion of human interleukin-2 into the milk of transgenic rabbits. *Biotechnology* **8:** 140.

Chung, C.S., T.D. Etherton, and J.P. Wiggins. 1985. Stimulation of swine growth by porcine growth hormone. *J. Anim. Sci.* **60:** 118.

Clark, A.J., H. Bessos, J.O. Bishop, P. Brown, S. Harris, R. Lathe, M. McClenaghan, C. Prowse, J.P. Simons, C.B.A. Whitelaw, and I. Wilmut. 1989. Expression of human anti-hemophilic factor IX in the milk of transgenic sheep. *Biotechnology* **7:** 487.

Conley, A.J., Y.C. Jung, N.K. Schwartz, C.M. Warner, M.F. Rothschild, and S.P. Ford. 1988. Influence of SLA haplotype on ovulation rate and litter size in miniature pigs. *J. Reprod. Fertil.* **82:** 595.

Crittendon, L.B. 1974. Two levels of genetic resistance to lymphoid leukosis. *Avian Dis.* **19:** 281.

Crowley, J.W. and R.P. Niedermeier. 1981. Dairy production 1955 to 2006. *J. Dairy Sci.* **64:** 971.

Cunningham, B.E., W.T. Magee, and H.D. Ritchie. 1987. Effects of using sires selected for yearling weight and crossbreeding with beef and dairy breeds: Birth and weaning weights. *J. Anim. Sci.* **64:** 1591.

Cuthbertson, R.A. and G.K. Klintworth. 1988. Biology of disease: Transgenic mice—a gold mine for furthering knowledge in pathobiology. *Lab. Invest.* **58:** 484.

Davis, M.E. 1987. Divergent selection for postweaning feed conversion in beef cattle: Predicted response based on an index of feed intake and gain vs feed:gain ratio. *J. Anim. Sci.* **65**:886.

Donaldson, L.E. 1968. The pattern of pregnancies and lifetime productivity of cows in a Northern Queensland beef cattle herd. *Aust. Vet. J.* **44:** 493.

Draincourt, M.A., P. Philipon, A. Locatelli, E. Jacques, and R. Webb. 1988. Are differences in FSH concentrations involved in the control of ovulation rate in Romanov and Ile-de-France ewes? *J. Reprod. Fertil.* **83:** 509.

Ebert, K.M. 1989. Gene transfer through embryo microinjection. In *Animal biotechnology: Comprehensive biotechnology*, 1st. supplement (ed. L.A. Babrick et al.), p. 233. Pergamon Press, England.

Ebert, K.M., T.E. Smith, F.C. Buonomo, E.W. Overstrom, and M.J. Low. 1990. Porcine growth hormone gene expression from viral promoters in transgenic swine. *Anim. Biotech.* **1:** 145.

Ebert, K.M., M.J. Low, E.W. Overstrom, F.C. Buonomo, C.A. Baile, T.M. Roberts, A. Lee, G. Mandel, and R.H. Goodman. 1988. A Moloney MLV-rat somatotropin fusion gene produces biologically active somatotropin in a transgenic pig. *Mol. Endocrinol.* **2**: 277.

Elzo, M.A., E.J. Pollak, and R.L. Quaas. 1987. Genetic trends due to bull selection and differential usage in the Simmental population. *J. Anim. Sci.* **64:** 983.

Etherton, T.D., J.P. Wiggins, C.S. Chung, C.M. Evock, J.F. Rebhun, and P.E. Walton. 1986. Stimulation of pig growth performance by porcine growth hormone and growth hormone-releasing factor. *J. Anim. Sci.* **63:** 1389.

Etherton, T.D., J.P. Wiggins, C.M. Evock, C.S. Chung, J.F. Rebhun, P.E. Walton, and N.C. Steele. 1987. Stimulation of pig growth performance by porcine growth hormone: Determination of the dose-response relationship. *J. Anim. Sci.* **64:** 433.

Feinberg, A.P. and B. Vogelstein. 1983. A technique for radiolabelling DNA restriction endonuclease fragments to high specific activity. *Anal. Biochem.* **132:** 6.

Fredeen, H.T. and B.G. Harmon. 1983. The swine industry: Changes and challenges. *J. Anim. Sci.* (suppl. 2) **57:** 100.

Frels, W.I., J.A. Bluestone, R.J. Hodes, M.R. Capecchi, and D.S. Singer. 1985. Expression of a microinjected porcine class I major histocompatibility complex gene in transgenic mice. *Science* **228:** 577.

Gaugler, H.R., D.S. Buchanan, R.L. Hintz, and R.K. Johnson. 1984. Sow productivity comparisons for four breeds of swine: Purebred and crossbred litters. *J. Anim. Sci.* **59:** 941.

Gertler, A., A. Weil, and N. Cohen. 1982. Hormonal control of casein synthesis in organ culture of the bovine lactating mammary gland. *J. Dairy Res.* **49:** 387.

Gibson, J.P. and C. Smith. 1989. The incorporation of biotechnologies into animal breeding strategies. In *Animal biotechnology: Comprehensive biotechnology*, 1st. supplement (ed. L.A. Babrick et al.), p.

203. Pergamon Press, England.

Goosens, M. and Y.W. Kan. 1981. DNA analysis in the diagnosis of hemoglobin disorders. *Methods Enzymol.* **76:** 805.

Gordon, J.W., G.A. Scangos, D.J. Plotkin, J.A. Barbosa, and R.H. Ruddle. 1980. Genetic transformation of mouse embryos by microinjection by purified DNA. *Proc. Natl. Acad. Sci.* **77:** 7380.

Gordon, K., E. Lee, J.A. Vitale, A.E. Smith, H. Westphal, and L. Hennighausen. 1987. Production of human tissue plasminogen activator in transgenic mouse milk. *Biotechnology* **5:** 1183.

Hammer, R.E., V.G. Pursel, C.E. Rexroad, Jr., R.J. Wall, D.J. Bolt, K.M. Ebert, R.D. Palmiter, and R.L. Brinster. 1985. Production of transgenic rabbits, sheep and pigs by microinjection. *Nature* **315:** 680.

Hanahan, D. 1989. Transgenic mice as probes into complex systems. *Science* **246:** 1265.

Johnson, I.D., I.C. Hart, and B.W. Butler-Hogg. 1985. The effects of exogenous bovine growth hormone and bromocriptine on growth, body development, fleece weight and plasma concentrations of growth hormone, insulin and prolactin in female lambs. *Anim. Prod.* **41:** 207.

Johnson, R.K., D.R. Zimmerman, and R.J. Kottock. 1984. Selection for components of reproduction in swine. *Livest. Prod. Sci.* **11:** 541.

Johnson, R.K., D.R. Zimmerman, W.R. Lamberson, and S. Sasaki. 1985. Influencing prolificacy of sows by selection for physiological factors. *J. Reprod. Fertil.* (suppl.) **33:** 139.

Koch, R.M. and J.W. Algeo. 1983. The beef cattle industry; Changes and challenges. *J. Anim. Sci.* (suppl. 2) **57:** 28.

Lacey, M., S. Alpert, and D. Hanahan. 1986. Bovine papillomavirus genome elicits skin tumours in transgenic mice. *Nature* **322:** 609.

Lahlou-Kassi, A., D. Schams, and P. Glatzel. 1984. Plasma gonadotrophin concentrations during the oestrous cycle and after ovariectomy in two breeds of sheep with high and low fecundity. *J. Reprod. Fertil.* **70:** 165.

Lasfargues, E.Y., J.C. Lasfargues, A.S. Dion, A.E. Greene, and D.H. Moore. 1976. Experimental infection of a cat kidney cell line with the mouse mammary tumor virus. *Cancer Res.* **36:** 67.

Lee, K.F., S.H. Atiee, and J.M. Rosen. 1989. Differential regulation of rat β-casein-chloramphenicol acetyltransferase fusion gene expression in transgenic mice. *Nucleic Acids Res.* **16:** 1027.

Legault, C. 1985. Selection of breeds, strains and individual pigs for prolificacy. *J. Reprod. Fertil.* (suppl.) **33:** 151.

LeMeur, M., P. Gerlinger, C. Benoist, and D. Mathis. 1985. Correcting an immunoresponse deficiency by creating Eα transgenic mice. *Nature* **316:** 38.

Lesmeister, J.L., P.J. Burfening, and R.L. Blackwell. 1973. Date of first calving in beef cows and subsequent calf production. *J. Anim.*

Sci. **36:** 1.
Long, C.R. 1980. Crossbreeding for beef production: Experimental results. *J. Anim. Sci.* **51:** 1197.
Machlin, L.J. 1973. Effect of growth hormone on milk production and feed utilization in dairy cows. *J. Dairy Sci.* **56:** 575.
Martinat-Botte', F., F. Bariteau, B. Badouard, and M. Terqui. 1985. Control of pig reproduction. *J. Reprod. Fertil.* (suppl.) **33:** 211.
McDevitt, H.O. and B. Benacerraf. 1969. Genetic control of specific immune responses. *Adv. Immunol.* **11:** 31.
McEvoy, T.G. and J.M. Sreenan. 1990. The efficiency of production, centrifugation, microinjection and transfer of one- and two-cell bovine ova in a gene transfer program. *Theriogenology* **33:** 819.
McLaren, D.G., D.S. Buchanan, and R.K. Johnson. 1987. Individual heterosis and breed effects for postweaning performance and carcass traits in four breeds of swine. *J. Anim. Sci.* **64:** 83.
Meade, H., L. Gates, E. Lacy, and N. Lonberg. 1990. Bovine α-casein gene sequences direct high level expression of active human urokinase in mouse milk. *Biotechnology* **8:** 443.
Miller, K.F., D.J. Bolt, V.G. Pursel, R.E. Hammer, C.A. Pinkert, R.D. Palmiter, and R.L. Brinster. 1989. Expression of human or bovine growth hormone gene with a mouse metallothionein-1 promoter in transgenic swine alters the secretion of porcine growth hormone and insulin-like growth factor-I. *J. Endocrinol.* **120:** 481.
Miller, P.D. 1981. Artificial insemination organizations. *J. Dairy Sci.* **64:** 1283.
Morris, C.A., G.L. Bennet, R.L. Baker, and A.H. Carter. 1986. Birth weight, dystocia and calf mortality in some New Zealand beef breeding herds. *J. Anim. Sci.* **62:** 327.
Murray, J.D., C.D. Nancarrow, J.T. Marshall, I.G. Hazelton, and K.A. Ward. 1989. Production of transgenic merino sheep by microinjection of ovine metallothionein-ovine growth hormone fusion genes. *Reprod. Fertil. Dev.* **1:** 147.
Nadarajah, K., D.R. Notter, T.J. Marlow, and A.L. Eller, Jr. 1987. Evaluation of phenotypic and genetic trends in weaning weight in Angus and Hereford populations in Virginia. *J. Anim. Sci.* **64:** 1349.
Niedermeier, R.P., J.W. Crowley, and E.C. Meyer. 1983. United States dairying: Changes and challenges. *J. Anim. Sci.* (suppl.) **57:** 44.
Nott, S.B., D.E. Kauffman, and J.A. Speicher. 1981. Trends on the management of dairy farms since 1956. *J. Dairy Sci.* **64:** 1330.
Nwakalor, L.N., J.S. Brinks, and G.V. Richardson. 1986. Selection in Hereford cattle. I. Selection intensity, generation interval and indexes in retrospect. *J. Anim. Sci.* **62:** 927.
Palmiter, R.D. and R.L. Brinster. 1986. Germ-line transformation of mice. *Annu. Rev. Genet.* **20:** 465.
Palmiter, R.D., R.L. Brinster, R.E. Hammer, M.E. Trumbauer, M.G. Rosenfield, N.C. Birnberg, and R.M. Evans. 1982. Dramatic

growth of mice that develop from eggs microinjected with metallothionein-growth hormone fusion genes. *Nature* **300:** 611.

Papaioannou, V.E. and K.M. Ebert. 1986. Development of fertilized embryos transferred to oviducts of immature mice. *J. Reprod. Fertil.* **76:** 603.

Parker, C.F. and A.L. Pope. 1983. The U.S. sheep industry: Changes and challenges. *J. Anim. Sci.* (suppl. 2) **57:** 75.

Parks, W.P., E.M. Scolnick, and E.H. Kozikowski. 1974. Dexamethasone stimulation of murine mammary tumor virus expression: A tiddus culture source of virus. *Science* **184:** 158.

Pinkert, C.A., T.J. Dyer, D.L. Kooyman, and D.J. Kiehm. 1990. Characterization of transgenic livestock production. *Domest. Anim. Endocrinol.* **7:** 1.

Pinkert, C.A., J. Manz, P.-J. Linton, N.R. Klinman, and U. Storb. 1989. Elevated PC responsive B cells and anti-PC antibody production in transgenic mice harboring anti-PC immunoglobulin genes. *Vet. Immunol. Immunopathol.* **23:** 321.

Pinkert, C.A., G. Widera, C. Cowing, E. Heber-Katz, R.D. Palmiter, R.A. Flavell, and R.L. Brinster. 1985. Tissue-specific, inducible and functional expression of the $E\alpha^d$ MHC class II gene in transgenic mice. *EMBO J.* **4:** 2225.

Pittius, C.W., L. Sankaran, Y.J. Topper, and L. Hennighausen. 1988a. Comparison of the regulation of the whey acidic protein gene with that of a hybrid gene containing the whey acidic protein gene promoter in transgenic mice. *Mol. Endocrinol.* **2:** 1027.

Pittius, C.W., L. Hennighausen, E. Lee, H. Westphal, E. Nicols, J. Vitale, and K. Gordon. 1988b. A milk protein gene promoter directs the expression of human tissue plasminogen activator cDNA to the mammary gland in transgenic mice. *Proc. Natl. Acad. Sci.* **85:** 5874.

Polge, E.J.C., S.C. Barton, M.A.H. Surani, J.R. Miller, T. Wagner, F. Rottman, S.A. Camper, K. Elsome, A.J. Davis, J.A. Goode, G.R. Foxcroft, and R.B. Heap. 1989. Induced expression of a bovine growth hormone construct in transgenic pigs. In *Biotechnology in growth regulation* (ed. R.B. Heap et al.), p. 279. Butterworths, London.

Pursel, V.G., R.J. Wall, L. Hennighausen, C.W. Pittius, and D. King. 1990. Regulated expression of the mouse whey acidic protein gene in transgenic swine. *Theriogenology* **33:** 302. (Abstr.)

Pursel, V.G., K.F. Miller, D.J. Bolt, C.A. Pinkert, R.E. Hammer, R.D. Palmiter, and R.L. Brinster. 1989a. Insertion of growth hormone genes into pig embryos. In *Biotechnology of growth regulation* (ed. R.B. Heap et al.), p. 181. Butterworths, London.

Pursel, V.G., C.A. Pinkert, K.F. Miller, D.J. Bolt, R.G. Campbell, R.D. Palmiter, R.L. Brinster, and R.E. Hammer. 1989b. Genetic engineering of livestock. *Science* **244:** 1281.

Pursel, V.G., C.E. Rexroad, Jr., D.J. Bolt, K.F. Miller, R.J. Wall, R.E.

Hammer, C.A. Pinkert, R.D. Palmiter, and R.L. Brinster. 1987. Progress on gene transfer in farm animals. *Vet. Immunol. Immunopathol.* **17:** 303.

Rexroad, C.E., Jr., R.E. Hammer, D.J. Bolt, K.E. Mayo, L.A. Frohman, and R.D. Palmiter. 1989. Production of transgenic sheep with growth-regulating genes. *Mol. Reprod. Dev.* **1:** 164.

Ross, S.R. and D. Solter. 1985. Glucocorticoid regulation of mouse mammary tumor virus sequences in transgenic mice. *Proc. Natl. Acad. Sci.* **82:** 5880.

Rothchild, M.F., D.W. Zimmerman, L.L. Christian, L. Venier, and C.M. Warner. 1984. Differences in SLA haplotypes in two lines of swine selected for high and low ovulation rates. *Anim. Blood Groups Biochem. Genet.* **15:** 155.

Seeburg, P.H., S. Siar, J. Adelman, H.A. deBoer, J. Hayflick, P. Jhurani, D.V. Goeddel, and H.L. Heyneker. 1983. Efficient bacterial expression of bovine and porcine growth hormones. *DNA* **2:** 37.

Selgrath, J.P., T.D. Bunch, W.C. Foote, and J.W. Call. 1983. *In vitro* culture and processing of sheep and goat embryos. *Encyclia* **60:** 43.

Selgrath, J.P., M.A. Memon, T.E. Smith, and K.M. Ebert. 1990. Collection and transfer of microinjectable embryos from dairy goats. *Theriogenology* **34:** 1195.

Short, R.E. and R.A. Bellows. 1971. Relationships among weight gains, age at puberty and reproductive performance in heifers. *J. Anim. Sci.* **32:** 127.

Simons, J.P., M. McClenaghan, and A.J. Clark. 1987. Alteration of the quality of milk by expression of sheep β-lactoglobulin in transgenic mice. *Nature* **328:** 530.

Simons, J.P., I. Wilmut, A.J. Clark, A.L. Archibald, J.O. Bishop, and R. Lathe. 1988. Gene transfer into sheep. *Biotechnology* **6:** 179.

Singer, D.A., R.D. Camerini-Otero, M.L. Satz, B. Osborne, D. Sachs, and S. Rodikoff. 1983. Characterization of a porcine genomic clone encoding a major histocompatibility antigen: Expression in mouse L cells. *Proc. Natl. Acad. Sci.* **79:** 1403.

Steele, N.C. and V.G. Pursel. 1990. Nutrient partitioning by transgenic animals. *Annu. Rev. Nutr.* **10:** 213.

Stewart, T.A., P.K. Pattengale, and P. Leder. 1984. Spontaneous mammary adenocarcinomas in transgenic mice that carry and express MTV myc fusion genes. *Cell* **38:** 627.

Storb, U. 1987. Transgenic mice with immunoglobulin genes. *Annu. Rev. Immunol.* **5:** 151.

Thonney, M.L. 1987. Growth, feed efficiency, and variation of individually fed Angus, Polled Hereford and Holstein steers. *J. Anim. Sci.* **65:** 1.

Thrift, F.A. and D.K. Aaron. 1987. The crossbred sire: Experimental results for cattle. *J. Anim. Sci.* **65:** 128.

Urick, J.J., O.F. Pahnish, W.L. Reynolds, and B.W. Knapp. 1986. Comparison of two- and three-way rotational crossing, beef x beef and beef x brown Swiss composite breed production. *J. Anim. Sci.* **62:** 344.

Vesely, J.A. and E.E. Swierstra. 1987. Reproductive traits of ewe lambs representing eight genetic types born in winter, spring, summer and fall. *J. Anim. Sci.* **65:** 1195.

Vilotte, J.L., S. Soulier, M.G. Stinnakre, M. Massoud, and J.C. Mercier. 1989. Efficient tissue-specific expression of bovine α-lactalbumin in transgenic mice. *Eur. J. Biochem.* **186:** 43.

Vize, P.D., A.E. Michalska, R. Ashman, B. Lloyd, B.A. Stone, P. Quinn, J.R.E. Wells, and R.F. Seamark. 1988. Introduction of a porcine growth hormone fusion gene into transgenic pigs promotes growth. *J. Cell Sci.* **90:** 295.

Voelker, D.E. 1981. Dairy herd improvement associations. *J. Dairy Sci.* **64:** 1269.

Wakelin, D. 1978. Genetic control of susceptibility and resistance to parasitic infection. *Adv. Parasitol.* **16:** 219.

Ward, K.A. 1982. Possible contributions of molecular genetics to animal improvement. In *Future developments in the genetic improvement of animals* (ed. J.S.F. Barker et al.), p. 17. Academic Press, New York.

Warner, C.M. 1986. Genetic manipulation of the major histocompatibility complex. *J. Anim. Sci.* **63:** 279.

Warner, C.M., D.L. Meerker, and M.F. Rothschild. 1987. Genetic control of immune responsiveness: A review of its use as a tool for selection for disease resistance. *J. Anim. Sci.* **64:** 394.

Warwick, E.J. 1958. Fifty years of progress in breeding beef cattle. *J. Anim. Sci.* **17:** 922.

Westphal, H. 1989. Molecular genetics of development studies in the transgenic mouse. *Annu. Rev. Cell Biol.* **5:** 181.

Wieghart, M., J.L. Hoover, M.M. McCrane, R.W. Hanson, F.M. Rottman, S.H. Holtzman, T.E. Wagner, and C.A. Pinkert. 1990. Production of transgenic swine harboring a rat phosphoenolypyruvate carboxykinase-bovine growth hormone fusion gene. *J. Reprod. Fertil.* (suppl.) **41:** 89.

Willham, R.L. 1982. Genetic improvement of cattle in the United States: Cattle, people and their interaction. *J. Anim. Sci.* **54:** 659.

Wilmut, I., A.L. Archibald, S. Harris, M. McClenaghan, J.P. Simons, C.B.A. Whitelaw, and A.J. Clark. 1990. Methods of gene transfer and their potential use to modify milk composition. *Theriogenology* **33:** 113.

Yamamura, K., H. Kikutsni, V. Folsom, L.K. Clayton, M. Kimoto, S. Akira, S. Kashiwamura, S. Tonegawa, and T. Kishimoto. 1985. Functional expression of a microinjected $E^d\alpha$ gene in C57BL/6 transgenic mice. *Nature* **316:** 67.

Yu, S.H., K.C. Deen, E. Lee, L. Hennighausen, R.W. Sweet, M. Rosen-berg, and H. Westphal. 1989. Functional human CD4 protein produced in milk of transgenic mice. *Mol. Biol. Med.* **6**: 255.

Prospects for the Establishment of Embryonic Stem Cells and Genetic Manipulation of Domestic Animals

C.L. Stewart

Department of Cell and Developmental Biology
Roche Institute of Molecular Biology
Nutley, New Jersey 07110

The genetic manipulation of embryos by the introduction of recombinant DNA, by either pronucleus injection or retroviral vector infection (for review, see Wagner and Stewart 1986; Palmiter and Brinster 1986), demonstrated the feasibility of using such techniques to circumvent the long tedious process of conventional breeding to introduce new traits into livestock.

This paper reviews the progress made in the last decade in the use of embryonic stem (ES) cells for genetic manipulation of mammals. These cells, which are in many ways analogous to having a hugely expanded population of embryos to work with, offer the prospect of being able to manipulate the genome in ways other than by DNA injection (for a review of pronuclear injection, see Ebert and Selgrath, this volume). It is now possible not only to add new genetic information into the germ line of an animal, but to add it with exquisite precision so that one may replace an endogenous gene by a modified version that was manipulated by conventional molecular biology methods in the test tube. Such technical achievement offers potentially unlimited prospects for the manipulation of livestock, so that within a relatively short time, it may be possible to readily introduce favorable traits or remove unfavorable traits. As discussed here, the molecular techniques do exist, but what is required is a better understanding of the cell biology and growth requirements of embryonic cells from

Animal Applications of Research in Mammalian Development

domesticated species before such prospects can be fully realized.

ISOLATION AND GROWTH OF ES CELLS FROM MOUSE AND OTHER MAMMALIAN SPECIES

Embryonal carcinoma (EC) cells, the forerunners of ES cells, were derived from embryonic tumors (teratocarcinomas) induced in adult mice (for review, see Graham 1977). However, it was the similarities between EC cells and cells of the inner cell mass (ICM) of mouse blastocysts, in their morphological, biochemical, and developmental characteristics, that stimulated investigation into whether cell lines could be directly established from embryos grown in vitro (Evans 1981), thus avoiding the requirement for producing teratocarcinomas from transplanted embryos.

ES (or EK) cells have been established from blastocysts that have been experimentally delayed from implanting into the uterus (Evans and Kaufman 1981); from ICM cells isolated from blastocysts by immunosurgery, with and without the use of EC-cell-conditioned medium (Martin 1981; Handyside et. al. 1989); and directly from day-4 blastocysts (day 1 = day of plug) (Axelrod 1984; Wagner et al. 1985; Doetschman et al. 1985; for review, see Robertson 1987).

The technique most frequently used, and the one we have employed to establish ES cells from a wide variety of different strains and of differing genetic constitutions, has been the culture of day-4 blastocysts. The general procedure involves explanting blastocysts with the zona pellucida removed onto a dish containing a layer of mitomycin-C-treated or -irradiated primary mouse embryo fibroblasts (PMEFs) onto which they attach. We prefer to use PMEFs (Wobus et al. 1984), although STO fibroblasts (a permanent mouse fibroblast line) are also widely employed (Evans and Kaufman 1981; Robertson 1987). There does, however, appear to be some variation in batches of STO fibroblasts in their ability to support ES cell growth, and it has been reported that PMEFs appear to be superior in sustaining growth of ES cells with regard to maintaining a stable karyotype and an undifferentiated stem cell phenotype (Suemori and Nakatsuji 1987).

Usually, blastocysts will attach and outgrow in normal tissue culture medium with 15% fetal calf serum (FCS) that has been carefully tested to maximize plating efficiency and to support the ability of an already established ES cell line to differentiate in vitro. The medium can, however, be supplemented by conditioned medium from either the BRL (Buffalo rat liver) cell line or the human bladder carcinoma line 5637, both of which are known to produce the cytokine leukemia inhibitory factor/differentiation activity (LIF/DIA) (Smith et al. 1988; Williams et al. 1988b). Attachment and outgrowth usually takes 3–4 days, with the trophoblast forming a flat plate of cells on which a knob of ICM cells sits. The ICM, before endoderm formation becomes visible, is then mechanically picked off the outgrowth using a fine mouth-controlled pipette and is disaggregated into single or small clumps of cells by a brief incubation in trypsin-EDTA. These are then replated onto a fresh layer of fibroblasts with conditioned medium. Within the next 3–10 days, colonies of cells within a typical ES morphology appear.

Other morphologically different cell types are also often visible; these do not, however, continue to proliferate. We routinely use 5637 conditioned medium in all experiments involving establishment of new ES lines, and once established, we switch to regular medium for maintaining ES lines. Under these conditions, between 30% and 100% of the picked ICM cells form ES lines.

ES cell lines have also been established in the absence of any feeders by explanting ICM cells into either 5637 conditioned medium or normal culture medium supplemented with recombinant LIF (Pease et al. 1990; C.L. Stewart, unpubl.). The presence of this cytokine, by either its secretion from fibroblasts or its addition to the medium, appears to be necessary since it allows proliferation of mouse ES cells without their differentiation. However, a totipotent EC cell line has been isolated in the absence of feeders or of any exogenously added LIF (Mintz and Cronmiller 1981; Stewart and Mintz 1981), and thus it is unclear whether there is an absolute requirement for this factor, although it does appear to make the process of isolation and establishment of ES cells more efficient.

Once ES cells have been established, they can then be expanded into mass cultures before freezing. It is also important, especially if the ES cells are to be used for genetic manipulation studies, that the karyotype of the lines be determined since many of the lines (25–50%) can contain a high proportion of cells with an aneuploid chromosome complement that will restrict their ability to form viable gametes.

At present, numerous ES cell lines have been established from a variety of different strains of mice, including lines from parthenogenetic, androgenetic, and haploid blastocysts, as well as from embryos carrying mutations in the T complex (Evans et al. 1985; Martin et al. 1987; Mann et al. 1990). The favored strain has been 129/Sv and with three cell lines, D3 (Doetschman et al. 1985b), CCE-1 (Bradley et al. 1984), and AB1 (McMahon and Bradley 1990) being principally used, although ES lines have also been established from other inbred strains such as C57BL6/J and C3H/He (C.L. Stewart, unpubl.), as well as from F_1 strain combinations (Evans et al. 1985).

The only other species from which cell lines have been established is the Syrian golden hamster (Doetschman et al. 1988). These lines were obtained under conditions identical to those used in the mouse, including the use of mouse embryo fibroblasts as a feeder layer. These ES cells have been reported to differentiate extensively in culture under conditions similar to those used for the mouse. However, a rigorous test as to whether they can form chimeras has not yet been reported, although attempts are being made (N. Maeda, pers. comm.). In domestic species, such as the sheep and pig, some preliminary reports describe the isolation of cells that morphologically resemble murine ES cells, some of which undergo differentiation in vitro. In no instance has there been any report of these cells being able to form chimeras (Handyside et al. 1987; Piedrahita et al. 1988; Evans et al. 1990).

It is not yet clear whether this poor success rate with domestic animals is due to some intrinsic difference between embryos of different species being able to form ES cells. The culture requirements for preimplantation stages do vary according to the species (Wright and O'Fallon 1987), and the blastocysts of domestic species have different growth characteristics, with delayed implantation as a common feature (Heyman and Menezo 1987). Furthermore, from cultures of their explanted blastocysts, it is often difficult to discern a recognizable ICM that is a characteristic of mouse embryo outgrowth (Prather et al. 1989).

It could thus be that the culture conditions necessary to support stem cell growth may differ between species. One report, describing the establishment of a pluripotent human EC cell line from a spontaneous testicular teratocarcinoma, demonstrated that, like pluripotent mouse EC cell lines isolated from tumors, undifferentiated growth of the EC cells was feeder-dependent (Pera et al. 1989). However, as was subsequently shown, this requirement for feeders to prevent differentiation of mouse lines was attributable to synthesis and secretion of LIF/DIA by the feeders (Smith and Hooper 1987; Williams et al. 1988b). Although the human EC cell lines are also dependent on a mouse fibroblast feeder cell line, this dependence is not attributable to LIF/DIA produced by BRL cells (Pera et al. 1989). The feeders are presumably supplying some other as yet unidentified factor(s). Therefore, it may be necessary to identify a unique set of conditions or additional factors other than LIF/DIA for establishment of ES cells from each species.

PRODUCTION OF CHIMERAS

Chimeras are generally made by injecting 10–15 trypsin-disaggregated ES cells into the blastocoel cavity of day-4 recipient blastocysts. They can also be produced by aggregating a small group (3–5 cells) of ES cells with one or two eight-cell-stage embryos or by injecting a similar number of cells under the zona pellucida of an eight-cell-stage embryo, where they will also aggregate with the blastomeres (Fuji and Martin 1980, 1983; Stewart 1980, 1982; Wagner et al. 1985; Bradley 1987; C.L. Stewart, unpubl.). An important observation concerning chimeras with regard to genetic modification of the germ line has been the influence of the strain combinations used. Thus, 129/Sv ES cells give a much higher frequency of germ line chimeras using C57BL6/J blastocyst recipients as compared to blastocysts at either outbred or certain inbred strains (Schwartzenberg et al. 1989; M. Taketo et al., in prep.). This presumably reflects competition and selection between donor and host cells that occur during development of the chimera, as recognized in chimeras produced by aggregation (Mullen and Whitten 1971; for review, see McLaren 1976). Therefore, in addition to selecting host blastocyst strains that carry the appropriate genetic markers for identifying an ES contribution to the mouse's tissues (i.e., coat color and/or glucose phosphate isomerase [Gpi-I] polymorphism), it is also necessary to identify combinations that result in a substantial ES cell contribution to both somatic and germ line lineages.

ES CELLS AND THE GENETIC MANIPULATION OF MAMMALS

An overwhelming interest in the use of ES cells for genetic manipulation was stimulated by the demonstration that ES cell lines could colonize the germ cell lineage and form functional gametes at frequencies much higher than those observed with established EC cell lines (Bradley et al. 1984; Evans et al. 1985). Variations in the techniques of making chimeras, and the discovery that certain strain combinations are better at generating germ line chimeras (Schwartzenberg et al. 1989; M. Taketo et al., in prep.), have made it practical to design experiments using ES cells to introduce new genetic traits into the germ line of mice with reasonable efficiency.

There are some advantages in using transfected ES cells instead of microinjected fertilized eggs. One is that transfected clones of ES cells carrying an introduced gene can be characterized for copy number and levels of expression of the introduced gene in vitro before the cells are transferred into the in vivo environment of the embryo. Such an approach has been used to study the constitutive expression of certain genes in the developing embryo (Stewart et al. 1985; Gossler et al. 1986), as well as to study the tissue specificity of expression of collagen and crystalline genes (Lovell-Badge et al. 1987; Takahashi et al. 1988). In addition, the relatively high frequencies with which ES cell chimeras can be generated has made these cells particularly useful for studies where the expression of the introduced gene may act in a dominant manner and have lethal consequences for the developing embryo. For example, such an approach was used in studying the expression of the polyomavirus middle T gene, which was found to induce hemangiomas in the yolk sac endothelium of chimeras (Williams et al. 1988a).

Another instance where ES cells have been particularly useful has been in the use of gene or enhancer trap constructs to identify genes whose pattern of expression in the developing embryo suggests that they might be of developmental importance (Allen et al. 1988; Gossler et al. 1989; Skarnes 1990). Not only is it easier to screen a large number of ES clones that give a potentially interesting pattern of expression in the chimeric embryo, but if a pattern is detected, the gene can then be isolated from the ES clone since the enhancer trap acts as a tag for the cloning of the gene. In addition, the gene or enhancer trap can potentially disrupt expression of the gene, thus making it possible to gain a further understanding of its function by determining what consequences such a mutated gene might have on embryonic development. However, it has been the ability to introduce mutations into specific endogenous genes, with the subsequent incorporation of the mutated gene into the germ line of the mouse, that has attracted the most interest.

ES CELLS AND THE INTRODUCTION OF TARGETED MUTATIONS INTO THE GERM LINE

Two approaches have been used to introduce mutations into specific genes. The first depends on the gene in question expressing a phenotype that can be selected. However, this strategy is limited to only a few genes expressed in ES cells for which there is a selection system. Using this approach, it was possible to isolate mutations in the purine nucleotide salvage pathway enzyme hypoxanthine phosphoribosyl transferase (HPRT). HPRT deficiency in humans results in Lesch-Nyhan syndrome, a disease characterized by mental retardation and early death (for review, see Stout and Caskey 1988). Mutations were introduced by homologous recombination (Doetschman et al. 1988a), by retroviral insertion (Kuehn et al. 1987), or by selection of spontaneous mutants (Hooper et al. 1987). The last approach is a relatively straightforward procedure with male ES lines because HPRT is located on the X chromosome. In all of these experiments, HPRT-deficient clones were isolated by their resistance to β-thioguanine, a toxic purine analog. Chimeras were produced from these clones, and individuals were identified among the offspring that carried the mutated gene. From these, in turn, male mice were detected that were completely deficient for HPRT activity. Ironically, none of the deficient mice appeared to exhibit any overtly abnormal phenotype, although subsequent analysis revealed that dopamine levels in the brain were reduced by 20%, a characteristic that also occurs in Lesch-Nyhan patients (Finger et al. 1988). The absence of a phenotype resembling Lesch-Nyhan syndrome may be due to the fact that mice possess urate oxidase activity and can thereby metabolize the toxic products that accumulate as a consequence of HPRT deficiency. These toxic products are thought to be responsible for the disease in humans, who lack urate oxidase activity (Wu et al. 1989). Attempts are in progress to mutate this gene and thus determine whether mice homozygous for both HPRT and urate oxidase deficiency exhibit a Lesch-Nyhan phenotype (X. Wu and C.T. Caskey, pers. comm.).

The possibility of mutating any gene, regardless of whether or not a selection system is available, became apparent when it was demonstrated that mammalian cells possess the biochemical mechanisms for mediating the homologous recombination between genes (Smithies et al. 1985; Thomas and Capecchi 1986; Thomas et al. 1986). This, coupled with the development of polymerase chain reaction (PCR) technology (Kim and Smithies 1988) and the design of vectors that increase the efficiency of detecting the occurrence of a homologous recombination event, has made it possible to screen for ES clones in which a particular endogenous gene is replaced with a mutated copy of the gene. The technology involved in this approach has already been extensively described in a number of reviews (Capecchi 1989; Camerini-Otero and Kucherlapati 1990) and thus is only briefly summarized here.

At present, two general approaches have been used to mutate endogenous genes. One of the approaches involves microinjection of DNA into ES cells. This technique was employed to disrupt the homeo-box gene, Hox 1.1, with a 20-bp oligonucleotide insertion in the coding region of the gene (Zimmer and Gruss 1989). After this construct was microinjected into ES cells, clones containing the homologous recombinant were initially identified by PCR screening. The frequency of homologous recombination (1/150) in cells injected with the construct was remarkably high. Because this approach has not yet been repeated, it is unclear whether this success was due to microinjection (which can result in frequencies of stable DNA incorporation in about 20% of injected cells), to limiting the alteration of the gene to approximately 20 bp, or to both reasons.

The other more widely used approach has been to construct a homologous recombination vector in which a genomic sequence of the gene to be targeted contains a selectable neomycin (*neo*) resistance expression cassette, usually integrated in an exon where it disrupts the coding sequence. This vector is introduced into the ES cells by electroporation or transfection, and selection for *neo* resistance colonies is initiated. The resulting clones are then screened by PCR and Southern analysis to identify clones in which the vector has recombined with the target gene homologously (Thomas and Capecchi 1987; Joyner et al. 1989). An increase in the frequency of homologous recombination (due to a reduction in the number of illegitimate or nonhomologous recombinants) can be obtained by incorporating another selectable cassette, herpes simplex virus thymidine kinase (HSV-TK), at one or both ends of the genomic sequence in the targeting vector (Mansour et al. 1988; Johnson et al. 1989). The use of this gene is based on the premise that those vectors which undergo homologous recombination will lose the HSV-TK cassettes, whereas those undergoing illegitimate recombination will retain it. When the antiviral drugs gancyclovir or FIAU (1-[2-deoxy-2-fluro-β-D-arabinofuranoscyl]-5-iodouracil) (McMahon and Bradley 1990), which are converted into toxic nucleotide analogs by the viral thymidine kinase, are added to the culture medium, clones that retain HSV-TK and express the thymidine kinase gene are killed. An alternative double-selection procedure utilizing a diphtheria toxin expression cassette instead of thymidine kinase has been described recently (Yagi et al. 1990). The principle underlying its use is the same as that for thymidine kinase, except that it is not necessary to use drugs to select against diphtheria toxin expression, because if homologous recombination does not occur, the diphtheria toxin gene is retained and its expression will result in cell death.

An alternative vector system that would only be of general use for obtaining recombinants for genes expressed in ES cells is the "promoter-less *neo*" construct, which is analogous to the enhancer trap vector described previously. In this vector, the *neo* resistance gene is incorporated into the targeting vector in such a way that its expression will only occur if it integrates close to a promoter. The presence of the targeted gene sequences in the vector greatly enriches the homologous recombination events with the consequent incorporation of the *neo* resistance gene into the endogenous gene (Schwartzenberg et al. 1989; Stanton et al. 1990).

Using these techniques, a number of genes have been disrupted when introduced into the germ line of mice. These mutations have all resulted in the absence of gene function, thereby producing novel and in some instances, unexpected, phenotypes. For example, deficiency in expression of the cell-surface–associated protein, β_2-microglobulin, resulted in apparently normal, viable mice whose only defect was the absence of a particular subset of T lymphocytes (Smithies et al. 1985; Zijlstra et al. 1990). Disruption of the Wnt-1 (*int*-1) proto-oncogene results in the homozygotes either lacking or having severe abnormalities in a substantial part of the midbrain (McMahon and Bradley 1990; Thomas and Capecchi 1990). In addition, these approaches have also described the complete restoration of a functional *hprt* gene from a clone containing a substantial deletion (Koller et al. 1989; Thompson et al. 1989), so demonstrating the feasibility of rescuing a mutated gene.

Other genes that have been targeted with the integration of the mutation into the germ line have included a modified c-*abl* proto-oncogene (Schwartzenberg et al. 1989), the N-*myc* proto-oncogene (Stanton et al. 1990), and the erythroid-specific transcription factor GATA, which, when mutated, results in the block of erythroid differentiation (Pevny et al. 1991). However, one of the more intriguing results has emerged from studies on the insulin-II-like growth factor (IGF-II), where disruption of this gene resulted in a distinct phenotype being apparent in the heterozygotes (DeChiara et al. 1990). Mice that inherited the mutated allele from their father attained only 60% of the size of their wild-type siblings, but apart from this, they were apparently normal and reached sexual maturity. It appears that the normal maternally derived allele either is not expressed or is at a tenfold lower level compared to the normal paternal allele. Thus, it has been proposed that IGF-II is an endogenous, imprinted gene, a category of developmentally important genes predicted from the aberrant development of parthenogenetic, gynogenetic, and androgenetic embryos (for review, see Surani 1987; Solter 1988), and of cells derived from them in chimeras with normal embryos (Thomson and Solter 1987; Nagy et al. 1989; Fundele et al. 1990; Mann et al. 1990).

Clearly, techniques for genetic manipulation can be used to introduce mutations in numerous genes, which should lead to a better understanding of their function. However, a full understanding of gene function as assessed by using these genetic approaches also requires the characterization of dominant and recessive alleles that result in different phenotypes. Current work represents the beginning in the application of such techniques to mammalian genetics, and it is already clear that a number of groups are developing strategies by which more subtle mutations, rather than the complete abolishment of a gene's expression, can be introduced.

CONCLUSIONS

In the decade since the first publication describing their establishment, ES cells have been developed as a major research tool for the genetic manipulation of mice and for the investigation of questions relevant to the study of mouse embryology.

Results from homologous recombination experiments have suggested that perhaps all genes may be mutated. This clearly offers important and far-reaching possibilities for the manipulation of animals that are of commercial importance. It is disappointing, however, that progress in the establishment of ES cell lines from other species has not progressed so rapidly. Cells that are equivalent to mouse EC or ES cells, or at least morphologically closely resemble them, have been observed from a number of mammalian species including humans (Pera et al. 1989), suggesting that with perseverance and a better understanding of the factors required for their establishment and growth, it is only a question of time before such cells can be isolated from the embryos of domestic species. The prospects for the genetic manipulation of domestic livestock should then be feasible.

ACKNOWLEDGMENTS

I thank Paul Wassarman and Roger Pedersen for helpful comments and Sharon Perry for her assistance in the preparation and typing of the manuscript.

REFERENCES

Allen, N.D., D.G. Cran, S.C. Barton, S. Heltle, W. Reik, and M.A. Surani. 1988. Transgenes as probes for active chromosomal domains in mouse development. *Nature* **333:** 852.

Axelrod, H.R. 1984. Embryonic stem cell lines derived from blastocysts by a simplified technique. *Dev. Biol.* **101:** 225.

Bradley, A. 1987. Production and analysis of chimeric mice. In *Teratocarcinomas and embryonic stem cells: A practical approach* (ed. E.J. Robertson), p. 113. IRL Press, Washington, D.C.

Bradley, A., M. Evans, M.H. Kaufman, and E. Robertson. 1984. Formation of germ-line chimeras from embryo derived teratocarcinoma cell lines. *Nature* **309:** 255.

Camerini-Otero, R.D. and R. Kucherlapati. 1990. Right on target. *New Biologist* **2:** 337.

Capecchi, M.R. 1989. The new mouse genetics: Altering the genome by gene targeting. *Trends Genet.* **5:** 70.

DeChiara, T.M., A. Efstratiadis, and E.J. Robertson. 1990. A growth deficiency phenotype in heterozygous mice carrying an insulin-like growth factor II gene disrupted by targeting. *Nature* **345:** 78.

Doetschman, T., N. Maeda, and O. Smithies. 1988a. Targeted mutation of the HPRT gene in mouse embryonic stem cells. *Proc. Natl. Acad. Sci.* **85:** 8583.

Doetschman, T., P. Williams, and N. Maeda. 1988b. Establishment of hamster blastocyst-derived embryonic stem (ES) cells. *Dev. Biol.* **127:** 224.

Doetschman, T.C., H. Eistetter, M. Katz, W. Schmidt, and R. Kemler. 1985. The *in vitro* development of blastocyst-derived embryonic stem cell lines: Formation of visceral yolk sac, blood islands and myocardium. *J. Embryol. Exp. Morphol.* **87:** 27.

Evans, M.J. 1981. Origin of mouse embryonal carcinoma cells and the possibility of their direct isolation into tissue culture. *J. Reprod. Fertil.* **62:** 625.

Evans, M.J. and M.H. Kaufman. 1981. Establishment in culture of pluripotent cells from mouse embryos. *Nature* **292:** 154.

Evans, M.J., A. Bradley, and E.J. Robertson. 1985. EK cell contribution to chimeric mice: From tissue culture to sperm. *Banbury Rep* **20:** 93.

Evans, M.J., E. Notarianni, S. Laurie, and R.M. Moore. 1990. Derivation and preliminary characterization of pluripotent cell lines from porcine and bovine embryos. *Theriogenology* **33:** 125.

Finger, S., R.P. Heavens, D.J.S. Sirinathsinghji, M.R. Kuehn, and S.B. Dunnett. 1988. Behavioural and neurochemical evaluation of a transgene mouse model of Lesch-Nyhan syndrome. *J. Neurol. Sci.* **86:** 9412.

Fujii, J. and G.R. Martin. 1980. Incorporation of teratocarcinoma stem cells into blastocysts by aggregation with cleavage stage embryos. *Dev. Biol.* **74:** 239.

———. 1983. Development potential of teratocarcinoma stem cells *in utero* following aggregation with cleavage state mouse embryos. *J. Embryol. Exp. Morphol.* **74:** 79.

Fundele, R.H., M.L. Norris, S.C. Barton, M. Fehlan, S.K. Howlett, W.E. Mills, and M.A. Surani. 1990. Temporal and spatial selection against parthenogenetic cells during development of fetal chimeras. *Development* **108:** 203.

Gossler, A., A.L. Joyner, J. Rossant, and W.C. Skarnes. 1989. Mouse embryonic stem cells and reporter constructs to detect developmentally regulated genes. *Science* **244:** 463.

Gossler, A., T. Doetschman, R. Korn, E. Senfling, and R. Kemler. 1986. Transgenesis by means of blastocyst-derived embryonic stem cell lines. *Proc. Natl. Acad. Sci.* **83:** 9065.

Graham, C.F. 1977. Teratocarcinoma cells and normal mouse embryogenesis. In *Concepts in mammalian embryogenesis* (ed. M.J. Sherman), p. 315. MIT Press, Cambridge, Massachusetts.

Handyside, A., M.L. Hooper, M.H. Kaufman, and I. Wilmutt. 1987. Towards the isolation of embryonal stem cell lines from the sheep. *Roux's Arch. Dev. Biol.* **196:** 185.

Handyside, A.H., G.T. O'Neill, M. Jones, and M.H. Hooper. 1989. Use of BRL-conditioned medium in combination with feeder layers to isolate a diploid embryonal stem cell line. *Wilhelm Roux's Arch.*

Dev. Biol. **198:** 48.

Heyman, Y. and Y. Menezo. 1987. Interaction of trophoblastic vesicles in bovine embryos developing *in vitro*. In *The mammalian preimplantation embryo* (ed. B. Bavister), p. 175. Plenum Press, New York.

Hooper, M., K. Hardy, A. Handyside, S. Hunter, and M. Monk. 1987. HPRT-deficient (Lesch-Nyhan) mouse embryos derived from germ line colonization by cultured cells. *Nature* **326:** 292.

Johnson, R.S., M. Sherry, M.E. Greenberg, R.D. Kolodner, V.E. Papaioannou, and B.M. Spiegelman. 1989. Targeting of nonexpressed genes in embryonic stem cells via homologous recombination. *Science* **245:** 1234.

Joyner, A.L., W.C. Skarnes, and J. Rossant. 1989. Production of a mutation in mouse En-2 gene by homologous recombination in embryonic stem cells. *Nature* **388:** 153.

Kim, H.S. and O. Smithies. 1988. Recombinant fragment assay for gene targeting based on the polymerase chain reaction. *Nucleic Acids Res.* **16:** 8887.

Koller, B.H., L.J. Hagerman, T. Doetschman, J.R. Hagerman, S. Huang, P.J. Williams, N.L. First, N. Maeda, and O. Smithies. 1989. Germ line transmission of a plasmid alteration made in a hypoxanthine phosphoriboxyl transferase gene by homologous recombination in embryonic stem cells. *Proc. Natl. Acad. Sci.* **86:** 8927.

Kuehn, M.R., A. Bradley, E.J. Robertson, and M.J. Evans. 1987. A potential animal model for Lesch-Nyhan syndrome through introduction of HPRT mutations into mice. *Nature* **326:** 295.

Lovell-Badge, R.H., A. Bygrave, A. Bradley, E.J. Robertson, R. Tilly, and K.S.E. Cheath. 1987. Tissue specific expression of the human type II collagen gene in mice. *Proc. Natl. Acad. Sci.* **84:** 2803.

Mann, J.R., I. Gadi, M.L. Harbison, S.J. Abbondanzo, and C.L. Stewart. 1990. Androgenetic mouse embryonic stem cells are pluripotent and cause skeletal defects in chimeras: Implications for genetic imprinting. *Cell* **62:** 251.

Mansour, S.L., K.R. Thomas, and M.R. Capecchi. 1988. Disruption of the proto-oncogene *int*-2 in mouse embryo-derived stem cells: A general strategy for targeting mutations to non-selectable genes. *Nature* **336:** 348.

Martin, G.R. 1981. Isolation of a pluripotent cell line from early mouse embryos cultured in medium conditioned by teratocarcinoma stem cells. *Proc. Natl. Acad. Sci.* **78:** 7634.

Martin, G.R., L.M. Silver, H.S. Fox, and A.L. Joyner. 1987. Establishment of embryonic stem cell lines from preimplantation mouse embryos homozygous for lethal mutations in the t-complex. *Dev. Biol.* **121:** 20.

McLaren, A. 1976. *Mammalian chimeras*. Cambridge University Press, England.

McMahon, A.P. and A. Bradley. 1990. The wnt-1 (int-1) proto-oncogene is required for development of a large region in the mouse brain. *Cell* **63:** 1073.

Mintz, B. and C. Cronmiller. 1981. METI-1; A. karyotypically normal *in vitro* line of developmentally totipotent mouse teratocarcinoma cells. *Somatic Cell Genet.* **7:** 489.

Mullen, R.J. and W. Whitten. 1971. Relationship of genotype and degree of chimerism in coat color to sex ratios and gametogenesis in chimeric mice. *J. Exp. Zool.* **178:** 165.

Nagy, A., M. Sarz, and M. Markulla. 1989. Systematic non-uniform distribution of parthenogenetic cells in adult mouse chimeras. *Development* **106:** 321.

Palmiter, R.D. and R.L. Brinster. 1986. Germ-line transformation of mice. *Annu. Rev. Genet.* **20:** 465.

Pease, S., P. Braghetta, D. Gearing, D. Grail, and R.L. Williams. 1990. Isolation of embryonic stem (ES) cells in media supplemented with recombinant leukemia inhibitory factor (LIF). *Dev. Biol.* **141:** 344.

Pera, M.F., S. Cooper, J. Mills, and J.M. Parrington. 1989. Isolation and characterization of a multipotent clone of human embryonal carcinoma cells. *Differentiation* **42:** 10.

Pevny, L., M. Celeste Simon, E. Robertson, W.H. Klein, S.F. Tsai, V. D'Agai, S.H. Orkin, and F. Costantini. 1991. A targeted mutation in the gene for transcription factor GATA-1 blocks erythroid differentiation in chimeric mice. *Nature* **349:** 257.

Piedrahita, J.A., G.B. Anderson, G.R. Martin, R.H. Bon Durant, and R.L. Pashen. 1988. Isolation of embryonic stem cell-like colonies from porcine embryos. *Theriogenology* **29:** 286.

Prather, R.S., L.J. Hagerman, and N.L. First. 1989. Preimplantation mammalian aggregation and injection chimeras. *Gamete Res.* **22:** 233.

Robertson, E.J. 1987. Embryo-derived stem cell lines. In *Teratocarcinomas and embryonic stem cells: A practical approach* (ed. E.J. Robertson), p. 71. IRL Press, Washington, D.C.

Schwartzenberg, P.L., S.P. Goff, and E.J. Robertson. 1989. Germ-line transmission of a c-*abl* mutation produced by targeted gene disruption in ES cells. *Science* **46:** 799.

Skarnes, W.C. 1990. Entrapment vectors: A new tool for mammalian genetics. *Biotechnology* **8:** 823.

Smith, A.G. and M.L. Hooper. 1987. Buffalo rat liver cells produce a diffusable activity which inhibits the differentiation of murine embryonal carcinomas and embryonic stem cells. *Dev. Biol.* **121:** 1.

Smith, A.G., J.K. Heath, D.D. Donaldson, G.G. Wong, J. Moreau, M. Stahl, and D. Rogers. 1988. Inhibition of pluripotent embryonic stem cell differentiation by purified polypeptides. *Nature* **336:** 688.

Smithies, O., R.G. Gregg, S.S. Boggs, M.A. Koralewski, and R.S. Kucherlapati. 1985. Insertion of DNA sequences into the human chromosomal β-globin locus by homologous recombination. *Nature*

317: 230.

Solter, D. 1988. Differential imprinting and expression of maternal and paternal genomes. *Annu. Rev. Genet.* **22:** 127.

Stanton, B.R., S.W. Reid, and L.F. Parada. 1990. Germ-line transmission of an inactive N-*myc* allele generated by homologous recombination in mouse embryonic stem cells. *Mol. Cell. Biol.* **10:** 6755.

Stewart, C.L. 1980. Aggregation between teratocarcinoma cells and preimplantation mouse embryos. *J. Embryol. Exp. Morphol.* **58:** 289.

———. 1982. Formation of viable chimeras by aggregation between teratocarcinomas and preimplantation mouse embryos. *J. Embryol. Exp. Morphol.* **67:** 167.

Stewart, C.L., M. Vanek, and E.F. Wagner. 1985. Expression of foreign genes from retroviral vectors in mouse teratocarcinoma chimeras. *EMBO J.* **4:** 3701.

Stewart, T.A. and B. Mintz. 1981. Successive generations of mice produced from an established culture line of euploid teratocarcinoma cells. *Proc. Natl. Acad. Sci.* **78:** 6314.

Stout, J.T. and C.T. Caskey. 1988. The Lesch-Nyhan syndrome: Classical molecular and genetic aspects. *Trends Genet.* **4:** 175.

Suemori, H. and N. Nakatsuji. 1987. Establishment of the embryo-derived stem (ES) cell lines from mouse blastocysts: Effects of the feeder cell layer. *Dev. Growth Differ.* **29:** 133.

Surani, M.A.H. 1987. Evidences and consequences of differences between maternal and paternal genomes during embryogenesis in the mouse. In *The mouse in experimental approaches to mammalian embryonic development* (ed. J. Rossant and R.A. Pedersen), p. 401. Cambridge University Press, England.

Takahashi, Y., K. Hamaoka, M. Hayasaka, K. Katoh, Y. Kato, T.S. Okado, and H. Kondoh. 1988. Embryonic stem cell-mediated transfer and correct regulation of the chicken σ-crystalline gene in developing mouse embryos. *Development* **102:** 259.

Thomas, K.R. and M.R. Capecchi. 1986. Introduction of homologous DNA sequences into mammalian cells induces mutations in the cognate gene. *Nature* **324:** 34.

———. 1987. Site-directed mutagenesis by gene targeting in mouse embryo derived stem cells. *Cell* **51:** 503.

———.1990. Targeted disruption of the murine int-1 proto-oncogene resulting in severe abnormalities in midbrain and cerebellar development. *Nature* **346:** 847.

Thomas, K.R., K.R. Folger, and M.R. Capecchi. 1986. High frequency targeting of genes to specific sites in the mammalian genome. *Cell* **44:** 419.

Thomson, J.A. and D. Solter. 1987. The development fate of androgenetic, parthenogenetic and gynogenetic cells in chimeric gastrulating mouse embryos. *Genes & Dev.* **2:** 1344.

Thompson, S., A.R. Clarke, A.M. Pow, M.L. Hooper, and D.W.

Shelton. 1989. Germ line transmission and expression of a corrected HPRT gene produced by gene targeting in embryonic stem cells. *Cell* **56:** 313.

Wagner, E.F. and C.L. Stewart. 1986. Integration and expression of genes introduced into mouse embryos. In *Experimental approaches to mammalian embryonic development* (ed. J. Rossant and R.A. Pedersen), p. 509. Cambridge University Press, England.

Wagner, E.F., G. Keller, E. Gilboa, U. Rüther, and C.L. Stewart. 1985. Gene transfer into murine stem cells and mice using retroviral vectors. *Cold Spring Harbor Symp. Quant. Biol.* **50:** 691.

Williams, R.L., S.A. Courtneidge, and E.F. Wagner. 1988a. Embryonic lethalites and endothelial tumors in chimeric mice expressing polyoma virus middle T oncogene. *Cell* **52:** 121.

Williams, R.L., D.J. Hilton, S. Pease, T.A. Willson, C.L. Stewart, D.P. Gearing, E.F. Wagner, D. Metcalf, N.A. Nicola, and N.M. Gough. 1988b. Myeloid leukemia inhibitory factor maintains the developmental potential of embryonic stem cells. *Nature* **336:** 684.

Wobus, A.M., H. Holzhauser, P. Jäkel, and J. Schoneich. 1984. Characterization of a pluripotent stem cell line derived from a mouse embryo. *Exp. Cell Res.* **152:** 212.

Wright, R.W. and J.V. O'Fallon. 1987. Growth of domesticated animal embryos *in vitro*. In *The mammalian preimplantation embryo* (ed. B. Bavister), p. 251. Plenum Press, New York.

Wu, X., C.C. Lee, D.M. Guzny, and C.T. Caskey. 1989. Urate oxidase: Primary structure and evolutionary implications. *Proc. Natl. Acad. Sci.* **86:** 9412.

Yagi, T., Y. Ikawa, K. Yoshida, Y. Shigetani, N. Takedo, I. Mabuchi, T. Yamamoto, and S. Aizawa. 1990. Homologous recombination at c-*fyn* locus of mouse embryonic stem cells with use of diptheria toxin A fragment gene in negative selection. *Proc. Natl. Acad. Sci.* **87:** 9918.

Zijlstra, M., M. Bix, N.E. Simister, J.M. Loring, D.H. Raulet, and R. Jaenisch. 1990. B_2-Microglobulin deficient mice lack CD4-8$^+$ cytolytic T-cells. *Nature* **344:** 742.

Zimmer, A. and P. Gruss. 1989. Production of chimeric mice containing embryonic stem (ES) cells carrying a homeobox Hox 1.1 allele mutated by homologous recombination. *Nature* **338:** 150.

Perspectives for Marker-assisted Selection and Velogenetics in Animal Breeding

M. Georges
GenMark Inc.
Salt Lake City, Utah 84108

Until recently, artificial selection has relied on the biometrical evaluation of breeding values from the performance of an individual animal and its relatives (Falconer 1989). This biometrical strategy is based on relatively simple genetic premises. The majority of economically important traits are complex or quantitative traits, meaning that the phenotype of an animal is determined by both environment and a large number of genes with individually small, additive effects. The heritability of the trait is the proportion of the phenotypic variation observed in a given population which is genetic in nature. Substantial genetic progress has been obtained using this approach. One of the powers of this biometrical approach is that it obviates the need for any detailed molecular knowledge of the underlying genes, termed economic trait loci.

It is believed that the molecular identification of these economic trait loci could increase genetic response by affecting both time and accuracy of selection through a procedure called marker-assisted selection (Soller and Beckmann 1982; Smith and Simpson 1986). One strategy toward the isolation of economic trait loci relies on the use of DNA sequence polymorphisms as genetic markers in linkage studies. This approach, paradoxically referred to as "reverse genetics" (Orkin 1986), is described in detail in this chapter. Moreover, we propose a new concept called "velogenetics" or the combined use of marker-assisted introgression and germ-line manipulations to shorten the generation interval of domestic species

Animal Applications of Research in Mammalian Development
 0-87969-333-9/91 $3.00 + 00

(especially cattle), which will allow the rapid and efficient introgression of mapped economic trait loci between genetic backgrounds.

DNA SEQUENCE POLYMORPHISMS

Types of DNA Sequence Polymorphisms

The typical mammalian genome is composed of approximately 3×10^9 base pairs of DNA, divided over a species-specific number of chromosomes, and contains all of the information required for the proper development and functioning of a normal being. An individual has two copies of each chromosome: one paternal and one maternal in origin. Although overall architecture and content are virtually identical, the paternal and maternal DNA sequences exhibit subtle allelic differences, hereafter referred to as DNA sequence polymorphisms (DSPs). DSPs that can be recognized in a given population are the molecular basis of the genetic component of the observed phenotypic variance. One can distinguish three types of DSPs.

Single-base-pair Polymorphisms

As the name implies, DSPs are due to single-base-pair differences between alleles. These can be either base-pair substitutions—transitions (purine to purine or pyrimidine to pyrimidine) and transversions (purine to pyrimidine and vice versa)—or the insertion/deletion of a single base pair. The frequency of a single-base-pair polymorphism is measured by the nucleotide diversity, π, the average heterozygosity per nucleotide site (Nei 1987). The nucleotide diversity has been estimated from restriction-fragment-length polymorphisms (RFLPs) at 0.002 for human (see, e.g., Kazazian et al. 1983) and at 0.0007 in cattle (Georges et al. 1987; Hilbert et al. 1989; Steele and Georges 1991). This means that, on the average, a human will be heterozygous for 1 out of every 500 nucleotides and a cow will be heterozygous for 1 out of every 1500 nucleotides.

One type of single-base-pair polymorphism deserves special attention: the CpG to TpG transition. The cytosine in the CpG dinucleotide sequence is known to be the substrate of a eukaryotic methylase, which will add a methyl group in position 5 of the pyrimidine ring if the cytosine of the complementary CpG dinucleotide is itself methylated. Deamination of a 5-methylcytosine generates a thymine, blurring the task of the DNA repair machinery, which may resolve the ensuing mismatch by replacing the original guanine instead of the mutated thymine. As a consequence, cytosines in the CpG doublet exhibit mutation rates at least ten times higher than other nucleotides and hence are rich sources of single-base-pair polymorphisms (Barker et al. 1984; Hilbert et al. 1989; Steele and Georges 1991).

DNA Sequence Rearrangements

In this kind of DSP, the difference between allelic variants involves DNA sequence rearrangements such as insertions, deletions, inversions, and duplications. Although there is a wide spectrum of molecular mechanisms by which such chromosomal rearrangements can be generated, it is well established that mobile genetic elements significantly contribute to this kind of DSP.

Rearrangements involving transposable elements account for a large proportion of new mutations detected in lower eukaryotes such as *Drosophila* and yeast (Rubin 1983). In the mouse, retrovirus-like sequences or retrotransposons have been shown to act as insertional mutagens (Hawley et al. 1982; Canaani et al. 1983; Ymer et al. 1985; Blatt et al. 1988; Stocking et al. 1988), and different strains of mice exhibit substantial heterogeneity with respect to the numbers and chromosomal sites of endogenous proviruses (Cohen and Varmus 1979). Variation in the distribution of endogenous retroviruses has been demonstrated in poultry as well.

In the human, at least 10% of the genome is known to be composed of retrotransposon-like sequences. Evidence for a role of these sequences in human genetic variability and disease stems from several reports of de novo mutations due to these sequences: (1) A mutation in the human low-density lipoprotein (LDL) receptor gene giving rise to familial hypercholesterolemia is caused by a deletion brought about by an intrastrand recombination event between two *Alu* sequences (Lehrman et al. 1987), (2) L1 (a human "long interspersed repeat" sequence) insertions were found to inactivate the factor VIII gene in hemophilia-A patients (Kazazian et al. 1988), (3) a c-*myc* rearrangement in a breast carcinoma was found to be due to insertion of an L1 element (Morse et al. 1988), (4) an *Alu* transposition event has been documented in human lung carcinoma cells (Lin et al. 1988), and (5) a homologous recombination between the long terminal repeats (LTRs) of a human retrovirus-like element was shown to cause a 5-kb deletion-polymorphism. Recently, Wong et al. (1990) reported evidence of human DNA polymorphism arising through DNA-mediated, rather than RNA-mediated, transfer between autosomes as well.

Expansion-Contraction-type Polymorphisms

A significant proportion of the eukaryotic genome is composed of sequences widely termed satellite DNA, sharing a common organization: a sequence motif varying in length between one and several thousand nucleotides, repeated in a head-to-tail (tandem) arrangement. Depending on the methodology originally used for their study (i.e., isopycnic centrifugation, pulsed-field gel electrophoresis, agarose gel electrophoresis, or polyacrylamide gel electrophoresis), satellite sequences were grouped into four size classes: macro-, midi-, mini-, and microsatellites. Minisatellites are also known as variable number of tandem repeats (VNTRs) (Jeffreys et al. 1985; Nakamura et al. 1987b; Vassart et al. 1987). Although macrosatellites seem to be confined to heterochromatic regions (Singer 1982), mini- and microsatellites have been found scattered throughout the genome (O'Connell et al. 1987, 1989; Lathrop et al. 1988a,b; Nakamura et al. 1988a,b,c; Litt and Luty 1989; Tautz 1989; Weber and May 1989; Georges et al. 1990b; Julier et al. 1990). In the human, however, minisatellite clusters seem to be particularly abundant in proterminal regions (Royle et al. 1988). The only midisatellite described to date has been mapped to the short arm of chromosome 1 (Nakamura et al. 1987a). In the human, the polydeoxyadenylate tract of *Alu* repetitive elements is also characterized by length variation and is thus an abundant source of genetic markers as well (Economou et al. 1990). The function of these satellite sequences is essentially unknown.

An important feature of all satellite sequences is that the maintenance of their tandemly repeated organization is dependent on the concerted evolution of the repeats. This evolution is thought to result from subsequent rounds of unequal crossing-over or related mechanisms (Doolittle 1985) that are favored by the tandemly repeated structure itself. The proposed unequal crossing-over mechanism, whether happening between sister chromatids or homologous chromosomes, explains the substantial degree of length polymorphism, referred to as expansion-contraction polymorphism, that characterizes those sequences. Moreover, the ensuing shuffling of slightly divergent repeat units or minisatellite variant repeats (Jeffreys et al. 1990) within the satellite generates additional internal site polymorphisms. These peculiar properties of satellite sequences have made them an invaluable source of highly informative genetic markers both in the human and in domestic species (for review, see Georges 1990).

Detection of DSPs

During the last 10 years, a multitude of methods have been developed for the detection of DSP. Two techniques, however, undoubtedly dominate this field: Southern blot hybridization (Southern et al. 1975) and the polymerase chain reaction (PCR) (Saiki et al. 1988), used either separately or in conjunction. A nonexhaustive list of methods presented here is grouped into four classes.

Restriction Pattern Analysis

DSPs may alter the restriction patterns of defined chromosomal regions, generating RFLPs. Depending on the size range of the explored restriction fragments, pulsed-field gel electrophoresis (Julier and White 1988), agarose gel electrophoresis (Jeffreys 1979), or polyacrylamide gel electrophoresis (Kreitman and Aguade 1986) can be used to study large, intermediate, or small fragments, respectively. RFLPs are classically detected by Southern blot hybridization. Alternatively, one can analyze the restriction patterns of defined DNA sequences amplified by PCR, generating amplified sequence polymorphisms (Skolnick and Wallace 1988). When studying chromosomal rearrangements or expansion-contraction-type polymorphisms, the use of PCR obviates the need for restriction enzyme digestion, as the DSP is reflected in the size of the amplified product.

Because of its simplicity, the detection of RFLPs has been by far the most popular approach toward DSP. The relative lack of power inherent to the method (only 20% of a given sequence is amenable to exploration using the most common restriction enzymes) can be compensated for by focusing on highly polymorphic sequences such as CpG dinucleotides (using enzymes such as *Taq*I and *Msp*I containing CpG in their recognition sequence) or hypervariable minisatellites. However, the discovery of microsatellites as a very abundant source of highly informative DSP in a broad taxonomic range, easily detectable by PCR, is likely to shift the focus toward these sequences for future marker development (Lift and Luty 1989; Tautz 1989; Weber and May 1989; Economou et al. 1990; Georges et al. 1990a).

Mismatch Analysis

Several methods for the detection of DSP are based on the study of mismatch analysis. The DNA to be analyzed is probed with a sequence corresponding to a defined genetic variant. The presence of a different variant in the target DNA generates a mismatched heteroduplex, which can be detected by various means as described below.

Detection of altered melting behavior. A mismatched heteroduplex will differentiate itself from the perfectly matched homoduplex by an altered melting behavior detected either as an all-or-none binary response that is positive for the homoduplex and negative for the heteroduplex or as a more graded response that allows the distinction between different heteroduplex variants.

The classic all-or-none test depends on the use of allele-specific oligonucleotides in hybridization experiments. With appropriate hybridization and washing conditions, the allele-specific oligonucleotide will only recognize a perfectly complementary sequence (Thein and Wallace 1986). With the advent of PCR, new variants of this approach have been described including reverse dot-blot (Saiki et al. 1989), the amplification refractory mutation system (Newton et al. 1989) or allele-specific PCR (Wu et al. 1989), and competitive oligonucleotide priming (Gibbs et al. 1989). The ligation amplification reaction, in which specific DNA sequences are amplified using sequential rounds of template-dependent ligation, can also be considered as a special application of the allele-specific oligonucleotide approach (Wu and Wallace 1989).

More discriminating is denaturing gradient gel electrophoresis, which explores the melting behavior of each heteroduplex as it is electrophoresed through an increasing gradient of DNA denaturants (Myers et al. 1988). The sensitivity of this method can also be improved by preamplifying the target sequence by PCR.

Ribonuclease and chemical mismatch detection. The presence of a mismatch in a heteroduplex makes those molecules susceptible to cleavage by various means, including chemical treatment with either hydroxylamine or osmium tetroxide (Roberts et al. 1989), as well as ribonucleases such as RNase A (for RNA:DNA heteroduplexes) (Myers et al. 1988). Electrophoretic analysis of the cleavage products allows the distinction between genetic variants. Again, implementing PCR will increase the sensitivity of the approach.

Single-stranded Conformation Polymorphisms

Under nondenaturing conditions, single-stranded DNA has a folded conformation that is stabilized by intrastrand interactions. Consequently, the conformation, and therefore the electrophoretic mobility, is dependent on the sequence. DNA variants exhibit mobility shifts when electrophoresed under such conditions, presumably resulting from conformational changes caused by sequence alterations. The altered mobility can be detected by blot hybridization analysis or by PCR (Orita et al. 1989a,b).

Direct Determination of the DNA Sequence

The most powerful approach toward DSP analysis is the direct determination of the DNA sequence. However, the need of a cloning step in classic sequencing protocols precluded the analysis of large samples. This limitation has been circumvented by the development of genomic sequencing techniques that allow the direct determination of defined DNA sequences from genomic DNA (Church and Gilbert 1984). More recently, the possibility to detect DSP by direct sequence determination of PCR-amplified products has been amply demonstrated in several independent studies (see, e.g., Wong et al. 1987).

Origin and Evolution of DSPs

DSPs encountered in a given population find their origins in mutational events (neomutations) occurring in the germ line and escaping the DNA repair machinery. The fate of these germ-line mutations in the population is dominated by two kinds of effects: stochastic and deterministic.

Stochastic effects

When a new mutation appears in the population, its initial survival depends largely on chance, regardless of its selective effect. This is easily illustrated by assuming an individual is heterozygous for a neomutation, inherited from its parent in whose germ line the mutation appeared. If this individual in turn has one, two, or three offspring the chance for the neomutation to be lost from the population because it is transmitted to none of the offspring is 0.5, 0.25, and 0.125, respectively. Even if it is inherited by some of the offspring, the same stochastic filter will operate in the next generation. In the course of this random drift, the overwhelming majority of mutant alleles are lost by chance. However, some will see their frequency increase in the population and, despite fluctuations over time, eventually become fixed in the population until substituted by the next mutant allele.

As demonstrated by Kimura (1986) in the framework of his neutral theory of molecular evolution, the probability for a selectively neutral neomutation to be fixed in a population of N

individuals (with $2N$ chromosomes) is equal to its initial frequency, $1/2N$. The average time for fixation is four times the "effective" population size, or $4Ne$. The rate κ of mutant substitution per generation is simply equal to the rate of mutation per gamete and per generation, μ, hence independent of the population size. According to this view, a polymorphism observed in a population at a given time is composed of transient alleles caught in their stochastic "odyssey" throughout the population.

Populations for which $4.Ne.\mu \leq 1$ are essentially monomorphic, and populations for which $4.Ne.\mu \approx 1$ are characterized by a substantial degree of transient polymorphism. The model predicts a steady-state level of heterozygosity, H:

$$H \approx \frac{4.Ne.\mu}{4.Ne.\mu + 1}$$

Deterministic Effects

Good evidence exists that the fate of a significant proportion of DSPs, especially those occurring in noncoding parts of the genome (composing the large majority of the genome), is essentially dominated by random drift. However, when a neomutation affects a DNA sequence expressed at the phenotypic level, the mutation may no longer be selectively neutral, and deterministic effects will be superimposed on the stochastic ones. Negative and positive selections will decrease or increase, respectively, the probability and rate of fixation, whereas "balancing selection" will maintain specific alleles in a population in an equilibrium state.

Negative selection. When comparing DNA sequences between taxa, it appears that the estimated number of mutant substitutions per nucleotide to account for the observed divergence is highest for noncoding sequences, such as pseudogenes and intronic sequences, and much lower for coding sequences. For the latter, however, a distinction must be made between first, second, and third positions of the codons. The third position, where only 28% of the substitutions are expected to cause an amino acid change (vs. 95% and 100% for first and second positions, respectively), exhibits the highest substitution rate (Nei 1983). When estimating that part of the substitutions at the third position which are "synonymous," rates very similar to those of the noncoding regions are observed (Kimura 1986). Moreover, DSPs are more prominent in noncoding sequences and at third codon positions within coding sequences (Nei 1983).

These observations are easily explained by assuming that the fate either of synonymous mutations or of neomutations arising in noncoding regions is dominated by stochastic effects, whereas the fate of mutations causing amino acid replacements will depend on whether or not they disrupt the function of the protein. In this case they will be eliminated from the population by negative, purifying selection. The higher the functional constraints imposed on a protein, the higher the proportion of neomutations expected to be harmful and hence the lower the substitution rate, expressed at the protein level as unit evolutionary time (average time required for one amino acid change to appear in a sequence of 100 amino acid residues).

Positive selection. According to the previous discussion, the major drive behind molecular evolution is mostly nonadaptive in nature, which is in conflict with the classic theory of adaptive, positive Darwinian selection. There is, however, evidence for positive and adaptive evolution at the molecular level in at least a few instances. Comparing DNA sequences from members of two gene families, the serine protease inhibitors in rat (Hill and Hastie 1987) and the pregnancy-specific β1 glycoprotein gene family in humans (Streydio et al. 1990), evidence for higher substitution rates at first and second codon positions in some protein regions has been found, indicating a positive selection mechanism. Moreover, there are a number of experimental data suggesting that some allelic differences identified by electrophoresis are associated with adaptation to different environments. In *Drosophila*, for instance, there is evidence for correlation between in vitro heat resistance of alcohol dehydrogenase (ADH) variants and the temperature characterizing their geographical origin (Nei 1983).

Balancing selection. The evolutionary forces described so far generate transient DSP, in the sense that the population frequencies of existing genetic variants will irrevocably change with time until either fixation or loss. In some cases, however, alleles may be maintained in a population at a steady-state level. Overdominance is one of the mechanisms susceptible to generate such a balanced polymorphism. For a two-allele system, this means that the heterozygous individuals benefit from a selective advantage when compared to both homozygous genotypes. This is expected to generate a steady state where both alleles are maintained in the population at respective equilibrium frequencies p and q, where $q = s/(s + t)$ and $p = t/(s + t)$, with s and t being the respective selection coefficients of the homozygotes.

The best-known examples of balanced polymorphism due to overdominance are the maintenance of the S α-globin allele (causing sickle-cell anemia in homozygotes) and thalassemia mutants (see, e.g., Flint et al. 1986) in populations subjected to malaria, because of the resistance exhibited by the heterozygotes toward the parasite. The high level of polymorphism observed at the major histocompatibility complex (MHC) locus is thought to result from overdominance selection as well (Hughes and Nei 1988). Frequency-dependent selection may be another cause of balanced polymorphism, as in the "rare mate advantage" observed in *Drosophila* (Ajala 1972).

CONSTRUCTION OF PRIMARY DNA MARKER MAPS

Linkage Strategies

Two loci are said to be genetically linked if during meiosis they recombine at a rate significantly lower than 50%, i.e., generate significantly more parental than recombinant gametes. The recombination rate between loci reflects the frequency of occurrence of an uneven number of crossovers between the loci. Because the probability for crossing-over is proportional to the distance separating the loci, the recombination rate can be used as a unit of chromosomal length. This length unit is known as the Morgan (M), with 0.01 M or 1 cM corresponding to the distance separating two loci exhibiting a 1% recombination rate. For small distances (≤30 cM), the relationship between centi-Morgan and recombination rate is essentially linear; for longer distances, however, the relationship is more complex, depending on the frequency of double crossovers, itself affected by possible interference (the effect of one crossover on the occurrence of an other one nearby). Parental and recombinant gametes will only be distinguishable for doubly heterozygous individuals, hence the need for highly polymorphic markers.

Due to the advent of the PCR, it has been possible to directly determine the genotype of individual gametes (Li et al. 1988). However, most often, the gametic contribution is inferred from the genotype of the offspring, and linkage studies are performed within families. Most modern linkage studies use the lodscore test for evaluation of linkage, a sequential test based on the method of maximum likelihood (Morton 1955). The lodscore is calculated as $\log_{10}$(LR), where LR (likelihood ratio) is the likelihood of genotypic observations under alternative hypothesis, $\Theta \leq 0.5$, divided by the likelihood of genotypic observations under null hypothesis of no linkage, $\Theta = 0.5$ (where Θ is the recombination rate). In human genetics, a lodscore greater than 3 is accepted as significant evidence for linkage. The prior probability of linkage between two loci has been used to justify this stringent critical value. Note that 2ln(LR) can be used as well, having a chi-square distribution with one degree of freedom under the null hypothesis of no linkage.

Recently, algorithms for multilocus linkage analysis have been developed, allowing an estimate of the most likely gene orders and genetic distances between several loci to be determined simultaneously (Lathrop et al. 1984, 1985; Lander and Green 1987).

Although usually determined within families, genetic linkage can also manifest itself at the population level, a phenomenon called linkage disequilibrium. According to the Hardy-Weinberg law, the equilibrium genotypic frequencies are reached in a single generation (except when the initial gene frequencies are not equal among sexes). For a diallelic system with alleles a1 and a2 and respective allelic frequencies p1 and p2, the equilibrium genotypic frequencies are $p1^2$, 2p1p2, and

$p2^2$ for a1a1, a1a2, and a2a2, respectively. This does not necessarily hold when considering two loci simultaneously. The genotypic equilibrium frequencies are reached only when the previous generation produces the four possible gametes at the expected frequencies: a1b1: p1q1; a2b1: p2q1; a1b1: p2q1; a2b2: p2q2. The difference between observed and expected gametic frequencies is the linkage disequilibrium, *D*. The value of *D* is reduced by Θ every generation, where Θ is the recombination rate between the two loci. For unlinked loci, *D* diminishes by one half every generation; for linked loci, however, the reduction of *D* per generation will be much smaller. The detection of a linkage disequilibrium is an indication of linkage between the corresponding loci.

Genetic Maps

Using this linkage approach combined with alternative mapping strategies such as in situ hybridization (see, e.g., Lichter et al. 1990), somatic cell hybrid panels, radiation hybrid mapping (for review, see Cox et al. 1990), and comparative mapping (Womack 1987), the location of large sets of DSPs can be determined in order to build a genetic marker map (see, e.g., White et al. 1985; Donis-Keller et al. 1987; Fries et al. 1989). Assuming a total map length of 30 M as for the human, and a desirable maximum distance of 20 cM between markers, a set of 150 DSPs could cover the entire genome. However, many more markers will be needed to generate reasonable maps for our domestic species essentially for two reasons. First, most of the time, we have no a priori information on the location of the characterized markers. Hence, some chromosomal regions will initially be overrepresented in our map and others will be underrepresented. This problem is expected to become critical in the later stages of the development of a map. Comparative data will then become critical, leading to a search for markers whose locations can be predicted from other species. Second, an individual will only be informative for the markers for which it is heterozygous; parts of its genome will thus not be explorable, because it will be homozygous for the corresponding markers. To compensate for this, one will have to identify more markers, the number required being inversely proportional to their heterozygosity. Hence, the importance of highly informative systems.

Once such a map is generated, any gene for which the appropriate segregating family material is available can be located. Assuming a maximum marker-target gene distance of 10 cM, the expected lodscore for a doubly informative, phase-known meiosis (alleles sorted by parental origin) is approximately 0.16 (Lander and Botstein 1986). Theoretically, 20 such meioses are sufficient to establish linkage with a lodscore of 3. In practice, however, the number of individuals it is necessary to analyze will be higher, function among other factors of the quality of the marker, expressed as its polymorphism information content (Botstein et al. 1980).

The efficiency of this approach has been illustrated by the recent mapping of a large number of genes involved in human single-gene disorders (see, e.g., White and Caskey 1988). The identification of DNA markers for a defined gene can be the first step toward its molecular cloning. Successful positional cloning, the isolation of a gene based on its map location, has been achieved in the human for chronic granulomatous disease (Royer-Pokora et al. 1986), Duchenne's muscular dystrophy (Monaco et al. 1986), retinoblastoma (Friend et al. 1986), cystic fibrosis (Kerem et al. 1989; Riordan et al. 1989; Rommens et al. 1989), type-1 neurofibromatosis (Cawthon et al. 1990; Viskochil et al. 1990), and the testis determining factor (Gubbay et al. 1990; Sinclair et al. 1990).

In domestic animals, genetic maps could be used to localize the genes underlying production traits, allowing for marker-assisted selection and constituting a first step toward their isolation, the understanding of their mechanism of action, and their manipulation by mutagenesis and gene transfer methods. Several laboratories worldwide are now involved in the development of markers and the construction of genetic maps for the main domestic species, cattle, pigs, and poultry.

Progress toward a Primary DNA Marker Map in Cattle

In the last 2 years, our laboratory has focused on the development of a primary DNA marker map for cattle. We have now developed more than 200 highly polymorphic DNA markers of three types.

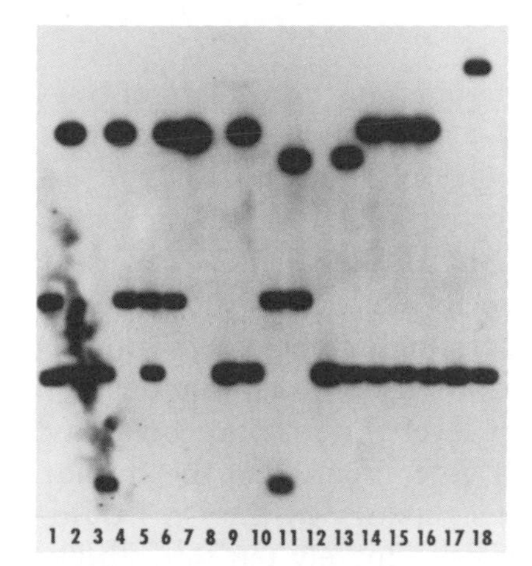

FIGURE 1 VNTR pattern obtained with probe GMBT-028 on 18 randomly selected Holsteins.

Variable Number of Tandem Repeat Markers

Hypervariable minisatellites are known to show significant cross-hybridization between species (Georges et al. 1988a, 1990b). We have exploited this to isolate bovine VNTRs using heterologous minisatellite probes (M. Georges et al., in prep.). By screening purpose-built libraries with minisatellite probes, we isolated 36 bovine VNTRs, characterized by a mean heterozygosity of 59.3% within the American Holstein breed. Matching probabilities and exclusion powers were estimated by Monte Carlo simulation, showing that the top five to ten probes could be used as a very efficient DNA-based system for individual identification and paternity diagnosis. The isolated VNTR systems should contribute significantly to the establishment of a bovine primary DNA marker map. Linkage analysis, use of somatic cell hybrids, and in situ hybridization demonstrate that these bovine VNTRs are organized as clusters scattered throughout the bovine genome, without evidence for proterminal confinement as in the human (Royle et al. 1988). Moreover, Southern blot analysis and in situ hybridization demonstrate conservation of sequence and map location of minisatellites. The patterns obtained with one of these VNTRs is shown in Figure 1.

Multisite Haplotypes

We used 110 random cosmids to probe Southern blots of DNA from nine unrelated cattle. Twelve restriction enzymes were used to analyze the DNA. Although only one third of the expected fragments could be detected, 85% of the cosmids revealed at least one polymorphism. The mean heterozygosity of the generated multisite haplotypes (White and Lalouel 1988) was estimated at 51.9%. A surprisingly high proportion of polymorphisms (≈25%) was attributed to insertion/deletion events, compensating for the lower level of nucleotide diversity, π, observed in cattle ($\pi \approx 0.0007$) as compared to the human ($\pi \approx 0.002$). The mutation rate at cytosines in the CpG dinucleotide was estimated as being approximately ten times higher as compared to other nucleotides. The generated markers should cover approximately 40% of the bovine genome when used in linkage studies (Steele and Georges 1991).

Microsatellites

Recently, microsatellites were proven to be an abundant source of highly polymorphic markers in the human (Litt and Luty 1989; Tautz 1989; Weber and May 1989). As the name implies, microsatellites are minute VNTR markers (Jeffreys et al. 1985; Nakamura et al. 1987b), characterized by tandem repetitions of very short repeats, one to four base pairs in length. Microsatellites exhibit levels of polymorphism comparable to those of VNTRs, but they are much more abundant and apparently evenly spread throughout the genome. We have estimated the frequency of $(CpA)_n$ dinucleotide repeats with $n \geq 9$ at ≥150,000 in the bovine genome. Because of their small size, their detection is greatly facilitated by PCR. Although this imposes the preliminary determination of flanking DNA sequences to design the appropriate primers, the subsequent PCR used for their analysis offers several advantages over Southern blotting: it is fast, requires less DNA, and is easier to automate.

As part of our effort to build a primary DNA marker map for cattle, we have isolated more than 100 bovine microsatellites and have amplified them in vitro. The majority of these micro-

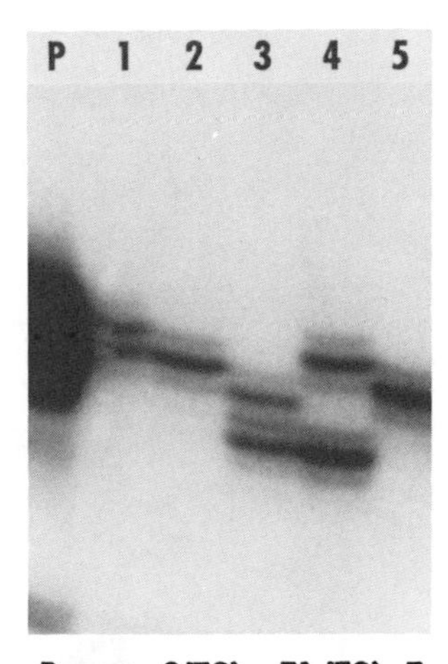

Repeat: G(TG)$_{13}$ TA (TG)$_6$ T
PCR-Product: 168bp

FIGURE 2 TGLA-9 microsatellite patterns from five randomly selected Holsteins (lanes *1–5*). (Lane P) Plasmid control.

satellites were found to be polymorphic in cattle (Fig. 2) (M. Georges, unpubl.), and several have been tentatively assigned to specific bovine chromosomes using a somatic cell hybrid panel (Fig. 3). Moreover, we have shown that approximately 50% of the bovine microsatellites can be successfully used in other Bovidae, particularly sheep, as well, which will greatly facilitate the construction of marker maps in these species (S.S. Moore et al., in prep.).

Magnetic solid-phase DNA-sequencing procedures (Hultman et al. 1989) are used for the massive generation of sequence information, and multiplex approaches, based on the simultaneous detection of molecules labeled with different fluorescent dyes using a laser-excited confocal fluorescence gel scanner (Quesada et al. 1991), are developed for genotype collection.

Map Construction

The relative location of the markers is determined by linkage analysis in pedigrees generated by multiple ovulation and embryo transfer. To assign linkage groups to specific chromo-

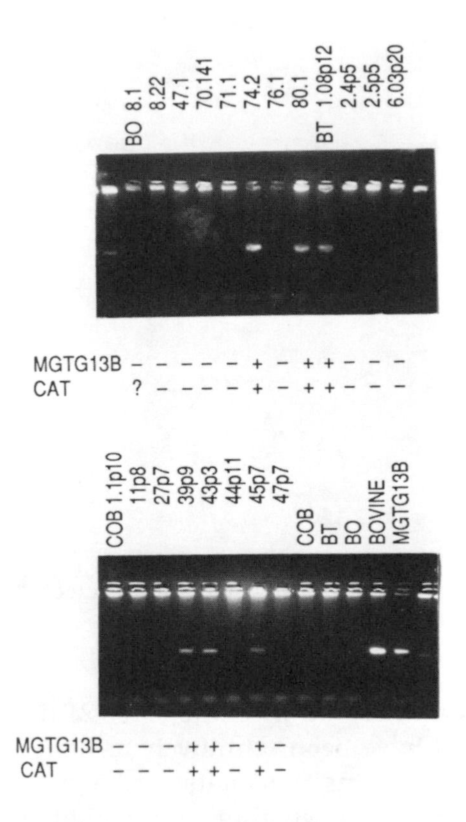

CAT = U19 = Chr. 15

FIGURE 3 PCR mapping of microsatellites using somatic cell hybrids. PCR amplifications were performed with primers specific for microsatellite MGTG-13B, on 50 ng of DNA from a somatic cell hybrid panel, as well as the appropriate plasmid, bovine, and rodent (COB, BT, BO) controls. The pattern obtained was perfectly concordant with the catalase enzyme, corresponding to bovine synteny group 19 or chromosome 15.

somes, highly polymorphic anchor markers are mapped using somatic cell hybrids (R. Fries et al.; J. Womack et al., both unpubl.) and in situ hybridization.

GENETIC MAPPING OF QUANTITATIVE TRAIT LOCI

The majority of the traits dealt with in animal production are quantitative traits, characterized by continuous variation. The phenotype of an animal with respect to a particular trait is the result of the effect of several polygenes, known as quantitative trait loci (QTL), combined with environmental effects. The number of polygenes involved in determining a QTL is essentially unknown, but is considered to be very large, with each gene contributing a very small part of the genetic variation. However, there is evidence from both the plant and animal worlds that QTL with significant effects are common (Martin et al. 1989; Paterson et al. 1989; Hoeschele and Meinert 1990). The most likely model is to assume that the polygene number is large, but that there is a broad distribution of effects, substantial in some cases. Polygenes with extreme effects, whose segregation in a population may cause skewness and bi- or trimodality, are known as major genes. Examples in animal breeding are double-muscling genes in both cattle and pigs, the White Shorthorn gene involved in the determinism of white heifer disease and the Booroola fertility gene in sheep (Hanset 1982). Even with significant effects on the trait of interest, however, the contribution of the polygenes to the total genetic variation may be limited if their frequency in the population is low.

When dealing with quantitative traits, direct determination of genotype for the corresponding QTL is impossible. Nevertheless, strategies have been designed to map QTL by linkage analysis. Within segregating populations, which is usually the case for domestic species, QTL mapping can be performed both within families and at the population level.

QTL Mapping Within Families

Traditionally, QTL mapping proceeds as follows: offspring from an individual that is heterozygous for both marker loci and QTL are grouped according to which marker allele they inherited; a statistically significant difference between the phenotypic means of the two groups indicates linkage between the marker and QTL. Tests for statistical significance are done by linear regression (i.e., one-way analysis of variance) under the assumption of normally distributed residual environmental variance. Markers are tested one at a time for possible linkage to a QTL affecting the trait of interest. One of the drawbacks of this approach is that it is impossible to estimate unequivocally both the map location of the QTL with respect to the marker and its effect on the considered trait; no distinction can be made between a closely linked QTL with a small effect and a loosely linked QTL with a major effect.

The lodscore method has been recently updated to deal with quantitative and other complex traits and to exploit fully the power of the nearly complete marker maps that have become available for different organisms. This approach is known as interval mapping. Interval mapping not only solves the problem of simultaneous estimation of location and effect, but also reduces the number of individuals to be tested to detect linkage with a QTL of given effect because of its increased power (Lander and Botstein 1989).

Assuming that the marker is the QTL, the number of individuals to be tested to detect an effect of given amplitude, δ, can be estimated from $n \geq 4(t_0 + t_1)^2 \cdot s^2/\delta^2$, where n gives the required sample size, s^2 is an estimate of the residual variance, t_0 is the t value associated with type I error, and t_1 is the t value associated with type II error; t_1 equals tabulated t for probability $2(1\text{-}p)$, where p is the required probability for detecting δ if such a difference exists (Soller et al. 1976).

In dairy production, for example, with linkage analysis using the daughter yield-deviations (DYD; $\sigma^2_{DYD} \approx 600$ lb) from paternal half-sibs (granddaughter design; Weller et al. 1990), one would have to study 1500, 378, and 168 individuals to detect QTL with differences of 200, 400, and 600 lb, respectively, between alternate alleles. Assuming a phenotypic variance of $(2500 \text{ lb})^2$, such effects correspond to 0.08, 0.16, and 0.24 standard deviations, respectively. These estimates assume a type I error of 5%, a type II error of 10%, and an absence of recombination between the marker and QTL.

If the tested marker and the QTL recombine at a rate Θ, the number of individuals to test increases by a factor $1/(1 - 2\Theta)^2$ for single marker analysis and by a factor $\approx(1 - \tau)/(1 - 2\Theta)^2$ in the case of interval mapping, with τ corresponding to the recombination rate between the flanking markers (Lander and Botstein 1989).

In view of the costs and time involved in genotyping, it is

important to minimize the required sample size. This can be achieved by various ways as described below.

1. *Identifying the individuals most likely to be heterozygous.* One way to achieve this is to cross highly divergent strains for the trait of interest. In plant breeding, where the use of exotic germ plasm is common practice, this is perfectly applicable. The identification of markers for interesting QTL from the exotic strains can then be used for their marker-assisted introgression in the commercial varieties. In animal breeding, however, introgression programs are very uncommon. Velogenetics (see below) may make the use of exotic germ plasm in introgression programs more attractive for animal breeders as well. An alternative approach is to identify individuals whose offspring show a higher variance for the trait of interest.
2. *Selective genotyping of the extreme progeny.* As discussed by Lander and Botstein (1986), individuals whose genotypes can be most clearly inferred from their phenotypes are the ones that will provide most of the linkage information for studying complex traits. For quantitative traits, these are the individuals whose phenotypic values deviate most from the mean; they are the ones at the tails of the distribution. Sample sizes could be reduced by 60–80% by focusing on individuals that deviate 1–2 standard deviations from the mean. Paradoxically, selective genotyping may be limited by the size of the studied population. Indeed, a larger sample will be required in order to find enough individuals 1 or 2 standard deviations from the mean.
3. *Decreasing environmental variance via progeny testing.* Weller et al. (1990) tested the effect of progeny testing to reduce the environmental variance by comparing the power of the daughter and granddaughter designs for the detection of QTL in dairy cattle. In the daughter design, marker genotype and quantitative trait values were assessed on daughters of heterozygous sires, and in the granddaughter design, marker genotypes were determined on sons of heterozygous sires; their breeding values were determined by progeny testing from the quantitative trait value measured on their daughters. It was demonstrated that for

equal power, the granddaughter design required half as many marker assays as the daughter design.

4. *Simultaneous search*. Just as environmental noise can be decreased by progeny testing, genetic noise can be reduced by searching for several unlinked QTL simultaneously (Lander and Botstein 1989).
5. *Using DNA pools*. Instead of genotyping all individuals separately, one can analyze DNA pools from individuals sorted by phenotype. Significant differences of allelic frequencies between pools point toward possible genetic linkage between the corresponding marker locus and a gene or genes affecting the trait of interest. This approach can be used both for within-family studies and for studies at the population level. The latter approach, however, requires linkage disequilibrium between QTL and marker locus. This method was first described by Arnheim et al. (1985) to study the role of human leukocyte antigen (HLA) class II loci in insulin-dependent diabetes mellitus. It was recently adopted by Plotsky et al. (1990) to study association between DNA fingerprint bands and abdominal fat deposition in poultry.
6. *Exploiting tagged QTLs*. The direct effect of selection for a production trait will be to increase the frequency of the favorable alleles at the segregating QTL. However, this selection pressure may indirectly affect loci in linkage disequilibrium by so-called "hitch-hiking." An example of this is seen in the genetic defect causing progressive degenerative myeloencephalopathy, known as "Weaver" in Brown Swiss cows, shown to be linked to a major gene for milk production (Hoeschele and Meinert 1990). Because of the deleterious effect of the Weaver-causing gene, it is the heterozygous carrier genotype that is selectively most advantageous, generating a balanced polymorphism with the Weaver-causing allele being maintained in the population at a relatively high frequency. This can be exploited to map the corresponding QTL by finding a marker linked to this single-gene disorder, a relatively easy exercise compared to QTL mapping. We have recently identified a marker linked to Weaver and presumably to the associated QTL (M. Georges, in prep.).

In addition to the Weaver phenotype, QTL for a variety of polygenic traits have been identified in both plants and animals. Using complete DSP maps in tomato, Paterson et al. (1989) identified at least six genes controlling fruit mass, four controlling soluble solids, and five controlling fruit pH. They accounted for 58%, 44%, and 48% of the phenotypic variance, respectively. Using a similar approach, Martin et al. (1989) identified at least three tomato genes controlling water-use efficiency. In cattle, Geldermann et al. (1985) found significant effects on milk yield (+200 kg) and fat content (+1%), especially for the β-lactoglobulin locus. More recently, Cowan et al. (1990) demonstrated significant effects on milk production traits using a prolactin DSP as marker.

QTL Mapping Within Populations

One can expect to find the effect of marker alleles linked to QTL at the population level if the two loci are in linkage disequilibrium. As reported by Hanset (1989), and assuming a diallelic marker (alleles M1 and M2 with respective frequencies p1 and p2) linked to a diallelic QTL, the phenotypic difference, δ, between the respective homozygotes at the marker loci is

$$\delta = 2a \frac{D}{p1p2}$$

with D measuring the linkage disequilibrium and 2a corresponding to the phenotypic difference between the two homozygotes for the QTL.

Markers for which evidence of linkage disequilibrium is highest are candidate genes, expected from their physiological role to be likely candidates for the QTL itself. DSPs at those loci, even if selectively neutral, can be expected to exhibit linkage disequilibrium with the hypothetical functional mutations because of their very tight linkage. As an example, the B allele of the K-casein gene has been shown in several studies to increase protein yield in milk by about 3%, and may improve cheese yield independent of the effect on protein yield (see, e.g., Grosclaude 1988; Gibson et al. 1990).

USE OF DNA MARKERS IN BREEDING PROGRAMS

Marker-assisted Selection

In classic selection programs, breeding values are estimated from an individual's own performances and performances of relatives (for review, see Ebert and Selgrath, this volume). The expected genetic progress is a function of the accuracy of selection, i.e., the correlation between estimated and true breeding values. All direct information on QTL can be used to increase the accuracy of selection and genetic response.

Soller and Beckmann (1982) proposed to exploit marker information for the preselection of young dairy sires before progeny testing. In cattle, marker-assisted selection is already used for sexing preimplantation embryos using Y-chromosome-specific probes (see, e.g., Herr et al. 1990) and for genotyping at the K-casein (see, e.g., Medrano and Aguillar-Cordova 1990) and prolactin loci (Cowan et al. 1990). In pigs, marker-assisted selection is used to reduce the frequency of the major gene causing porcine stress syndrome (PSS). Susceptibility to PSS correlates with halothane sensitivity or malignant hyperthermia (Hal). This condition has been mapped to a linkage group on pig chromosome 6, encompassing the following markers: S(A-O)-GPI-Hal-H-A1BG-PGD (for review, see Archibald and Imlah 1985). These markers are used for selection against the PSS condition. Recently, the ryanodine receptor gene has been identified as a good candidate for the Hal gene (MacLennan et al. 1990).

As shown by Smith and Simpson (1986), the gain to be made with marker-assisted selection increases with the proportion of QTL identified and is highest for low heritability traits. Unfortunately, the QTL determining the latter traits are also the hardest ones to identify. It should be noted that the increase in accuracy is subordinate to the accurate estimation of the QTL effects. This may require larger samples than those needed for the detection of linkage. Once a QTL is mapped by within-family linkage studies, it may be more effective to identify supplementary flanking markers and to determine accurately the effect of the generated haplotypes at the population level. Selection can then focus on the best haplotype, instead of spending initial selection efforts on intermediate ones.

The use of genetic markers in selection programs may as well reveal dominance deviation (particularly overdominance) and interaction deviation at defined QTL. Specific programs may be required to exploit these QTL fully. In the case of overdominance, for instance, two lines each homozygous for the different alleles at each QTL could be developed and crossed to produce multiple heterozygotes.

Velogenetics: The Synergistic Use of Marker-assisted Introgression and Germ-line Manipulation

There is widespread interest in resolving quantitative traits into their Mendelian components by mapping the underlying QTL. The implementation of marker-assisted selection into breeding schemes, however, has not always been received with enthusiasm. Part of the skepticism expresses the doubt that the genetic gains obtainable by marker-assisted selection will justify expensive and tedious large-scale genotyping. Although the costs of genotyping will drop substantially in the near future, due to the rapid pace at which automation and robotics are being applied to DNA technology, this objection remains very pertinent.

Another major limitation of marker-assisted selection under its present form is its limitation to the exploitation of genetic variation preexisting within the commercial breed of interest, and only that present in a "high-merit," superior genetic background. Favorable mutations appearing within a mediocre background or in exotic germ plasm would be difficult to exploit, even with markers.

We have therefore proposed a scheme, designated as velogenetics, combining marker-assisted introgression and germline manipulations to reduce the generation interval, which might drastically increase the power of marker-assisted selection (Georges and Massey 1991).

Marker-assisted Introgression

The basic principle underlying marker-assisted introgression is well known. A gene responsible for a favorable attribute can

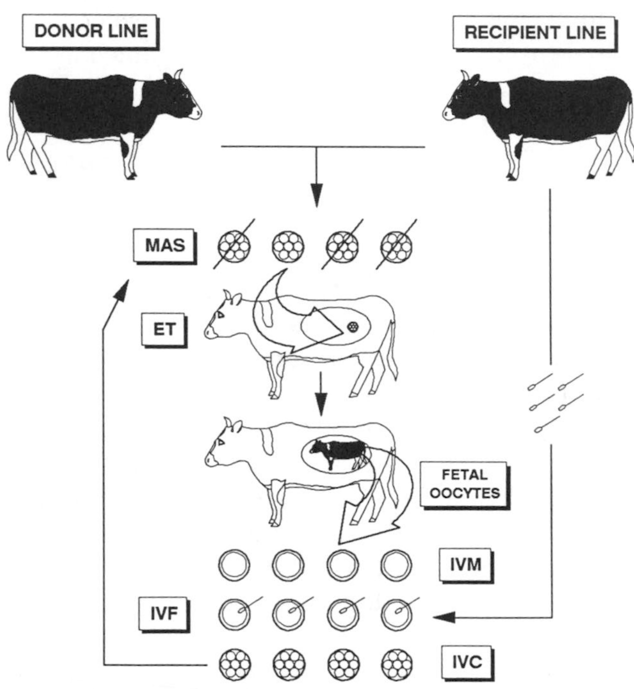

FIGURE 4 Schematic representation of velogenetics, the synergistic use of marker-assisted introgression and germ-line manipulations to reduce the generation interval in cattle. (MAS) Marker-assisted selection; (ET) embryo transfer; (IVM) in vitro maturation; (IVF) in vitro fertilization; (IVC) in vitro culture.

be introgressed from a donor strain into a recipient strain by repeated backcrossing. During the introgression process, the retention of the favorable gene is monitored in the backcross products, with linked flanking DNA markers. This latter aspect is particularly important for traits involving multiple genes and/or characterized by sex- or age-limited expression.

Classic genetic theory tells us that, with the exception of the marked segment whose retention is desired, the genomic contribution of the donor line is diluted by half after each backcross. Hence, after four backcrosses, the recipient genome is reconstituted to ±90% of the original. At the marked locus, however, the backcross retains one copy of the desired donor variant. If required, one intercross will then generate 25% of offspring homozygous for the favorable donor variant.

The net result is a graft of an advantageous gene within a recipient background. The procedure respects organization and chromosomal localization of the grafted gene and avoids aberrant expression patterns, which too often characterize transgenes (see Ebert and Selgrath, this volume). Significantly, the gene to be transferred does not need to be cloned. Only its genetic map location is required, as defined by the availability of linked markers, ideally flanking the gene of interest on each side. Hence, this procedure is perfectly applicable for the introgression of QTL identified through the previously described mapping strategies.

Marker-assisted introgression can be applied simultaneously to several genes, which is of particular interest for complex traits involving several genes. Introgressing more than one gene from a donor to a recipient line, however, increases the selection intensity at each backcross: with 1 marker, 1/2 of the offspring have the favorable genotype; with 2 markers, 1/4; and with *n* markers, $(1/2)^n$.

Selecting for the retention of defined donor genes will hamper the recovery of the recipient background genotype in adjacent chromosomal regions. This can be compensated for by increasing the number of backcrosses or by monitoring the fate of additional adjacent markers to identify the backcross products resulting from recombinations as close to the grafted gene as possible.

Shortening the Generation Interval by Velogenesis

Introgression by repeated backcrossing, with or without assistance from genetic markers, is a common practice in a variety of organisms, but it is essentially unfeasible in domestic animals such as cattle because of their prohibitively long generation time. The generation interval of such species could, however, be reduced using schemes such as the one illustrated in Figure 4, based on the in vitro maturation and fertilization of fetal oocytes, hereafter referred to as velogenesis. An overview of female gametogenesis (Russe 1983; Betteridge et al. 1989) indicates that the feasibility of such a scheme may not be that unlikely.

Oogenesis begins with the formation of primordial germ cells in the region of the allantois. These precursor cells migrate to the developing gonads where, after a period of mitotic proliferation, they enter meiosis. Meiosis is arrested at the diplotene stage of prophase I by the poorly understood meiotic division I arrest system, after which the primary oocyte enters a resting phase. During the lifetime of the animal, small numbers of resting primary oocytes are successively recruited into a pool of growing oocytes, within the environment of a developing follicle. These activated oocytes grow in size, acquire the competence to resume meiosis if appropriately stimulated, and accumulate the required material to sustain the early stages of the subsequent embryonal development. Resumption of meiosis and oocyte maturation is triggered by a hypothetical maturation-inducing signal produced by granulosa cells in response to gonadotropins. At least in rodents, oocyte maturation seems to be mediated by a drop in cAMP in the oocyte and subsequent inactivation of a type-A protein kinase. Evidence for the role of this pathway in oocyte maturation is, however, much more controversial in ruminants. Note that in the granulosa cells, gonadotropins act, among other pathways, through the activation of adenylate cyclase with subsequent increase in cAMP concentrations (Eppig 1989). In the oocyte, a cascade of events then probably leads to the phosphorylation and activation of a phosphatase, probably homologous to the *Schizosaccharomyces pombe cdc25* gene (Gould et al. 1990). This phosphatase activates the M-phase promoting factor (MPF), now known to be a complex of a $p34^{cdc2}$ protein kinase subunit with a B-type cyclin (for review, see Hunt 1989). The maturing oocyte completes the first meiotic division and enters the second (becoming a secondary oocyte) and becomes arrested at metaphase II until fertilization. This meiotic division II arrest system is thought to reflect the stabilization of MPF mediated by the kinase activity of $pp39^{mos}$ either on cyclin itself or on the ubiquitin-dependent pathway of cyclin degradation (for review, see Hunt 1991). Fertilization relieves this block by increasing the intracellular Ca^{++} concentration, triggering calcium-dependent protease activity (for review, see Hunt 1989).

In cattle, primordial germ cells reach the genital ridge at about 40 days of gestation. After a period of mitotic proliferation, they differentiate into oogonia starting at about 60 days of gestation. Mitotic proliferation of the germ line ceases by about day 170 of gestation, fixing the maximum number of oocytes the female will ever have. Meiosis starts at about 80 days, and the first primordial follicles are discernible at 90 days of gestation. Remarkably, activation of resting primordial follicles starts already in utero, about day 140, and secondary and tertiary follicles can be seen at 210 and 230 days, respectively. It is estimated that two to four resting primordial follicles are recruited daily into the pool of activated, developing follicles. These activated fetal oocytes, however, are irrevocably committed to follicular atresia. Indeed, spontaneous oocyte maturation and ovulation do not begin until puberty. Submitted to the appropriate hormonal stimulus, however, prepubertal oocytes can resume meiosis, be fertilized, and produce viable offspring. Indeed, offspring have been obtained from gonadotropin-stimulated calf oocytes, transferred to postpubertal recipient animals (for review, see Seidel 1981). The purpose of velogenesis would be to attempt to obtain similar results with fetal oocytes at the earliest stage possible, maybe as early as 90–180 days of gestation.

The development of culture systems in mice that support the growth of primary follicles, yielding mature oocytes capable of fertilization in vitro and development to term (Eppig and Schroeder 1989; Carroll et al. 1990), is very encouraging. It is reasonable to anticipate that similar conditions supporting development of bovine oocytes will become available in a species where primary oocytes from relatively small antral follicles can already be successfully matured and fertilized in vitro.

The impact of velogenesis on breeding programs has been discussed by Betteridge et al. (1989). In dairy breeding, for instance, annual responses in milk yield could be doubled when compared to yields from conventional progeny testing. With the added power of marker-assisted introgression, the approach becomes much more powerful. Velogenetics and velogenesis can be viewed as procedures for the rapid and efficient intraspecies transfer of desirable genes between genetic backgrounds. By analogy with the term transgene, the manipulated genes would be referred to as velogenes. In particular, desirable traits identified outside commercial breeding stock could be quickly introgressed into high-merit genetic backgrounds. Examples would include disease resistance, genes affecting milk and meat composition, polled hornless, and coat color genes. Moreover, the possibility of exploiting exotic genetic variation identified outside the breed of interest is particularly attractive because it greatly facilitates the mapping of the genes of interest (see QTL mapping).

INDIVIDUAL IDENTIFICATION AND PATERNITY DIAGNOSIS

Methods to estimate the breeding value of a particular animal consider information from its relatives. Keeping track of familial relationships has always been one of the major concerns of animal breeders, and parentage control is now a widely used procedure for several domestic species. Parentage control relies on the use of polymorphic systems within the studied population. The alleles that characterize an individual originate from the mother or the father. If one of the parents is known (usually the mother), the alleles necessarily transmitted by the other parent can be easily deduced. Paternity testing consists of scoring the existence or lack of those obligate paternal alleles in the genotype of the putative parent. Lack of one or more of these alleles points toward incorrectly assigned paternity. Nevertheless, one must always consider the possibility of fortuitous coincidence. The higher the variation of the genetic markers used, the higher the probability of detecting incorrectly assigned paternity, thus the higher the exclusion power.

Until now, the systems used most often for paternity testing were blood group systems, biochemical polymorphisms, or the MHC system. The availability of DSPs, however, opens new perspectives for paternity diagnosis. Hypervariable minisatellites in particular, characterized by their remarkably high degree of polymorphism, have proven especially useful in this respect. Multilocus DNA fingerprints, based on the simultaneous detection of related minisatellite loci, have been shown to be extremely powerful for paternity diagnosis, both in humans (Jeffreys and Wilson 1985) and in animals (Morton et al. 1987; Georges et al. 1988b). Exclusion powers as high as 99.999996% have been obtained with as few as two probes in humans (Jeffreys and Wilson 1985). With such high exclusion powers, absence of exclusion can be considered proof for truc biological parentage. Another corollary is that very high exclusion powers can be obtained even when a single parent is available and tested for parenthood. Multilocus DNA fingerprints, however, tend to be replaced by the combined use of a limited number of locus-specific VNTR markers (Nakamura et al. 1987b), giving equally powerful, but more reproducible, sensitive, and easily interpretable patterns. With the advent of locus-specific VNTR markers and PCR-amplifiable microsatellites in animal species (Georges et al. 1990a), the same results will probably be seen in this field. DNA markers can also be used for individual identification. Using expansion/contraction-type polymorphisms, individual-specific DNA "bar codes" can literally be generated (see Jeffreys and Wilson 1985; Georges et al. 1988a).

CONCLUSIONS

Primary DNA marker maps will soon become available for several domestic animal species. This will allow animal geneticists to map the Mendelian components that underlie production traits and to obtain insight into the molecular organization of the "black box," so far blindly manipulated by animal breeders. This increased understanding will result in improved, more efficient breeding schemes. The prospects of velogenetics, in particular, seems very promising. At least in cattle, the two components required for this approach are within reach: (1) genetic mapping of cconomic trait loci and (2) velogenesis, i.e., germ-line manipulations to reduce the generation interval. Velogenetics will allow the engineering of new animal strains by the rapid marker-assisted introgression of mapped genes without the need for the cloning of the corresponding genes and without disrupting their chromosomal organization.

We have no doubt that marker-assisted selection will have

its place in future breeding programs, not only because it enhances the power of the present-day breeding strategies, but also because, in combination with the manipulation of the germ line, it offers entirely new possibilities as well.

ACKNOWLEDGMENTS

The continuous interest and support of Dr. J.M. Massey, Professors R. White, R. Gesteland, and J.-M. Lalouel, and Mr. J. Fordyce are sincerely acknowledged. We thank Professor Pedersen for critically reviewing this manuscript.

REFERENCES

Ajala, F.J., 1972. Frequency-dependent mating advantage in *Drosophila. Behav. Genet.* **2:** 85.

Archibald, A.L., and P. Imlah. 1985. The halothane sensitivity locus and its linkage relationships. *Anim. Blood Groups Biochem. Genet.* **16:** 253.

Arnheim, N., C. Strange, and H. Erlich. 1985. Use of pooled DNA samples to detect linkage disequilibrium of polymorphic restriction fragments and human disease: Studies of the HLA class II loci. *Proc. Natl. Acad. Sci.* **82:** 6970.

Barker, D., M. Schafer, and R. White. 1984. Restriction sites containing CpG show a higher frequency of polymorphism in human DNA. *Cell* **36:** 131.

Betteridge, K.J., C. Smith, R.B. Stubbings, K.P. Xu, and W.A. Kang. 1989. Potential genetic improvement of cattle by fertilization of fetal oocytes in vitro. *J. Reprod. Fertil.* (suppl.) **38:** 87.

Blatt, C., D. Aberdam, R. Schwartz, and L. Sachs. 1988. DNA rearrangement of a homeobox gene in myeloid leukaemic cells. *EMBO J.* **7:** 4283.

Botstein, D., R.L. White, M. Skolnick, and R.W. Davis. 1980. Construction of a genetic linkage map in man using restriction fragment length polymorphism. *Am. J. Hum. Genet.* **32:** 314.

Canaani, E., O. Dreazen, A. Klar, G. Rechavi, D. Ram, J.B. Cohen, and D. Givol. 1983. Activation of the c-*mos* oncogene in a mouse plasmocytoma by insertion of an endogenous intracisternal A-particle genome. *Proc. Natl. Acad. Sci.* **80:** 7118.

Carroll, J., D.G. Whittingham, M.J. Wood, E. Telfer, and R.G. Gosden. 1990. Extra-ovarian production of mature viable mouse oocytes from frozen primary follicles. *J. Reprod. Fertil.* **90:** 321.

Cawthon, R.M., R. Weiss, G. Xu, D. Viskochil, M. Culver, J. Stevens, M. Robertson, D. Dunn, R. Gesteland, P. O'Connell, and R. White.

1990. A major segment of the neurofibromatosis type 1 gene: cDNA sequence, genomic structure, and point mutations. *Cell* **62:** 193.

Church, G.M. and W. Gilbert. 1984. Genomic sequencing. *Proc. Natl. Acad. Sci.* **81:** 991.

Cohen, J.C. and H.E. Varmus. 1979. Endogenous mammary tumor virus DNA varies among wild mice and segregates during inbreeding. *Nature* **278:** 418.

Cowan, C.W., M.R. Dentine, R.L. Ax, and L.A. Schuler. 1990. Structural variation around prolactin gene linked to quantitative traits in elite Holstein sire family. *Theor. Appl. Genet.* **79:** 577.

Cox, D.R., M. Burmeister, E.R. Price, S. Kim, and M. Myers. 1990. Radiation hybrid mapping: A somatic cell genetic method for constructing high-resolution maps of mammalian chromosomes. *Science* **250:** 245.

Donis-Keller, H. et al. (32 additional authors). 1987. A genetic linkage map of the human genome. *Cell* **51:** 319.

Doolittle, W.F. 1985. RNA mediated gene conversion? *Trends Genet.* **1:** 64.

Economou, E.P., A. Bergen, A.C. Warren, and S.E. Antonorakis. 1990. The polydeoxyadenylate tract of Alu repetitive elements is polymorphic in the human genome. *Proc. Natl. Acad. Sci.* **87:** 2951.

Eppig, J.J. 1989. The participation of cyclic adenosine monophosphate (cAMP) in the regulation of meiotic maturation of oocytes in the laboratory mouse. *J. Reprod. Fertil.* (suppl.) **38:** 3.

Eppig, J.J. and A.C. Schroeder. 1989. Capacity of mouse oocytes from pre-antral follicles to undergo embryogenesis and development to live young after growth, maturation and fertilization in vitro. *Biol. Reprod.* **41:** 268.

Falconer, D.S. 1989. *Introduction to quantitative genetics*, 3rd. edition. Longman Scientific, England.

Flint, J., A.V. Hill, D.K. Bowden, S.J. Oppenheimer, P.R. Sill, S.W. Serjeantson, J. Bana-Koiri, K. Bhatia, M.P. Alpers, A.J. Boyce, D.J. Weatherall, and J.B. Clegg. 1986. High frequencies of α-thalassaemia are the results of natural selection by malaria. *Nature* **321:** 744.

Friend, S.H., R. Bernards, S. Rogelj, R.A. Weinberg, J.M. Rapaport, D.M. Albert, and T.P. Dryja. 1986. A human DNA segment with properties of the gene that predisposes to retinoblastoma and osteosarcoma. *Nature* **323:** 643.

Fries, R., J.S. Beckmann, M. Georges, M. Soller, and J. Womack. 1989. The bovine gene map. *Animal Genet.* **20:** 3.

Geldermann, H., U. Pieper, and B. Roth. 1985. Effects of marked chromosome sections on milk performance in cattle. *Theor. Appl. Genet.* **70:** 138.

Georges, M. 1990. Hypervariable minisatellites and their use in animal breeding. In *Gene mapping: Strategies, techniques and applications* (ed. L.B. Schook et al.). Marcel Dekker, New York. (In press.)

Georges, M. and J.M. Massey. 1991. Velogenetics or the synergistic use of marker assisted selection and germ-line manipulation. *Theriogenology* **35:** 151.

Georges, M., A.S. Lequarré, R. Hanset, and G. Vassart. 1987. Genetic variation of the bovine thyroglobulin locus studied at the DNA level. *Animal Genet.* **18:** 41.

Georges, M., A.S. Lequarré, M. Castelli, R. Hanset, and G. Vassart. 1988a. DNA fingerprinting in domestic animals using four different minisatellite probes. *Cytogenet. Cell Genet.* **47:** 127.

Georges, M., A. Mishra, L. Sargeant, M. Steele, and M. Zhao. 1990a. Progress towards a primary DNA marker map in cattle. In *Proceedings of the 4th World Congress on Genetics Applied to Livestock Production* (ed. W. Hill et al.), vol. XIII, p. 107. Edinburgh, Scotland.

Georges, M., P. Hilbert, A. Lequarré, V. Leclerc, R. Hanset, and G. Vassart. 1988b. Use of DNA bar codes to resolve a canine paternity dispute. *J. Am. Vet. Med. Assoc.* **193:** 9.

Georges, M., M. Lathrop, P. Hilbert, A. Marcotte, A. Schwers, S. Swillens, G. Vassart, and R. Hanset. 1990b. On the use of DNA fingerprints for linkage studies in cattle. *Genomics* **6:** 461.

Gibbs, R.A., P.-N. Nguyen, and C.T. Caskey. 1989. Detection of single DNA base differences by competitive oligonucleotide priming. *Nucleic Acids Res.* **17:** 2437.

Gibson, J.P., G.B. Janson, and P. Rozzi. 1990. The use of K-casein genotypes in dairy cattle breeding. In *Proceedings of the 4th World Congress on Genetics Applied to Livestock Production* (ed. W. Hill et al.), vol. XIV, p. 163. Edinburgh, Scotland.

Gould, K.L., S. Moreno, N.K. Tonks, and P. Nurse. 1990. Complementation of the mitotic activator, $p80^{cdc25}$, by a human protein-tyrosine phosphatase. *Science* **250:** 1573.

Grosclaude, F. 1988. Le polymorphisme génétique des principales lactoprotéines bovines. *INRA Product. Anim.* **1:** 5.

Gubbay, J., J. Collignon, P. Koopman, B. Capel, A. Economou, A. Munsterberg, N. Vivian, P. Goodfellow, and R. Lovell-Badge. 1990. A gene mapping to the sex-determining region of the mouse Y chromosome is a member of a novel family of embryonically expressed genes. *Nature* **346:** 245.

Hanset, R. 1982. Major genes in animal production, examples and perspectives: Cattle and pigs. In *Proceedings of the 2nd World Congress on Genetics Applied to Livestock Production*, vol. 6, p. 439. Madrid, Spain.

———. 1989. Les perpectives d'une selection assistee par marqueurs. *Ann. Med. Vet.* **133:** 465.

Hawley, R.G., M.J. Shulman, H. Murialdo, D.M. Gibson, and N.

Hozumi. 1982. Mutant immunoglobulin genes have repetitive DNA elements inserted into their intervening sequences. *Proc. Natl. Acad. Sci.* **79:** 7425.

Herr, C.M., K.I. Matthaei, M.P. Bradley, and K.C. Reed. 1990. Rapid, accurate sexing of livestock embryos. In *Proceedings of the 4th World Congress on Genetics Applied to Livestock Production* (ed. W. Hill et al.), vol. XVI, p. 334. Edinburgh, Scotland.

Hilbert, P., A. Marcotte, A. Schwers, R. Hanset, G. Vassart, and M. Georges. 1989. Analysis of genetic variation in the Belgian Blue cattle breed using DNA sequence polymorphism at the growth hormone, low density lipoprotein receptor, α-subunit of glycoprotein hormones and thyroglobulin loci. *Anim. Genet.* **20:** 383.

Hill, R.E. and N.D. Hastie. 1987. Accelerated evolution in the reactive centre regions of serine protease inhibitors. *Nature* **236:** 96.

Hoeschele, I. and T.R. Meinert. 1990. Association of genetic defects with yield and type traits: A major production gene is linked to Weaver. *J. Dairy Sci.* **73:** 2503.

Hughes, A.L. and M. Nei. 1988. Pattern of nucleotide substitution at major histocompatibility complex class I loci reveals overdominant selection. *Nature* **335:** 167.

Hultman, T., S. Staahl, E. Hornes, and M. Uhlen. 1989. Direct solid phase sequencing of genomic and plasmid DNA using magnetic beads as solid support. *Nucleic Acids Res.* **17:** 4837.

Hunt, T. 1989. Under arrest in the cell cycle. *Nature* **342:** 483.

———. 1991. Destruction's our delight. *Nature* **349:** 100.

Jeffreys, A.J. 1979. DNA sequence variants in the $^{G}\gamma$–, $^{A}\gamma$–, δ– and β-globin genes of man. *Cell* **18:** 1.

Jeffreys, A.J. and V. Wilson. 1985. Individual-specific "fingerprints" of human DNA. *Nature* **316:** 76.

Jeffreys, A.J., V. Wilson, and S.L. Thein. 1985. Hypervariable "minisatellite" regions in human DNA. *Nature* **314:** 67.

Jeffreys, A.J., R. Neumann, and V. Wilson. 1990. Repeat unit sequence variation in minisatellites: A novel source of DNA polymorphism for studying variation and mutation by single molecular analysis. *Cell* **60:** 473.

Julier, C. and R. White. 1988. Detection of a NotI polymorphism with the pmetH probe by pulsed-field gel electrophoresis. *Am. J. Hum. Genet.* **42:** 45.

Julier, C., B. De Gouyon, M. Georges, J.L. Guenet, P. Avner, and G.M. Lathrop. 1990. Minisatellite linkage maps in the mouse by cross-hybridization with human VNTR probes. *Proc. Natl. Acad. Sci.* **87:** 4585.

Kazazian, H.H., A. Chakravarti, S.H. Orkin, and S.E. Antonorakis. 1983. DNA polymorphisms in the human β globin gene cluster. In *Evolution of genes and proteins* (ed. M. Nei and R.K. Koehn), p. 137. Sinauer, Sunderland, Massachusetts.

Kazazian, H.H., C. Wong, H. Youssoufian, A.F. Scott, D.G. Phillips, and S.E. Antonorakis. 1988. Haemophilia A resulting from de novo insertion of L1 sequences represents a novel mechanism for mutation in man. *Nature* **332:** 164.

Kerem, B.-S., J.M. Rommens, J.A. Buchanan, D. Markiewicz, T.K. Cox, A. Chakravarti, M. Buchwald, and L.-C. Tsui. 1989. Identification of the cystic fibrosis gene: Genetic analysis. *Science* **245:** 1073.

Kimura, M. 1986. *The neutral theory of molecular evolution*. Cambridge University Press, England.

Kreitman, M. and M. Aguade. 1986. Genetic uniformity in two populations of *Drosophila melanogaster* as revealed by filter hybridization of four-nucleotide-recognizing enzyme digests. *Proc. Natl. Acad. Sci.* **83:** 3562.

Lander, E.S. and D. Botstein. 1986. Mapping complex genetic traits in humans: New methods using complete RFLP linkage maps. *Cold Spring Harbor Symp. Quant. Biol.* **51:** 49.

———. 1989. Mapping Mendelian factors underlying quantitative traits using RFLP linkage maps. *Genetics* **121:** 185.

Lander, E.S. and P. Green. 1987. Construction of multilocus linkage maps in humans. *Proc. Natl. Acad. Sci.* **84:** 2363.

Lathrop, G.M., J.M. Lalouel, C. Julier, and J. Ott. 1984. Strategies for multilocus linkage analysis in humans. *Proc. Natl. Acad. Sci.* **101:** 3443.

———. 1985. Multilocus linkage analysis in humans: Detection of linkage and estimation of recombination. *Am. J. Hum. Genet.* **37:** 482.

Lathrop, M., Y. Nakamura, P. O'Connell, M. Leppert, S. Woodward, J. Lalouel, and R. White. 1988a. A mapped set of genetic markers for human chromosome 9. *Genomics* **3:** 361.

Lathrop, M., Y. Nakamura, P. Cartwright, P. O'Connell, M. Leppert, C. Jones, H. Tateishi, T. Bragg, J. Lalouel, and R. White. 1988b. A mapped set of genetic markers for human chromosome 10. *Genomics* **2:** 157.

Lehrman, M.A., J.L. Goldstein, D.W. Russell, and M.S. Brown. 1987. Duplication of seven exons in LDL receptor gene caused by Alu-Alu recombination in a subject with familial hypercholesterolemia. *Cell* **48:** 827.

Li, H., U.B. Gyllensten, X. Cui, R.K. Saiki, H.A. Erlich, and N. Arnheim. 1988. Amplification and analysis of DNA sequences in single human sperm and diploid cells. *Nature* **335:** 414.

Lichter, P., C.-J. Chang Tang, K. Call, G. Hermanson, G.A. Evans, D. Housman, and D.C. Ward. 1990. High-resolution mapping of human chromosome 11 by in situ hybridization with cosmid clones. *Science* **247:** 64.

Lin, C.S., D.A. Goldthwait, and D. Samols. 1988. Identification of Alu transposition in human lung carcinoma cells. *Cell* **54:** 153.

Litt, M. and J.A. Luty. 1989. A hypervariable microsatellite revealed by in vitro amplification of a dinucleotide repeat within the cardiac muscle actin gene. *Am. J. Hum. Genet.* **44:** 397.

MacLennan, D.H., C. Duff, F. Zorzato, J. Fujii, M. Phillips, R.G. Korneluk, W. Frodis, B.A. Britt, and R.G. Worton. 1990. Ryanodine receptor gene is a candidate for predisposition to malignant hyperthermia. *Nature* **343:** 559.

Martin, B., J. Nienhuis, G. King, and A. Schaeffer. 1989. Restriction fragment length polymorphisms associated with water use efficiency in tomato. *Science* **243:** 1725.

Medrano, J.F. and E. Aguilar-Cordova. 1990. Genotyping of bovine K-casein loci following DNA sequence amplification. *Biotechnology* **8:** 144.

Monaco, A.P., R.L. Neve, C. Colletti-Feener, C.J. Bertelson, D.M. Kurnit, and L.M. Kunkel. 1986. Isolation of candidate cDNA for portions of Duchenne muscular dystrophy gene. *Nature* **323:** 646.

Morse, B., P.G. Rotherg, V.J. South, J.M. Spandorfer, and S.M. Astrin. 1988. Insertional mutagenesis of the myc locus by a LINE-1 sequence in a human breast carcinoma. *Nature* **333:** 87.

Morton, D.B., R.E. Yaxley, I. Patel, A.J. Jeffreys, S.J. Howes, and P.G. Debenham. 1987. Use of DNA fingerprint analysis in identification of the sire. *Vet. Rec.* **121:** 592.

Morton, N.E. 1955. Sequential tests for the detection of linkage. *Am. J. Hum. Genet.* **7:** 277.

Myers, R.M., V.C. Sheffield, and D.R. Cox. 1988. Detection of single base changes in DNA: Ribonuclease cleavage and denaturing gradient gel electrophoresis. In *Genome analysis* (ed. K.E. Davies). IRL Press, England.

Nakamura, Y., M. Lathrop, P. O'Connell, M. Leppert, J.-M. Lalouel, and R. White. 1988a. A mapped set of DNA markers for human chromosome 15. *Genomics* **3:** 342.

———. 1988b. A primary map of ten DNA markers and two serological markers for human chromosome 19. *Genomics* **3:** 67.

Nakamura, Y., C. Julier, R. Wolff, T. Holm, P. O'Connell, M. Leppert, and R. White. 1987a. Characterization of a human "midisatellite" sequence. *Nucleic Acids Res.* **15:** 2537.

Nakamura, Y., M. Lathrop, P. O'Connell, M. Leppert, D. Barker, E. Wright, M. Skolnick, S. Kondoleon, M. Litt, J.-M. Lalouel, and R. White. 1988c. A mapped set of DNA markers for human chromosome 17. *Genomics* **2:** 302.

Nakamura, Y., M. Leppert, P. O'Connell, R. Wolff, T. Holm, M. Culver, C. Martin, E. Fujimoto, M. Hoff, E. Kumlin, and R. White. 1987b. Variable number of tandem repeat (VNTR) markers for human gene mapping. *Science* **235:** 1616.

Nei, M. 1983. Genetic polymorphism and the role of mutation in evolution. In *Evolution of genes and proteins* (ed. M. Nei and R.K. Koehn), p. 165. Sinauer, Sunderland, Massachusetts.

———. 1987. *Molecular evolutionary genetics*. Columbia University

Press, New York.

Newton, C.R., A. Graham, L.E. Heptinstall, S.J. Powell, C. Summers, N. Kalsheker, J.C. Smith, and A.F. Markham. 1989. Analysis of any point mutation in DNA. The amplification refractory mutation system (ARMS). *Nucleic Acids Res.* **17:** 2503.

O'Connell, P., G.M. Lathrop, Y. Nakamura, M.L. Leppert, R.H. Ardinger, J.L. Murray, J.M. Lalouel, and R. White. 1989. Twenty-eight loci form a continuous linkage map of markers for human chromosome 1. *Genomics* **4:** 12.

O'Connell, P., G.M. Lathrop, M. Law, M.L. Leppert, Y. Nakamura, M. Hoff, E. Kumlin, W. Thomas, T. Elsner, L. Ballard, P. Goodman, E. Azen, J. Sadler, G. Cai, J.M. Lalouel, and R. White. 1987. A primary genetic linkage map for human chromosome 12. *Genomics* **1:** 93.

Orita, M., Y. Suzuki, T. Sekiya, and K. Hayashi. 1989a. Rapid and sensitive detection of point mutations and DNA polymorphisms using polymerase chain reaction. *Genomics* **5:** 874.

Orita, M., H. Iwahana, H. Kanazawa, K. Hayashi, and T. Sekiya. 1989b. Detection of polymorphism of human DNA by gel electrophoresis as single-strand conformation polymorphism. *Proc. Natl. Acad. Sci.* **86:** 2766.

Orkin, S.H. 1986. Reverse genetics and human disease. *Cell* **47:** 845.

Paterson, A.H., E.S. Lander, J.D. Hewitt, S. Peterson, S.E. Lincoln, and S.D. Tanksley. 1989. Resolution of quantitative traits into Mendelian factors by using a complete linkage map of restriction fragment length polymorphisms. *Nature* **335:** 721.

Plotsky, Y., A. Cahaner, A. Haberfeld, U. Lavi, and J. Hillel. 1990. Analysis of genetic association between DNA fingerprint bands and quantitative traits using DNA mixes. In *Proceedings of the 4th World Congress on Genetics Applied to Livestock Production* (ed. W. Hill et al.), vol. XIII, p. 133. Edinburgh, Scotland.

Quesada, M.A., H.S. Rye, J.C. Gingrich, A.N. Glazer, and R.A. Mathies. 1991. High-sensitivity DNA detected with a laser-excited confocal fluorescence gel scanner. *Biotechniques* (in press).

Riordan, J.R., J.M. Rommens, B.-S. Kerem, N. Alon, R. Rozmahel, Z. Grzelczak, J. Zielenski, S. Lok, N. Plavsic, J.L. Chou, M.L. Drumm, M.C. Ianuzzi, F.S. Collins, and L.-C. Tsui. 1989. Identification of the cystic fibrosis gene: Cloning and characterization of complementary DNA. *Science* **245:** 1066.

Roberts, R.G., A.J. Montandon, M. Bobrow, and D.R. Bentley. 1989. Detection of novel genetic markers by mismatch analysis. *Nucleic Acids Res.* **17:** 5961.

Rommens, J.M., M.C. Iannuzzi, B.-S. Kerem, M.L. Drumm, G. Melmer, M. Dean, R. Rozmahel, J.L. Cole, D. Kennedy, N. Hidaka, M. Zsiga, M. Buchwald, J.R. Riordan, L.-C. Tsui, and F. Collins. 1989. Identification of the cystic fibrosis gene: Chromosome walking and jumping. *Science* **245:** 1059.

Royer-Pokora, B., L.M. Kunkel, A.P. Monaco, S.C. Goff, P.E. Newburger, R.L. Baehner, F.S. Cole, J.T. Curnutte, and S.H. Orkin. 1986. Cloning the gene for an inherited human disorder–chronic granulomatous disease–on the basis of its chromosomal location. *Nature* **322:** 32.

Royle, N.J., R.E. Clarkson, Z. Wong, and A.J. Jeffreys. 1988. Clustering of hypervariable minisatellites in the proterminal regions of human autosomes. *Genomics* **3:** 352.

Rubin, G.M. 1983. Dispersed repetitive DNAs in *Drosophila*. In *Mobile genetic elements* (ed. J.A. Shapiro), p. 329. Academic Press, New York.

Russe, I. 1983. Oogenesis in cattle and sheep. *Bibl. Anat.* **24:** 77.

Saiki, R.K., P.S. Walsh, C.H. Levenson, and H.A. Erlich. 1989. Genetic analysis of amplified DNA with immobilized sequence-specific oligonucleotide probes. *Proc. Natl. Acad. Sci.* **86:** 6230.

Saiki, R.K., D.H. Gelfland, S. Stoffel, S.S. Scharf, R. Higuchi, G.R. Horn, K.B. Mullis, and H.A. Erlich. 1988. Primer directed enzymatic amplification of DNA with a thermostable DNA polymerase. *Science* **239:** 487.

Seidel, G.E. 1981. Superovulation and embryo transfer in cattle. *Science* **211:** 351.

Sinclair, A.H., P. Berta, M.S. Palmer, J.R. Hawkins, B.L. Griffiths, M.J. Smith, J.W. Foster, A.-M. Frischauf, R. Lovell-Badge, and P.N. Goodfellow. 1990. A gene from the human sex-determining region encodes a protein with homology to a conserved DNA-binding motif. *Nature* **346:** 240.

Singer, M.F. 1982. Highly repeated sequences in mammalian genomes. *Int. Rev. Cytol.* **76:** 67.

Skolnick, M.H. and R.B. Wallace. 1988. Simultaneous analysis of multiple polymorphic loci using amplified sequence polymorphisms (ASPs). *Genomics* **2:** 273.

Smith, C. and S.P. Simpson. 1986. The use of genetic polymorphism in livestock improvement. *J. Anim. Breed. Genet.* **103:** 205.

Soller, M. and J.S. Beckman. 1982. Restriction fragment length polymorphisms and genetic improvement. In *Proceedings of the 2nd World Congress on Genetics Applied to Livestock Production*, vol. 6. p. 396. Madrid, Spain.

Soller, M., T. Brody, and A. Genizi. 1976. On the power of experimental designs for the detection of linkage between marker loci and quantitative loci in crosses between inbred lines. *Theor. Appl. Genet.* **47:** 35.

Southern, E.M. 1975. Detection of specific sequences among DNA fragments separated by gel electrophoresis. *J. Mol. Biol.* **98:** 503.

Steele, M. and M. Georges. 1991. Generation of bovine multisite haplotypes using random cosmids. *Genomics* (in press).

Stocking, C., C. Loliger, M. Kawai, S. Suciu, N. Gough, and W. Ostertag. 1988. Identification of genes involved in growth factor autonomy of hematopoietic cells by analysis of factor independent mutants. *Cell* **53:** 869.

Streydio, C., S. Swillens, M. Georges, C. Szpirer, and G. Vassart. 1990. Structure, evolution and chromosomal localization of the human pregnancy specific β1 glycoprotein gene family: Evidence in favor of positive Darwinian selection. *Genomics* **6:** 579.

Tautz, D. 1989. Hypervariability of simple sequences as a general source for polymorphic DNA markers. *Nucleic Acids Res.* **17:** 6463.

Thein, S.L. and R.B. Wallace. 1986. The use of synthetic oligonucleotides as specific hybridization probes in the diagnosis of genetic disorders. In *Human genetic diseases: A practical approach* (ed. K.E. Davies), p. 33. IRL Press, England.

Vassart, G., M. Georges, R. Monsieur, H. Brocas, A.S. Lequarre, and D. Christophe. 1987. A sequence in M13 phage detects hypervariable minisatellites in human and animal DNA. *Science* **235:** 683.

Viskochil, D., A.M. Buchberg, G. Xu, R.M. Cawthon, J. Stevens, R.K. Wolff, M. Culver, J.C. Carey, N.G. Copeland, N.A. Jenkins, R. White, and P. O'Connell. 1990. Deletions and a translocation interrupt a cloned gene at the neurofibromatosis type 1 locus. *Cell* **62:** 187.

Weber, J.L. and P.E. May. 1989. Abundant class of human DNA polymorphisms which can be typed using the polymerase chain reaction. *Am. J. Hum. Genet.* **44:** 388.

Weller, J.I., Y. Kashi, and M. Soller. 1990. Power of "daughter" and "granddaughter" designs for genetic mapping of quantitative traits in dairy cattle using genetic markers. *J. Dairy Sci.* **73:** 2525.

White, R. and C.T. Caskey. 1988. The human as an experimental system in molecular genetics. *Science* **240:** 1483.

White, R. and J.M. Lalouel. 1988. Linked sets of genetic markers for human chromosomes. *Annu. Rev. Genet.* **22:** 259.

White,. R., M. Leppert, D. Bishop, J. Berkowitz, C. Brown, P. Callahan, T. Holm, and L. Jerominski. 1985. Construction of linkage maps with DNA markers for human chromosomes. *Nature* **313:** 101.

Womack, J.E. 1987. Comparative gene mapping: A valuable tool for mammalian developmental studies. *Dev. Genet.* **8:** 281.

Wong, C., C.E. Dowling, R.K. Saiki, R.G. Higuchi, H.A. Erlich, and H.H. Kazazian. 1987. Characterization of β-thalassaemia mutations using direct genomic sequencing of amplified single copy DNA. *Nature* **330:** 384.

Wong, Z., N.J. Royle, and A.J. Jeffreys. 1990. A novel human DNA polymorphism resulting from transfer of DNA from chromosome 6 to chromosome 16. *Genomics* **7:** 222.

Wu, D.Y. and R.B. Wallace. 1989. The ligation amplification reaction (LAR)—Amplification of specific DNA sequences using sequential rounds of template-dependent ligation. *Genomics* **4:** 560.

Wu, D.Y., L. Ugozzoli, B.K. Pal, and R.B. Wallace. 1989. Allele-specific enzymatic amplification of β-globin genomic DNA for diagnosis of sickle cell anemia. *Proc. Natl. Acad. Sci.* **86:** 2757.

Ymer, S., W.Q.J. Tucker, C.J. Sanderson, A.J. Hapel, H.D. Campbell, and I.G. Young. 1985. Constitution synthesis of interleukin-3 by leukaemia cell line WEHI-3B is due to retroviral insertion near the gene. *Nature* **317:** 255.

Index